# Minnesota's Natural Heritage

*Publication of this book is made possible by:*

*David R. and Elizabeth P. Fesler through their generous addition to the John K. and Elsie Lampert Fesler Fund,*

*and by an appropriation from the Minnesota Legislature, ML 1993, Chapter 172, Sec. 14, Subd. 7 (g), as recommended by the Legislative Commission on Minnesota Resources from the Minnesota Futures Resources Fund.*

•

*The John K. and Elsie Lampert Fesler Fund was established in 1972 to support publication by the University of Minnesota Press of books significant to the people of Minnesota.*

*The Legislative Commission on Minnesota Resources (LCMR) was established in 1963 to provide the legislature with the background necessary to evaluate programs proposed to preserve, develop, and maintain the natural resources of the state of Minnesota. The information presented in* Minnesota's Natural Heritage: An Ecological Perspective *must be credited in part to the data that are available from the many surveys and inventory projects funded by the LCMR over the past thirty years.*

*John R. Tester*

# MINNESOTA'S
# NATURAL HERITAGE

## An Ecological Perspective

*Mary Keirstead • Developmental Editor*

University of Minnesota Press

*Minneapolis & London*

Published by the University of Minnesota Press
111 Third Avenue South, Suite 290, Minneapolis, MN 55401-2520
Third printing 2013

Printed in China on acid-free paper
Title page photo by Bob Firth, Firth PhotoBank, Minneapolis
Book and jacket design by Diane Gleba Hall

Library of Congress Cataloging-in-Publication Data

Tester, John R.
    Minnesota's Natural Heritage / John R. Tester ; Mary Keirstead, developmental editor.
       p.   cm.
    Includes bibliographical references and index.
    ISBN 0-8166-2133-0 (alk. paper)
    1. Biotic communities—Minnesota. 2. Natural history—Minnesota.
    I. Keirstead, Mary. II. Title.
    QH105.M55T47   1995
    574.5'247'09776—dc20                        94–45153

# CONTENTS

*v*

# PREFACE

The line of images that runs through my life reflects my involvement with and love of Minnesota. Our state's natural heritage is complex and rich. I have lived with this heritage most of my life, growing up in a farming community in south-central Minnesota and then working in nearly every county for the Minnesota Department of Natural Resources and the University of Minnesota, first in the James Ford Bell Museum of Natural History and now in the Department of Ecology, Evolution and Behavior.

I have written this book because of my deep interest in the natural history and ecology of Minnesota and with the hope that you, the reader, will develop admiration and respect for your environment. This will come, I am confident, through knowledge of Minnesota's physical and biological attributes and how they interact ecologically.

The purpose of this book is not to tell people what they should or should not do to their environment, nor to expound on philosophical aspects of life or nature. Rather, the book explains how ecological systems are structured, how they work, and how they respond to what is done to them and around them. This knowledge provides a basis for understanding Minnesota's ecosystems. Such understanding will ultimately lead to their wise use.

Because this book is written primarily to provide a general introduction for people interested in Minnesota, I have made no attempt to include all that is known about the ecology of the state. Furthermore, no one can know all of the kinds of living plants and animals that live in or migrate

through Minnesota. Consider that the state is or has been the home of about 81 mammals, 403 breeding and migrant birds, 29 reptiles, 19 amphibians, 144 fish, thousands of kinds of insects and other invertebrates, 2,010 trees, shrubs, and herbs, and thousands of kinds of algae, fungi, and other nonflowering plants. Thus, only the most conspicuous features of the landscape and some of the more ecologically significant and interesting plants and animals are discussed, and little is said about such things as microscopic organisms in the soil or about many of the plants and animals that we may see every day. The information is scientifically accurate, and references are cited to enable readers to go to the original sources to explore ideas in depth or to interpret findings in their own way. We all have the capability to see the same things in the natural world. It is the interpretation and understanding of these things, however, that give us insight into what we observe.

The common and scientific names of the plants mentioned in this book are listed in appendixes B, C, and D. All of Minnesota's mammals, breeding birds, reptiles, and amphibians are listed in appendixes E, F, and G.

I fully recognize that it is not possible to document the complete ecology of Minnesota, because so much is unknown. For example, little is written and, in fact, we know little about the ecology of corn fields, wheat fields, cities, clearcuts, roadsides, sewage lagoons, dogs, cattle, and especially people. Many aspects of Minnesota's ecology are still to be explored and explained.

Although most of the information presented deals with Minnesota, this book has relevance to a large region of central North America. Tallgrass prairie extends westward into the Dakotas and south and southeast to Kansas and Illinois. Northern coniferous forests cover much of southeastern Canada and parts of the northeastern states. Deciduous forests extend east and southeast to the Atlantic Ocean. Species differences occur across each of these vegetation types, but their ecology is similar in many ways. As a result, some aspects of the structures of these ecosystems and how they function in Minnesota have broad applications.

This book begins with a summary of geologic history and a description of how glaciation determined the surface geology, or landforms, in Minnesota. A description of vegetation and soil development from the time of the retreat of the continental glaciers about 10,000 years ago to the time of European exploration and settlement follows. Chapter 1 concludes with a brief look at the present landscape. Climate and weather are discussed in chapter 2; these physical factors are extremely important in determining the community of plants and animals that exists in an area.

Chapter 3 briefly explains general principles of ecology, such as energy flow, nutrient cycling, and population dynamics. The role of physical factors in influencing the distribution and abundance of plants and animals in a given area is considered in terms of the natural vegetation pattern in Minnesota.

Chapters 4, 5, and 6 discuss the structure and functioning of the three major natural ecosystems of Minnesota, deciduous forest, northern coniferous forest, and tallgrass prairie. The content of these and subsequent chapters is arranged so that the reader moves from familiar and easily observed species and ideas to more conceptual aspects, which may be difficult to observe. The influences of post-settlement human activities, such as logging and agriculture, are discussed to provide both historical and present-day perspective.

Lakes, wetlands, rivers, and streams are abundant throughout Minnesota. To many people, these features make our state unique and special. The ecology of wetlands is presented in chapter 7. Lakes are discussed in chapter 8, and the ecology of rivers and streams is discussed in chapter 9.

In chapter 10, I consider the future of the ecosystems in Minnesota viewed in the context of their history and present status. Harvesting of timber, reforestation, agriculture, drainage of wetlands, preservation of natural areas, acid rain, climate warming, and many other activities and impacts will have a profound influence on Minnesota's ecology. Problems will be raised by many of these activities and impacts, and solutions will not always come easily. I hope that the information presented in this book will assist the reader in evaluating alternatives and in making wise decisions.

# ACKNOWLEDGMENTS

The subject of ecology is too broad for one person to be expert in all of its aspects. Each chapter manuscript was read critically by one or more scientists with special competence in the subject matter. I am extremely grateful to the following reviewers for their corrections, suggestions, and encouragement: John C. Almendinger (chapters 4, 5), Donald G. Baker (2), David B. Czarnecki (8), Robert Dana (6), Eville Gorham (7), David Grigal (1, 5), Steve Heiskary (8), Jeffrey Lang (4, 5, 6), Robert O. Megard (8), Patrice Morrow (3), Katherine Ralls (10), Donald L. Rubbelke (5), Richard Rust (1), Kurt A. Rusterholz (4, 5), Joseph Shapiro (8), Gregory Spoden (2), Ed Swain (8), Tony Thompson (6), Thomas F. Waters (9), Herbert E. Wright Jr. (1), and James A. Zandlo (2). I thank Edward J. Cushing, Billy Goodman, and Dorothy K. Bromenshenkel for reading the entire manuscript and for making many helpful suggestions.

Gilbert Ahlstrand, Edward J. Cushing, Robert O. Megard, Artis Orjala, Jon Ross, William Schmid, David Tilman, and Herbert E. Wright Jr. generously contributed use of their photographs. Jim LaVigne photographed numerous subjects for this book and provided valuable counsel regarding selection of illustrations. I am especially grateful to the Minnesota Department of Natural Resources for providing photographs and for making unpublished data available to me.

Don Luce, artist, created the original illustrations and the inserts in many of the figures. Patti Isaacs, cartographer, created the excellent color graphics and drawings. I thank them for their special efforts.

Mary Keirstead deserves special recognition for her skillful editing of the entire text. Her ability to clarify my writing, organize concepts, and recognize the need for additional explanations has greatly improved the book. I am most grateful for her willing and insightful contributions.

Barbara Coffin, editor at the University of Minnesota Press, generously provided insights on many ecological issues and on the organization of the book. Her detailed comments are greatly appreciated.

The staff of the University of Minnesota Press assisted in many ways, and I thank them for their patience and consideration.

Lastly, and most importantly, I thank my family for their support. My wife, Joyce, provided patience and encouragement through many years of writing and editing. My sons, Hans and Peter, were both inspirational and critical in their reviews, and I thank them for their efforts.

# Minnesota's Natural Heritage

# THE LANDSCAPE     **I**

Imagine flying in a small plane westward across Minnesota. We begin in Pine County, about half-way between the Twin Cities and Duluth. Wetlands with clumps of black spruce and tamarack dominate the landscape for the first few miles near the Wisconsin border (figure 1.2). A bull moose, which has strayed south of its usual range, walks in a wetland, moving as easily as we walk on a path. Soon the vegetation changes, and mixed hardwood and conifer forests of aspen, birch, white spruce, and balsam fir become more common. Occasionally we pass over giant white and red pines, sticking out like sentinels above the rest of the forest (figure 1.3). These trees are remnants of the great pine forests that once covered extensive areas of the state. The small streams and occasional lakes contain a few wood ducks and mallards. A soaring red-tailed hawk watches us as we pass each other in the sky. We cross roads, towns, and farms, and in about an hour we see Mille Lacs Lake in the distance.

Here we notice a marked change in the forest. Tall pines stand out clearly, and a few spruce and fir trees can still be seen, but aspen, birch, maple, basswood, and other deciduous trees are becoming more and more common (figure 1.4). After crossing the Mississippi River, we see no more spruce and fir, and clearings in the forest are more frequent. Some clearings are covered with grass, and others.

**Figure 1.1** *The northern coniferous forest is a mosaic of stands of white spruce, balsam fir, and pines interspersed with stands of aspen and birch. (Photo by Blacklock Nature Photography.)*

are farmland. Wood ducks, ring-necked ducks, and mallards flush as we fly over. Three deer jump quickly from their beds, race across a field, jump in unison over a fence, and disappear into the forest.

Gradually the deciduous forest gives way to farmland as we cross into the hilly landscape of Otter Tail County. Now streams, lakes, and wetlands seem to be everywhere we look (figure 1.5), and most are dotted with mallards, blue-winged teal, redheads, and numerous other ducks. The trees that are present are along streams and lakes and around farmsteads. Here and there a tract of undisturbed tallgrass prairie stands out from the agricultural fields. A red fox raises its head to watch

**Figure 1.2** *(left) Black spruce and tamarack characterize wetlands near the Wisconsin border in Pine County and throughout the northern coniferous forest.*
**Figure 1.3** *(below) Large red and white pines, several hundred years old, rise above the forest, suggesting that they dominated the landscape before the logging era. (Photos by Richard Hamilton Smith.)*

**Figure 1.4** *Deciduous forests exist in a narrow band stretching diagonally across Minnesota from northwest to southeast. Mille Lacs Lake can be seen in the distance. (Photo by Richard Hamilton Smith.)*

the plane pass but does not get up from its resting spot in the sun on the south side of a hill.

As we continue to fly west, wetlands and lakes become much less common. We cross the flat, fertile Red River valley, with its neat geometric pattern of fields of wheat, potatoes, and sugar beets. In a few minutes, we reach the Bois de Sioux River, which forms the boundary with North Dakota.

From just this one transect of Minnesota, we can see the diversity of topography, vegetation, and wildlife in the state. If we were to continue flying, we would become more aware of the varied nature of the landscape, a result of geologic history, within the state's 84,068 square miles. Although most of Minnesota's topography is characterized by gently rolling hills, the northwestern corner contains

**Figure 1.5** *The combination of numerous small, shallow and large, deep prairie wetlands provides ideal habitat for ducks and other wildlife. (Photo by Bob Firth, Firth PhotoBank, Minneapolis.)*

broad expanses of very flat land, and the northeast contains rugged hills. These hills drop to the shoreline of Lake Superior, which gives Minnesota a coastline. Water is a prominent feature of the landscape throughout the state. Lakes and ponds number over 15,000. Wetlands, including thousands of acres of peatlands, occur in most areas. Three of North America's major drainage systems arise within the state, containing a variety of streams and rivers.

The diversity of ecological communities is especially apparent from the dramatic changes in vegetation that occur across the state (figure 1.6). The location of Minnesota near the center of North America results in a climate characterized by extremes of high and low temperatures and moder-

ate precipitation. This combination of temperature and moisture determined the natural pattern of vegetation in the state present before European settlement and still present, although in altered form, today. Tallgrass prairie with its rolling waves of grasses and myriad flowers covered the southern and western parts of Minnesota. In the southeast and extending in a narrow corridor northwestward nearly to the Canadian border were deciduous forests composed mainly of sugar maple, basswood, and elm. In the northeastern third of the state, forests were a mosaic of white pine, red pine, spruce, fir, aspen, and birch (figure 1.1).

These three vegetation types are part of large areas of central North America. The spruce-fir forests extend northward and eastward to the tree line at the tundra border and to the Atlantic Ocean in the northeast. The deciduous forests extend eastward from Minnesota to the Atlantic Ocean. Tallgrass prairies extend into the Dakotas, south into Kansas and Oklahoma, and southeast as far as Illinois and Indiana. These major ecological regions are called biomes (figure 1.6). The presence of three biomes in one nonmountainous state is unusual and accounts for the diversity of ecological communities.

These ecological communities and the landscape in which they occur are a result of thousands of years of interactions between geology, soils, and climate. Thus, we begin our study of the state's ecology with the story of Minnesota's geology, as summarized from Ojakangas and Matsch (1982).

## Geologic history

Since the beginning of planet Earth, powerful forces of mountain building and erosion, movement of tectonic plates deep beneath the surface of the earth, flooding by continental seas, and a continuing saga of bitter cold, blazing heat, howling winds, and torrential rains have all contributed to what we know today as Minnesota. Although this history is very complex, a simplified story can be told using the geologic time scale as a guide (table

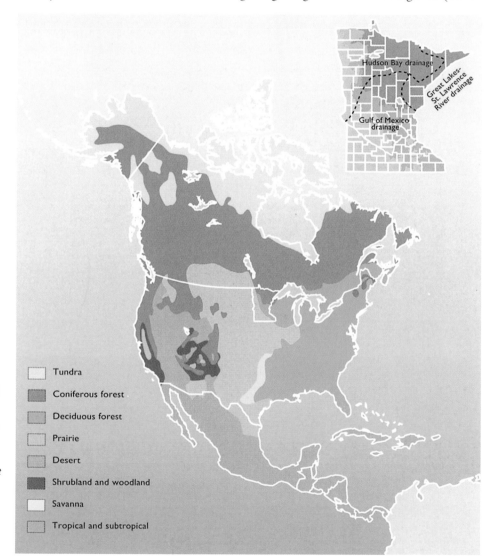

**Figure 1.6** *Major ecological regions of North America. Climate, soil, and topography, or landform, determine the location of the ecological regions, called biomes. The fact that three of the largest biomes meet in Minnesota makes the ecology of our state especially interesting. (Modified from Vankat 1979, by permission.)*

Tundra

Coniferous forest

Deciduous forest

Prairie

Desert

Shrubland and woodland

Savanna

Tropical and subtropical

7

I.I). Four eras—Precambrian, Paleozoic, Mesozoic, and Cenozoic—make up the scale of geologic time, dating from the origin of the earth about 4.5 billion years ago to the present.

Most of our knowledge about the ancient history of our planet comes from studies of the layers of rocks that make up the earth's crust, with the oldest rocks on the bottom and the youngest on top. The most precise measures of the real age of the materials making up the crust are based on measurements of the radioactive decay of isotopes of uranium, thorium, potassium, carbon, and other elements. Fossils, the remains of plants and animals embedded in the rocks, aid in determining age and in interpreting what the environment was like in prehistoric times.

Nearly 90% of the earth's past, from about 4.5 billion years to 600 million years ago, is included in the Precambrian era, when the earth and the first continents were formed. At about 2.7 billion years

ago and again at about 1.8 billion years ago, the area that was to become Minnesota had sizable mountain ranges (Ojakangas and Matsch 1982; Schwartz and Theil 1954). About 1.1 billion years ago, volcanoes and lava flows were common in Minnesota. These mountain ranges and volcanic cones have long since been eroded, and evidence of their existence, the massive deposit of basalt, is only available in the deeply buried rocks.

In a few places, Precambrian rocks can be observed on the landscape. Morton Gneiss, the oldest type of rock in Minnesota and perhaps the oldest rock on the planet, is visible on a hill in the Minnesota River valley northwest of Morton and at other sites near Granite Falls. This rock is estimated to be about 3.5 billion years old. A massive layer of rock about 1.0 billion years younger, called the Canadian Shield, underlies much of northeastern North America. The southwestern part of the shield is exposed in northeastern Minnesota near Lake Superior and in the Boundary Waters Canoe Area Wilderness, where the rocks provide an exceptional view of ages long past (figure 1.7). Because of their great age and potential to provide information about the formation of our planet, these rocks have been extensively studied by geologists from throughout the world.

Following long periods of erosion, which resulted in the leveling of the mountain ranges and volcanoes, large seas repeatedly covered central North America. The presence of these seas is documented in Minnesota by the flat beds of limestone, sandstone, and shale containing shark teeth and fossils of long-extinct marine invertebrates, such as trilobites, corals, and sponges.

About 2.0 billion years ago, when most of Minnesota was covered by a shallow sea, iron-bearing sediments combined with oxygen to form extensive deposits of iron ore. These deposits, located in northeastern Minnesota, have been mined and have thus been important to the state's economy. Glaciation occurred during the Precambrian era, and ice probably covered all of Minnesota. Scraping and scouring by the ice combined with the erosional

**Table 1.1** *Eras representing geologic time from the beginning of planet Earth to the present*

| Era | Years ago | Characteristics and events |
|---|---|---|
| Cenozoic | 65 million to present | present landscape established<br>Age of Mammals<br>Rocky Mountains uplifted<br>glaciation |
| Mesozoic | 225 to 65 million | continental seas<br>Age of Dinosaurs<br>first birds and mammals<br>advanced flowering plants |
| Paleozoic | 600 to 225 million | continental seas<br>subtropical and tropical vegetation<br>coal formation |
| Precambrian | 4.5 billion to 600 million | earliest life developed in the oceans<br>glaciation<br>mountain building<br>iron deposited<br>volcanism<br>formation of oceans and continents<br>formation of planet Earth |

*Source:* Modified from Ojakangas and Matsch (1982).

8

**Figure 1.7** *Cracks in the rock surfaces of the Canadian Shield in northeastern Minnesota hold just enough soil for a few trees to grow. These rocks are about 2.5 billion years old. (Photo by Bob Firth, Firth PhotoBank, Minneapolis.)*

forces of wind and water leveled any remnants of Minnesota's mountains.

During the Paleozoic era, from 600 to 225 million years ago, continental seas again flooded the interior of North America. These shallow seas repeatedly receded and flooded over much of Minnesota, depositing layers of sediment each time they flooded. Barn Bluff in Goodhue County provides an excellent example of layered limestone and sandstone representing the advances and retreats of the Paleozoic seas. Limestone layers were deposited in deep, calm seas, whereas sandstone layers were deposited in shallow, turbulent seas (Sansome 1983).

Reconstruction of the positions of ancient continents reveals that Minnesota was located very close to the equator during the Paleozoic era. Over millions of years, the movement of the underlying plates within the earth gradually shifted the continents to their present positions. Thus, during the Paleozoic, Minnesota had a subtropical or tropical

climate, and lush vegetation grew in forests and marshes. In the early Mesozoic, which began about 225 million years ago and continued to about 65 million years ago, the equator still cut through Minnesota. But North America was moving slowly northward, and gradual climatic cooling occurred. This period is often called the Age of Dinosaurs. The first birds and mammals and many kinds of flowering plants also developed during this era. Seas invaded Minnesota several times during the Mesozoic, most recently about 100 million years ago. These warm and shallow waters probably created subtropical conditions over much of North America. Although no dinosaur remains have been located in Minnesota, shark teeth and the skull of a marine crocodile provide proof of the existence of tropical seas long ago (Ojakangas and Matsch 1982).

As the Mesozoic seas slowly subsided, the earth's crust started to uplift, resulting in forma-

**Figure 1.8** *Glaciers, like this one in the Beartooth Mountains in Montana, some probably thousands of feet thick, covered Minnesota during the Great Ice Age. Deposits of rock and other material carried by the ice formed moraines along the margins of the glacier. (Photo by Richard Hamilton Smith.)*

tion of the Rocky Mountains. This event marks the beginning of the next era, the Cenozoic. Little evidence from the early Cenozoic has been discovered in Minnesota, but it is probable that a slow uplift of the land and erosion of the older rocks continued under a moderate but cooling climate. During the Cenozoic, from about 65 million years ago to the present, the continents continued to move slowly toward their present locations. Erosion and weathering continued, and the diversity of plants and animals increased.

### Glaciation

During the last two million years of the Cenozoic, the climate became colder and colder, and the Great Ice Age, known as the Quaternary period, began. Glaciers (figure 1.8) formed in the polar regions and began to flow toward the equator. Glaciers form when the snow that accumulates in a given area does not melt during the summer. Each year the snow layer becomes thicker, eventually compressing the bottom layers into ice. If the layers of snow continue to accumulate, their weight exerts pressure on the ice, causing it to flow away from the places of greatest pressure. Thus, glaciers are simply large masses of ice that move under the control of gravity and topography. Glacier margins expand as snow accumulation continues. The edges of glaciers recede through evaporation and melting when snowfall decreases. Glacial ice does not move backward but appears to do so as the ice edge melts.

During the Great Ice Age, Minnesota experienced the effects of four major periods of glaciation. Each of these periods of glaciation lasted for thousands of years, and the warmer interglacial periods had similar durations. During the first three advances, the ice mass, which was centered near Hudson Bay, advanced to the south as far as Illinois, Missouri, and Kansas, probably completely covering Minnesota. Most of the features formed by the glaciers during these three advances are either buried beneath more recent glacial deposits or have been eroded away.

We know that the lobes of ice advanced and retreated in a complex pattern in many parts of Minnesota because the glacial till, the material scraped up and carried by the ice, was deposited in distinct layers. The characteristics of each layer reflect the path over which the lobe of ice moved. A cross section of the glacial deposits along Hawk Creek in Renville County shows layers deposited by four different glacial advances from different source areas. During the intervals between these advances, which lasted for thousands of years, soils developed on the uplands, lakes were present in depressions, and erosion was common, as shown by the presence of buried soils, lake sediments, and a stone line. These buried remnants of the past are archives documenting part of Minnesota's geologic history.

The most recent advance, referred to as the Wisconsin Glaciation, covered all of Minnesota except for small areas in the southeastern and southwestern corners. During the Wisconsin Glaciation, which began about 75,000 years ago, lobes of ice pushed into Minnesota from a number of directions. These lobes were often channeled or directed by lowlands, such as the Lake Superior basin or those associated with the Red River along the Minnesota-Dakota border. The maximum extent of the ice during the Wisconsin Glaciation is shown in figure 1.9.

As the lobes or fingers of ice advanced and melted, they left various land forms, such as moraines, drumlins, eskers, and a variety of lakes, streams, and wetlands (figure 1.10). Most of Minnesota's hills and valleys, prominent features of the landscape, are due to moraines. Large moraines,

such as the Itasca Moraine in Becker and Hubbard Counties, formed when the edge of the glacier remained in about the same location for a long period of time. In these locations the ice acted like a conveyor belt, carrying rock, sand, and clay scooped up by the ice and depositing them at the edge of the glacial front as the moving ice melted.

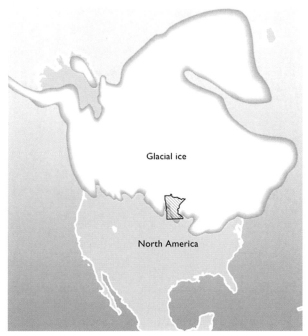

**Figure 1.9** *Maximum extent of the ice coverage in North America during the last glacial advance. (Modified from Ojakangas and Matsch 1982, by permission.)*

Prominent moraines in Minnesota are identified by the names of towns or other geographic features. The St. Croix Moraine, on which St. Paul and Minneapolis are situated, is an example of a moraine formed at the front of a glacier, and the Alexandria Moraine, which extends from Douglas County northward into Mahnomen County, is a good example of a lateral moraine formed along the glacier's edge.

Movement of the glacial ice was unrelenting and powerful. As the ice advanced, it scraped and scoured the landscape, destroying all vegetation, pulverizing softer rocks into silt and sand, and carrying boulders larger than houses. The rocks over which the ice moved were often scratched by

**Figure 1.10** *Landforms characteristic of glacier margins. Water from melting ice formed lakes and rivers, and deposits of till formed moraines, drumlins, eskers, kames, and other landforms. (Modified from Ojakangas and Matsch 1982, by permission.)*

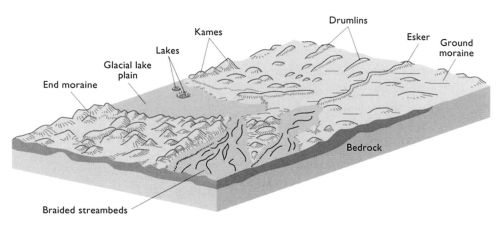

the rocks frozen in the ice, like grit in sandpaper (figure 1.11). The direction of the scratches indicates the direction of movement of the ice. Meltwater from the glaciers formed rivers under the ice, which eroded tunnel valleys beneath the ice sheet. The water carried fine particles of clay, silt, and sand away from the ice front and dropped them in large, flat, sometimes fan-shaped deposits known as outwash plains. Scouring of the land by the ice and melting of blocks of ice buried in the glacial till caused depressions in the landscape in which lakes and wetlands later formed.

During the final retreat of the Wisconsin Glacier, about 10,000 to 12,000 years ago, the ice mass in Canada prevented normal drainage of the Red River through Hudson Bay and into the Atlantic Ocean. The Continental Divide in western Minnesota prevented the meltwater from draining southeastward into the Mississippi River. As a result, a huge lake called Glacial Lake Agassiz formed along the Minnesota-Dakota border and

extended eastward and northward many hundreds of miles (figure 1.12). Lake Agassiz fluctuated in size and depth, covering a total area of about 123,500 square miles and having a maximum depth of nearly 400 feet (Elson 1983).

When the lake level was stable, beaches of sand and gravel developed along the shoreline, except on the northern side, which was glacial ice. These beach lines can be seen today as ridges of sand and gravel, often broken by gravel pits, running northward through the nearly flat Red River valley.

Lake Agassiz eventually rose high enough that it began draining over the top of a moraine dam near Browns Valley on the Minnesota–South Dakota border. The Glacial River Warren, which formed from this drainage, carved what is now the valley of the Minnesota River. The huge volume of water, filling the valley from bank to bank, must have quickly lowered the water level in the lake. At the same time that Glacial Lake Agassiz was draining through Glacial River Warren, the ice mass was receding northward. Finally the natural drainages through Lake Superior and Hudson Bay opened, and Glacial Lake Agassiz began draining to the east and north. Glacial River Warren ceased to exist, and today the Minnesota River winds its way through the valley following the path of the once mighty glacial river.

During the final stages of the Wisconsin Glaciation, when the ice mass was located in the western region of the state, a lobe of ice, known as the Grantsburg Sublobe, began moving in a northeasterly direction. This ice lobe crossed the Mississippi River over a broad area between Minneapolis and St. Cloud. The waters in the Mississippi were forced to flow northeastward along the edge of the ice. Eventually the ice pushed into Wisconsin near Grantsburg, and the Mississippi flowed around the tip of the ice and then back to the southwest, eventually reaching the St. Croix River valley and flowing southward through the St. Croix River to join the original Mississippi River. As the Grantsburg Sublobe melted, the waters of the Mississippi continued to flow across its tip, washing small particles from the till and leaving a deposit of fine uniform sand. Eventually, the Grantsburg Sublobe receded to the southwest of the old Mississippi River channel, and the river again resumed its old course, draining through the area now occupied by the Twin Cities. The deposit of fine sand, known as the Anoka Sand Plain, covers an area approximately 60

**Figure I.11** *As the glacier moved, rocks frozen in the glacial ice cut the grooves and scratches in this slab of the Canadian Shield in the Boundary Waters Canoe Area Wilderness. (Photo by Blacklock Nature Photography.)*

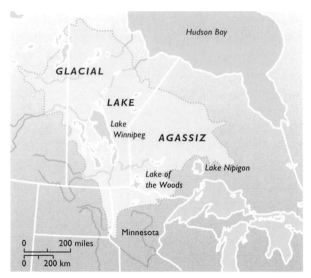

**Figure 1.12** *The total area covered by Glacial Lake Agassiz was about 123,500 square miles. The lake was never this large at any one time, however, because its borders changed as the ice margin changed. (Modified from Elson 1983, by permission.)*

miles from west to east and 50 miles from north to south and shows as outwash plain on figure 1.13 (Cooper 1935).

Glaciation sculpted the surface of Minnesota into a diverse landscape (figure 1.13), characterized mainly by glacial moraines and hills, by flat glacial lake beds, and by the presence of many rivers, lakes, and wetlands. The Arrowhead region in the northeast has the state's lowest and highest elevations and, therefore, its greatest relief (figure 1.14). Lake Superior's shore, the lowest elevation in Minnesota, is at 602 feet above sea level, and Eagle Mountain, the highest point, is at 2,301 feet. Much of the land in the northeast is rocky and ice-scoured. Beautiful lakes have formed in the many troughs in the rock formations (figure 1.15). Northwestern Minnesota is characterized by the low-lying, flat topography of the basin of Glacial Lake Agassiz. To the south of Upper and Lower Red Lakes, glacial moraines are prominent features of the landscape, continuing in an irregular pattern all the way to the Iowa border (figure 1.13). In the southwest and extending into South Dakota, bedrock forms a wedge-shaped plateau known as the Coteau des Prairies. The southeast corner of the state was not covered by the Wisconsin Glaciation. This area is characterized by a surface layer of wind-deposited tiny uniform particles, called loess, which covers the older till and bedrock. The deposits in the southeast have been deeply eroded, giving the landscape a hill and valley aspect (figure 1.16).

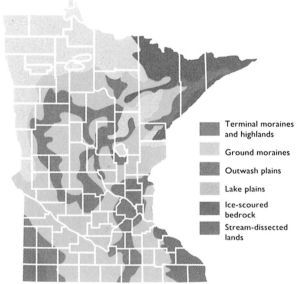

Terminal moraines and highlands

Ground moraines

Outwash plains

Lake plains

Ice-scoured bedrock

Stream-dissected lands

**Figure 1.13** *The major landforms in Minnesota were formed or influenced by glacial activity. (Modified from Wright 1972, by permission.)*

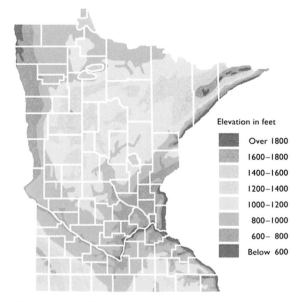

Elevation in feet

Over 1800

1600–1800

1400–1600

1200–1400

1000–1200

800–1000

600– 800

Below 600

**Figure 1.14** *Elevation, or topographic relief, in Minnesota. (Modified from Borchert and Yaeger 1968, by permission.)*

**Figure 1.15** *The landscape in northeastern Minnesota is characterized by mixed coniferous forests, steep rock ridges, and many lakes and rivers. (Photo by Richard Hamilton Smith.)*

**Figure 1.16** *The landscape in southeastern Minnesota is characterized by deciduous forests and cropland, gently rolling hills, and eroded valleys. (Photo by Blacklock Nature Photography.)*

## Vegetation and soil development since glaciation

Ecologists have been especially interested in the history of Minnesota since the final retreat of the ice sheet, because it is during this time that the present landscape developed. The major landforms left by glaciation were continually acted upon by wind and water, the primary forces of erosion. On these landforms, vegetation and soils developed under the influence of climate.

### Vegetation

No written history exists for most of the last 10,000 years. Scientists have therefore developed a variety of methods to determine what the landscape of Minnesota was like from the time the glaciers receded until the availability of written records,

which began about 200 years ago. One of these methods involves the study of the plant fragments and pollen grains that accumulated in the sediments of lakes and peatlands. How these remains serve as archives of the past is a fascinating but little known aspect of the study of Minnesota history.

Immediately after the last ice receded, the surface of most of Minnesota was barren of vegetation and had many streams formed by the water from melting ice. Depressions in the moraines filled with water to form lakes. Imagine standing at the edge of a newly formed lake in central Minnesota about 10,000 years ago. The shoreline, and as far as one can see, is bare. No plants or animals are present. In all probability the water is crystal clear, and the bottom of the lake is covered with sand or gravel. The temperature is cold, because the receding glacier is only a short distance away. Within a very short time, probably a few months or years, the first plants invade this barren environment. Tiny plants called algae, carried by the wind or on the feet of birds, are among the first living things to grow in the lake. On the upland surrounding the lake, light, fluffy seeds carried by winds from places beyond the farthest reaches of the glaciers become the first plants to grow on the till.

As plants grow on land and reach the flowering stage, they produce pollen, some of which is carried by wind or runoff to the lake. Most of these pollen grains became waterlogged and settle to the bottom. Sand and silt also wash into the lake and settle to the bottom. At the end of the first growing season, a thin layer of sediment composed of silt, sand, pollen, and perhaps small bits of stems or leaves exists on the bottom of the lake. Each year an additional layer is added to the sediments, a process that is still occurring today. As the vegetation surrounding the lake develops and as the plant species change, the kinds of pollen and plant fragments that accumulate in the sediments also change. It is this record in the layers of lake sediments that ecologists use today to interpret the history of vegetation following glaciation in Minnesota.

To obtain the pollen record, as it is called, a core several inches in diameter is extracted from the lake sediments. Usually ecologists obtain cores through the ice in winter so that they have a solid place to stand while pulling up the sampling tube containing the core. In shallow water, a framework may be used to obtain the core (figure 1.17). Layers in the core are frequently visible when the core is examined because of the presence of insect parts, charred wood, ashes, clumps of algae, and the varying nature of the material comprising the sediment (figure 1.18). Small samples are then taken throughout the length of the core. Pollen grains present in these samples are examined under a microscope after they have been concentrated by a chemical treatment.

Pollen grains, although only a few thousandths of an inch in diameter, are characteristic of certain

**Figure 1.17** *A core containing pollen grains is extracted from the bottom of a lake or wetland by pushing a sampling tube through the sediment until it hits the sand and gravel at the bottom. The tube is then pulled up, and the core is extracted and carefully preserved. (Photo courtesy of H. E. Wright Jr.)*

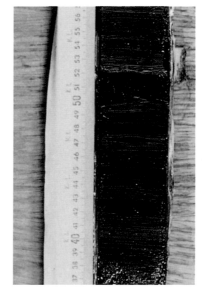

**Figure 1.18** *This sediment core, taken from Elk Lake in Clearwater County, shows layers, or varves, representing seasonal changes in the pattern of deposition (scale in centimeters). (Photo courtesy of Robert Megard.)*

plant species or at least groups of species, called genera (figure 1.19). For example, it is possible to differentiate white pine from red pine pollen, but it is not possible to differentiate pollen of bur oak from white oak or red oak. The oak pollen grains, however, are easily distinguished from pollen of pine or other genera. After the pollen grains are identified, a diagram, like the one shown in figure 1.20, is produced by graphing the relative abundance of different kinds of pollen grains. This diagram can then be interpreted to provide the history of the vegetation surrounding the lake or pond where the core was obtained.

Pollen data were obtained by John McAndrews (1966) from Bog D Pond in Itasca State Park in Hubbard County . On the left side of the diagram (figure 1.20) are numbers representing the depth from which samples were taken from the core. Note that the core was about 26 feet long. The date at the bottom of the right side of the diagram reads 11,000 ± 90 years. Thus, the sediments in the pond accumulated to a depth of 26 feet during the past 11,000 years.

Note also that other dates appear at several places along the right margin. These dates were obtained through use of a technique known as radiocarbon dating, another tool used to determine the history of vegetation. Growing plants incorporate carbon in their tissues through the process of photosynthesis. A known proportion of this carbon is the naturally occurring radioactive $C_{14}$, which has always been present in the atmosphere. The ratio of normal carbon, $C_{12}$, to $C_{14}$ has remained relatively constant over time. From the moment a plant dies, no more carbon is added to its tissues, and the $C_{14}$ begins to disintegrate at a known rate. Thus, the older a sample is, the less $C_{14}$ it contains and the greater is the ratio of $C_{12}$ to $C_{14}$.

Using this ratio, scientists are able to determine the age of plant fragments incorporated into the lake sediments. The best materials for radiocarbon dating are fragments of wood that have remained well preserved in the sediments. McAndrews obtained radiocarbon dates for four such samples from the core from Bog D Pond.

Now let us look at the section labeled total pollen on the diagram and examine in a general way the nature of the vegetation around the pond. Tree pollen comes from trees and shrubs, and herb pollen comes from grasses and other herbs. The full width of total pollen in the diagram represents 100% of the pollen grains identified and counted. Note that at the bottom of the diagram the percentage of tree pollen is increasing. Most likely few

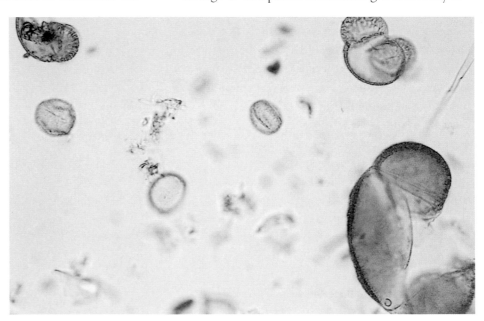

**Figure 1.19** *Microscopic view of pollen grains of pine (upper right), fir (lower right), oak (right center), and elm (lower left). (Photo courtesy of E. J. Cushing.)*

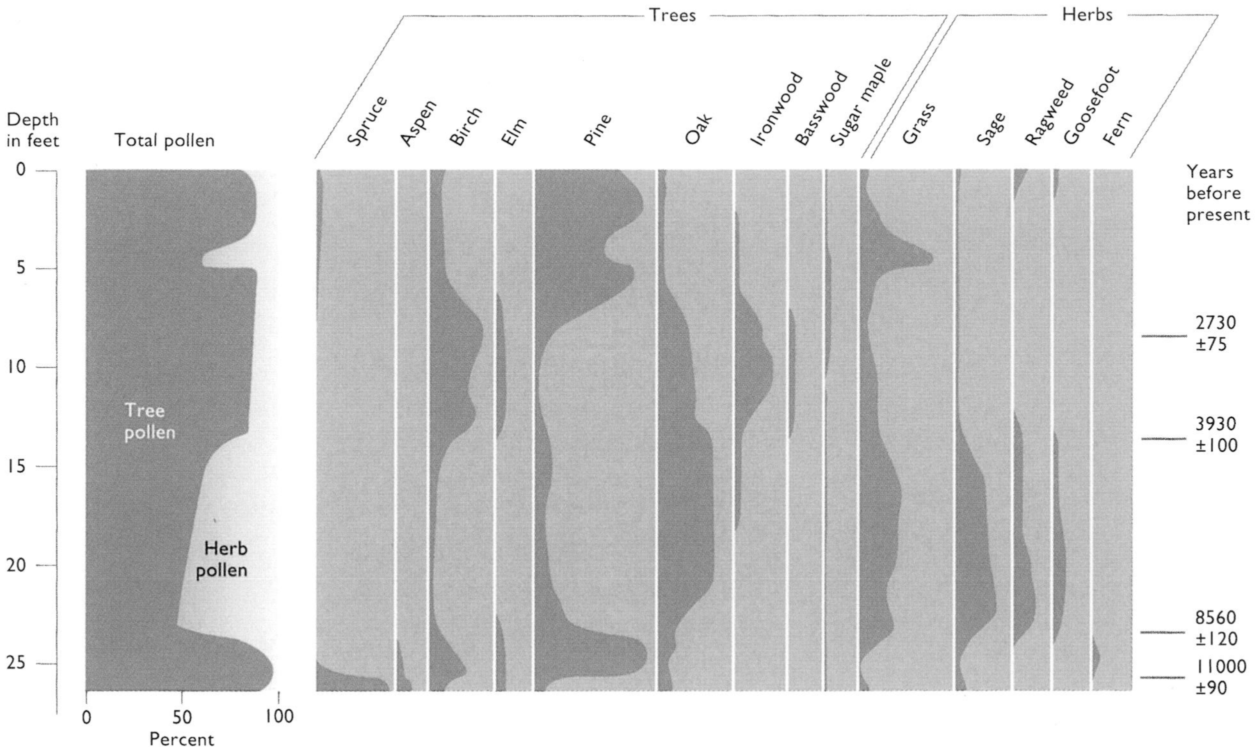

**Figure 1.20** *This pollen diagram from Bog D Pond in Hubbard County shows the percentage of pollen for each plant species or group plotted against depth in the sediment core. Time before present is shown by the radiocarbon dates on the right. (Modified from McAndrews 1966, by permission.)*

trees were present in the vegetation around the pond during the first years after glaciation. Then the percentage of tree pollen increases, indicating that the pond was surrounded by forests. The diagram shows that these forests existed for several thousand years. About 8,000 years ago the amount of tree pollen decreases, indicating the presence of fewer trees and more grasses and other herbs. We interpret this to mean that the vegetation around the pond changed to a savanna, defined as a habitat with scattered trees and large open areas of prairie vegetation.

Ecologists generally believe that this change from forest to savanna was caused by a major change in the earth's climate. Immediately following glaciation the climate was cool and moist. For unexplained reasons, probably related to the position of the earth and the sun, the climate became warmer and drier. This change led to the death of many of the forest trees and to an increase in prairie vegetation. Studies of pollen records from other lakes in Minnesota indicate that this warm, dry climate probably varied considerably, from years of drought to years of abundant precipitation. This variability was discovered from studies of water levels in lakes in west-central Minnesota. Some lakes may have fluctuated as much as 20 feet in depth during this warm, dry period (Almendinger 1988). How much warmer was this climate? Recent investigations using pollen records and climate models suggest that the temperature was about 2° to 4°F warmer than today (Webb, Bartlein, and Kutzbach 1987).

Several thousand years later the climate again changed, becoming cooler and wetter. This change led to expansion of the forest and a decrease in the grasses and herbs. This condition has lasted up to the present time.

The right side of the pollen diagram provides more detail by showing the percentage of pollen

**Figure 1.21** *This sparse stand of black spruce, with its mosses and low shrubs, is probably similar in appearance to the vegetation that covered much of Minnesota shortly after the glaciers disappeared from the state. (Photo by Richard Hamilton Smith.)*

grains of each species or closely related group of species. Not all of the plant species that were believed to have been present are shown in the pollen diagram. Only those plants that were abundant in the surrounding area and whose pollen is well preserved are illustrated. The width of the green bar indicates the abundance of that particular kind of pollen over the 11,000-year time interval. The first vegetation following the retreat of the ice was probably somewhat like that in communities composed mainly of mosses and sedges that exist today in northern Canada (figure 1.21). This community was followed by a forest of spruce and fir, and then of pine, which existed until the beginning of the warm, dry period. Oaks then became the predominant tree, and prairie grasses and herbs were abundant. As the climate became cooler and wetter, both red and white pine increased in abundance while the prairie species decreased. The oak savanna was thus replaced by a forest with many red and white

pines and, of course, a variety of other trees, shrubs, and herbs. Pine forests existed for hundreds of years around the pond, and today an increase in hardwoods such as elm, basswood, and sugar maple is occurring, although the increase does not show in the pollen diagram. Forests around the pond today are a mixture of pines and hardwoods.

The top layers in the pollen diagram represent deposits made during recent years. The column in the herbs section of the diagram labeled ragweed indicates human activity. Ragweed commonly grows on disturbed areas such as fields and roadsides. The first agriculture and road construction probably occurred in the area around Itasca State Park in the late 1800s, and the increase in ragweed pollen shown at the top of the pollen diagram clearly reveals this activity.

The history of vegetation in other parts of Minnesota was very likely similar to that just described. Studies show that the sequence of tundra vegetation followed by coniferous forests followed by an expansion of prairie and savanna eastward occurred over much of the state. The location of the border between forest and prairie shifted along its northwest to southeast axis at various times during the period from 9,000 to 500 years ago. The areas in the south and west were probably somewhat drier and warmer than the areas in the north and east, and the vegetation in these areas would have therefore reflected these differences in climate.

In north-central Minnesota, at the eastern side of the basin of Glacial Lake Agassiz, drainage was poor. Here, high water tables inhibited the decomposition of dead vegetation, and peat accumulated. The extensive peatland that formed north and east

of Upper and Lower Red Lakes is one of the largest continuous peat deposits in the world. This area is marked by stringlike patterns of vegetation that reflect the slow movement of water along the gently sloping peat surface (Heinselman 1970; Wright, Coffin, and Aaseng 1992). From studies of peat cores similar to the lake sediment cores described earlier, H. E. Wright Jr. (1972) estimated that the peatland began to form about 3,000 years ago and that the depth of the peat averages about 10 feet.

### Soils

After the retreat of the glaciers, till, rock fragments in an unsorted matrix of sand and finer particles deposited on the land by glaciers, covered nearly all

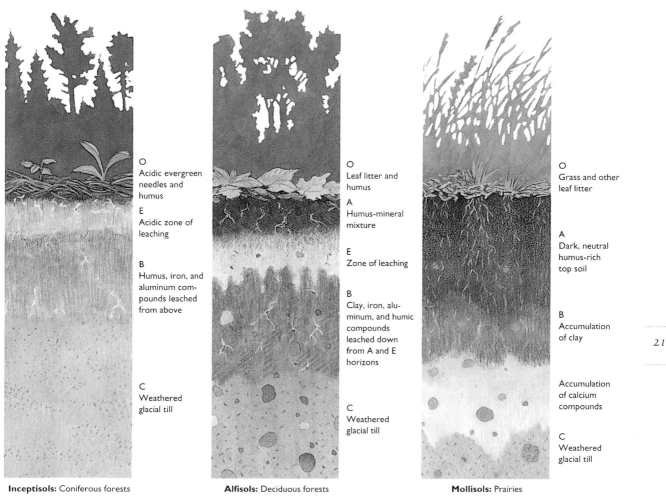

O
Acidic evergreen needles and humus

E
Acidic zone of leaching

B
Humus, iron, and aluminum compounds leached from above

C
Weathered glacial till

**Inceptisols:** Coniferous forests

O
Leaf litter and humus

A
Humus-mineral mixture

E
Zone of leaching

B
Clay, iron, aluminum, and humic compounds leached down from A and E horizons

C
Weathered glacial till

**Alfisols:** Deciduous forests

O
Grass and other leaf litter

A
Dark, neutral humus-rich top soil

B
Accumulation of clay

Accumulation of calcium compounds

C
Weathered glacial till

**Mollisols:** Prairies

**Figure 1.22** *Profiles of soils found in the three major biomes in Minnesota. Texture, color, thickness, and composition of the horizons, identified by letters, are used in the soil classification system. (Illustration by Don Luce.)*

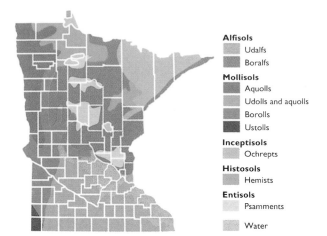

**Figure 1.23** *Orders and suborders of Minnesota soils (Data from Anderson and Grigal 1984; D. Grigal and R. Rust, personal communication.)*

of Minnesota. Since its deposition, this till has been modified by freezing and thawing, by chemical weathering, and by accumulation of organic matter from plants and animals. The results of these actions are the soils that we see today.

Particles of clay, silt, and sand from the till and loess, called parent material, living organisms, organic matter from dead plants and animals, water, and mineral nutrients make up the mixture we call soil. Most soils are arranged in layers, called horizons, each with its characteristic composition, texture, and color (figure 1.22). Horizons differ depending on the history of the formation of the soil, and few soils have all horizons. The soil profiles shown in figure 1.22 illustrate differences in the horizons of the soils in the northern coniferous forest, the deciduous forest, and the tallgrass prairie regions of Minnesota.

Topography influences soil formation in large part through its effect on water movement. For example, small particles are carried down hillsides in runoff, leaving the larger sand and gravel particles on the hilltops. The small particles form deep, fine-grained soils in valleys and depressions.

Climate also influences soil formation, both through direct actions such as freezing, thawing, and leaching and indirectly through its effects on vegetation. For example, the warm, dry climate in

southern and western Minnesota favored the growth of prairie. The dense root systems in the prairie sod and the litter produced each year by the grasses and other plants added much organic matter to the soil, making it rich and black. Nutrients from decomposition remained in the soil because precipitation was low and little leaching occurred.

The wide variety of initial conditions that resulted from Minnesota's geologic and climatic history have resulted in soils that are highly variable and have a complex distribution pattern (figure 1.23). The many kinds of soils are classified based on characteristics such as particle size, color, depth to bedrock, carbon and mineral content, and the presence of hardpans and other horizons (Soil Survey Staff 1975). Orders, the highest level of classification, will be described in the following general information about Minnesota soils. Suborders, the next lower level of classification, are shown in figure 1.23 and are described in J. L. Anderson and D. F. Grigal (1984).

The process of soil formation can be described from the standpoint of the type of vegetation that is growing in a region. We will examine the soils in Minnesota with reference to the four major vegetation types: northern coniferous forest, deciduous forest, prairie, and wetlands.

Most soils in the northeast belong to the order Inceptisols. These light-colored to brownish soils formed slowly, probably because of the rocky nature of the parent material, glacial till, combined with cool temperatures. Low nutrients and cool temperatures resulted in slow growth of vegetation. Most are low in lime and contain many boulders and little clay. Organic matter is sparse in the upper layer. Because Inceptisols are immature soils, they have weakly developed horizons and closely resemble the parent material.

In areas with a warmer, drier climate, deciduous forests predominated and influenced soil formation. The leaves of many deciduous trees, especially birch, maple, and basswood, are rich in calcium and other bases. These bases are quickly returned to the soil as the leaf litter decomposes. Downward per-

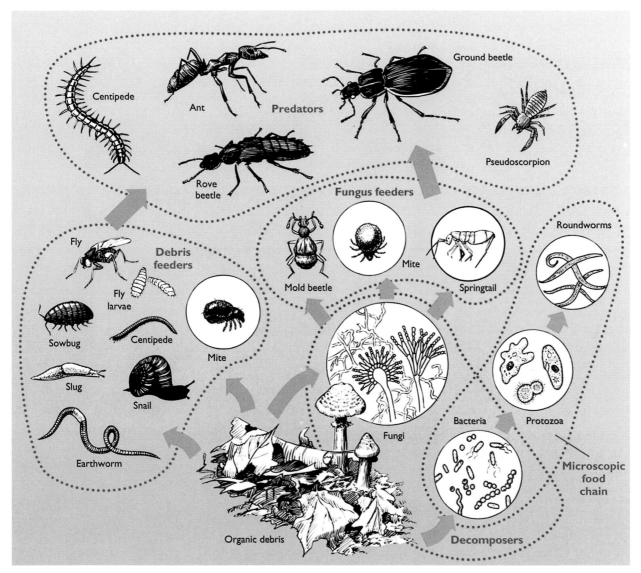

**Figure 1.24** *The many kinds of organisms living in a soil influence its fertility and texture. Organisms within circles are microscopic in size. (Modified from Miller 1992, by permission.)*

colation is much slower because of reduced infiltration. As a result, the upper layers of the soil become enriched rather than depleted. Forest soils produced by this process belong to the order Alfisols, whose name reflects the accumulation of aluminum (Al) and iron (Fe). These fertile soils are usually formed in loam or clay and have a gray to brown color near the surface. In the north, wet soils in this order support forests of aspen, black ash, and alder. In the north-central counties, aspen and red and white pine occur on drier sites. In the

southeast, sugar maple, basswood, elm, oak, and hickory are the common trees.

The prairies of southern and western Minnesota produced still other kinds of soils belonging to the order Mollisols. Calcification, an intense accumulation of bases, takes place in the upper layers of soil under grassland vegetation, and the decomposition of the fibrous roots of grasses results in a large amount of organic matter and nutrients being incorporated into the soil. The rich, black layer of the prairie soil may be from 6 to 40

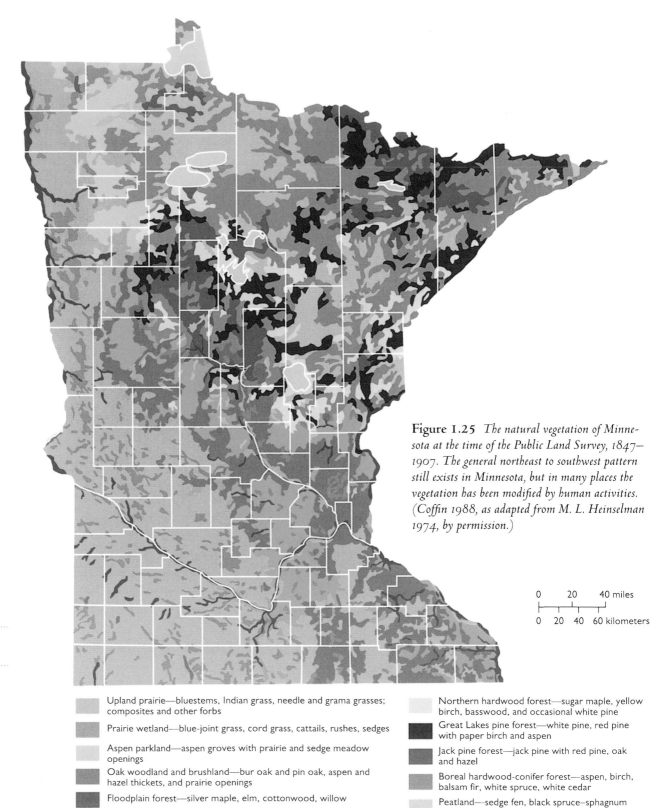

**Figure 1.25** *The natural vegetation of Minne-sota at the time of the Public Land Survey, 1847–1907. The general northeast to southwest pattern still exists in Minnesota, but in many places the vegetation has been modified by human activities. (Coffin 1988, as adapted from M. L. Heinselman 1974, by permission.)*

0    20    40 miles

0    20    40    60 kilometers

Upland prairie—bluestems, Indian grass, needle and grama grasses; composites and other forbs

Prairie wetland—blue-joint grass, cord grass, cattails, rushes, sedges

Aspen parkland—aspen groves with prairie and sedge meadow openings

Oak woodland and brushland—bur oak and pin oak, aspen and hazel thickets, and prairie openings

Floodplain forest—silver maple, elm, cottonwood, willow

Maple-basswood forest—elm, basswood, sugar maple, red oak, white oak

Northern hardwood forest—sugar maple, yellow birch, basswood, and occasional white pine

Great Lakes pine forest—white pine, red pine with paper birch and aspen

Jack pine forest—jack pine with red pine, oak and hazel

Boreal hardwood-conifer forest—aspen, birch, balsam fir, white spruce, white cedar

Peatland—sedge fen, black spruce–sphagnum bog, white cedar–black ash swamp inches thick. High nutrient levels and the excellent loose, soft texture of the prairie soil make it ideally suited for agriculture.

Two other soil orders, Entisols and Histosols, occur in Minnesota. Entisols are young soils such as those formed on river-deposited material or on sand, where development is extremely slow because of the weather-resistant nature of the parent material. Entisols occur on rock outcrops and steep slopes in the northeast and wherever sandy outwash deposits are found. These soils support jack pine or oak forests and dry prairie communities.

Histosols are formed in wet areas from the remains of plants. These organic soils are often called peat or muck. They occur throughout Minnesota but are most prevalent in the north. Peatlands are vegetated by sphagnum moss and shrubs in some areas and by black spruce and tamarack forest in other areas.

Soil provides us with a logical link between the physical factors of climate and geology and the biological world. We have already briefly mentioned the most conspicuous plant communities associated with Minnesota's soil orders. Less conspicuous, but also very important, are the animals associated with these plant communities and with the soil itself. The variety and number of organisms in the soil, including algae, fungi, bacteria, and nematodes are astonishingly high (figure 1.24). These organisms play a vital role in maintaining soil texture and fertility through their role in decomposition and nutrient cycling. Chapter 3, "Principles of Ecology," describes how these and other processes function in ecosystems.

## The present landscape

Minnesota's ecological communities are clearly evident in figure 1.25, an adaptation of the map titled "The Original Vegetation of Minnesota," which was compiled from the U. S. General Land Office survey notes by F. J. Marschner in 1930. Deciduous forests interspersed with brushlands,

oak savannas, and clearings occupy a narrow strip in the northwest and a broader region in the southeast. Since European settlement in the 1800s, however, much of the deciduous forest has been cleared for agriculture. Nearly all of the 3,000-square-mile Big Woods, which occupied the south-central part of the state, has been converted to farmland.

Northeastern Minnesota can be characterized as coniferous forest, a mosaic of stands of white spruce and balsam fir interspersed with stands of aspen and birch. Most of the original stands of white and red pine, however, have been logged and to a large extent have been replaced by aspen and birch forests. The peatlands are still relatively intact despite efforts in the early 1900s to ditch and drain large tracts north of the Red Lakes.

Southern and western Minnesota are now predominantly agricultural land. Less than 1% of the tallgrass prairie remains. Most of the prairie wetlands have been drained.

Despite these changes, many of the conditions that led to the development of Minnesota's ecological communities—the geologic setting, vegetation, soils, and climate (see chapter 2)—are present today. The variation in these conditions from the northeast to the southwest, to which I will refer often, still exists and is central to understanding the ecology of the state. The present and future landscape, whether the outcome of human activities or of nature's own design, is still bound by the topographic and climatic setting of the landscapes that occur in mid-continent North America.

Thus, the three biomes provide a framework for examining the ecology of the terrestrial communities in Minnesota in chapters 4, 5, and 6. Wetlands, lakes, and rivers occur throughout Minnesota and are greatly influenced by the ecology of the biome in which they are located or through which they flow. The ecology of these aquatic communities is discussed in chapters 7, 8, and 9. Finally, in chapter 10, I offer some thoughts about what the future may hold for the ecology of Minnesota.

25

# CLIMATE & WEATHER ❷

The Armistice Day blizzard of 1940, the tornadoes of 1965, extreme cold, ice storms (figure 2.1), hot summer drought—these are among the many topics Minnesotans mention when the weather is the subject of conversation. Indeed, memorable events and weather extremes might be considered trademarks of Minnesota weather. These events are often frightening. Forty-nine people and thousands of animals died in the Armistice Day blizzard in 1940. The tornadoes in Fridley in 1965 killed 14 people, injured 683, and caused $57 million in damage (Douglas 1990; Kuehnast 1972). In light of these extremes we may forget about the balmy spring days with cool breezes and fluffy clouds (figure 2.2) and the crisp, clear autumn nights, which are just as much a part of Minnesota's weather.

Day by day the weather influences our activities, our moods, our economy, and even our survival. Weather also influences the growth and survival of plants and animals. But weather, the atmospheric conditions at a given time, is only a manifestation of climate, which is the pattern of rain, snow, temperature, wind, radiation, and cloudiness over many years. Data about climate are often expressed in terms of monthly, seasonal, or yearly averages and extremes.

Past climatic conditions influenced the geologic history of Minnesota, the development of the soils

**Figure 2.1** *Vegetation coated with ice from freezing rain and sleet is beautiful but subject to extensive damage from the weight of the ice. (Photo by Bob Firth, Firth PhotoBank, Minneapolis.)*

**Figure 2.2** *Cumulus clouds rolling over western Minnesota form a soft background for the long, silky flowers of prairie smoke. (Photo by Richard Hamilton Smith.)*

and vegetation, and the distribution and abundance of living things. Today, averages of temperature and precipitation as well as extremes of heat, cold, wet, and dry continue to play a large role in determining the nature and location of Minnesota's ecological communities.

Minnesota's climate, characterized by warm, humid summers and cold, dry winters, is determined by the state's geographical location in North America, in particular by its location with respect to the north-south ranges of mountains in the West and to the low-relief plains extending from the Gulf of Mexico northward to the Canadian Arctic (Bryson and Hare 1974). These geographic features influence the movement of air masses over the state. The mountains obstruct the westerly

winds and, by forcing the moving air masses upward, cause cooling and precipitation in the mountains. The air masses that reach western Minnesota are therefore relatively dry. In contrast, the plains east of the Rocky Mountains allow the flow of warmer, more humid air from the Gulf of Mexico over southern Minnesota and the flow of cooler, drier air from the north over northern Minnesota. When the warmer air masses from the south are forced upward by denser polar air masses, the air cools and drops its moisture in the form of rain or snow. Air masses from the Pacific and from the northwest, guided eastward by the high-altitude jet stream, interrupt these periods of clouds and precipitation with calm days and bright sunshine.

Because of Minnesota's location, the state may

have the steepest spatial and temporal climatic gradients, that is, the greatest changes in the shortest distances and times, in North America except for mountainous regions. Average temperatures increase from northeast to southwest, and average precipitation increases from northwest to southeast. These climatic gradients, which will be discussed later, were a major factor in the development of Minnesota's diverse ecological communities. The prairies of northwestern Minnesota developed under the combination of cool temperatures and low precipitation, whereas the southwestern prairies developed under warm temperatures and low precipitation. The deciduous forests in the southeast developed under the combination of warm temperatures and high precipitation, and the northern coniferous forests developed under cool temperatures and high precipitation.

Before discussing these communities in the following chapters, let us first examine some details of Minnesota's climate and weather. The main source of this information is the series Climate of Minnesota by D. G. Baker and colleagues (1963, 1978, 1985) and a climatic summary by E. L. Kuehnast (1972).

## Solar radiation

Radiation from the sun provides the energy that heats the earth and "powers" every ecological system. Solar radiation does not heat the earth's surface evenly, a fact that is obvious from the experience of walking barefoot on a green lawn and then stepping onto a concrete sidewalk baking in the July sun. Such differential heating on a large scale makes air masses rise and fall at different rates, causing the wind to blow. The prevailing winds bring the different air masses over the state, influencing temperature and precipitation.

The amount of radiation reaching a given spot on the earth depends on the tilt of the earth's axis away from the sun. This tilt, or angle, determines the day length, the area over which a beam of radiation is spread when it reaches the earth's surface,

and the amount of atmosphere the radiation must pass through. For example, during a Minnesota winter the northern hemisphere is tilted away from the sun. As a result, the sun appears lower in the sky, the day length is short, the sun's rays are spread over a larger area, and the sun's rays pass through a thicker portion of the atmosphere than in summer, resulting in lower temperatures. In addition, more heat escapes through the atmosphere in winter than in summer because cold winter air holds less water vapor, which absorbs heat, than warm summer air.

Total solar radiation and net radiation, the difference between all incoming and outgoing radiation, are at the maximum in July and at the minimum in January in St. Paul. Net radiation is negative from November through February; in other words, no energy is available for heating the air, which of course leads to cooling and winter. During summer, the positive net radiation is dissipated through heating of the air and soil, by evaporation of water, and by photosynthesis (Baker and Crookston 1980).

A direct impact of solar radiation on ecological systems is evident in the explosion of flowers, such as trillium and bloodroot, in deciduous forests in early spring before leaves have developed on the shrubs and trees. These flowers complete their growth and seed production rapidly while shade is minimal and light intensity is high. A similar impact can be observed on burned prairies, where blackened soil surfaces warm quickly in the spring from direct solar radiation. This warming appears to stimulate prairie plants to grow earlier than on unburned prairies.

## Temperature

Minnesota's cold winters have long been noted as a distinguishing feature of the state. In 1855, Governor Gorman, in his state-of-the-territory message, said, "During the past year I have received almost innumerable letters from the middle states propounding a variety of questions about our territory, especially desiring to know if our winters are not very long, and so exceedingly cold that stock freezes to death, and man hardly dare venture out of his domicil" (as quoted in Lass 1984, 149).

To counter such inquiries, Governor Gorman initiated a public relations campaign that promoted Minnesota's variety of seasons and its beautiful, enjoyable, and healthful winters. The campaign pointed out that Minnesota lay at the same latitude that spawned the great civilizations in Europe. Further, winter was lauded for contributing greatness to the character of Minnesota's people. Winter was a time for rest and relaxation after the toil of summer and a time for contemplation and sharing of ideas, which led to an enlightened population (Lass 1984).

Temperature is certainly discussed more frequently than any other aspect of climate and weather, partly because extreme temperatures in Minnesota cover a wide range, as much as 173°F, from a maximum of +114°F at Beardsley on July 29, 1917, and at Moorhead on July 6, 1936, to a minimum of -59°F at Leech Lake on February 9, 1899 (Baker and Crookston 1980). The temperature extremes in Minnesota come from the hot, dry air pushed up from the deserts in the southwest and the frigid arctic air pushed down from Canada by the jet stream. Daily extremes can also range widely. Temperatures at a given location frequently vary by more than 30°F in 24 hours on spring and fall days.

Average temperatures in Minnesota are about 2° to 3°F cooler for every 100 miles of northward travel. Monthly average temperatures for International Falls, the Twin Cities, and Rochester show the wide variation that exists in the state (table 2.1). Temperatures at the Twin Cities are warmer

than at Rochester because urban areas produce a large amount of heat. Weather station records from 159 observers throughout the state and along the Minnesota border in adjacent states show a 14°F difference in average winter temperatures increasing from north to south (Baker, Kuehnast, and Zandlo 1985). In summer the north-south variation is nearly as great because of the cooling influence of Lake Superior. Monthly north-south variation in average temperatures is smallest in fall, the difference being 8°F in September and 7°F in October.

Another factor influencing temperature is the reflection and absorption of solar energy by land and water. In summer, most of the incoming energy is absorbed by the surface, and it heats up. During the growing season, about 20% to 25% of solar energy is reflected, and minor amounts are used by plants in photosynthesis. When snow covers the ground, however, about 75% of solar energy is reflected.

Solar energy is also lost as heat during the evaporation of water. Consequently, summer air temperatures are lower near lakes. Such effects on temperature are quite local in nature, however, even near very large bodies of water. Although temperatures along Lake Superior are cooler in summer because of evaporation and are warmer in winter because the lake water takes longer to cool than the surrounding land, these effects do not extend very far inland because of the steep slope of the North Shore and the prevailing westerly winds.

Topography has a marked influence on temperature in mountainous regions because of differences in air pressure at different altitudes, and even the comparatively small-scale glacial features that characterize Minnesota can alter temperature. As air

**Table 2.1** *Average monthly temperatures in °F, 1961–90*

|  | Jan. | Feb. | Mar. | Apr. | May | June | July | Aug. | Sept. | Oct. | Nov. | Dec. |
|---|---|---|---|---|---|---|---|---|---|---|---|---|
| International Falls | 2.2 | 8.0 | 21.5 | 38.8 | 51.5 | 61.0 | 66.4 | 63.6 | 53.1 | 43.0 | 24.9 | 8.3 |
| Twin Cities | 13.1 | 17.1 | 30.0 | 46.0 | 58.2 | 67.9 | 73.3 | 70.7 | 61.5 | 49.6 | 32.9 | 19.2 |
| Rochester | 12.1 | 16.6 | 29.4 | 44.8 | 56.8 | 66.2 | 71.4 | 68.9 | 59.9 | 48.5 | 32.6 | 18.5 |

*Source:* G. Spoden, personal communication.

masses rise or fall as a result of topography or other forces, temperature decreases or increases at 5.5°F per 1,000 feet, due respectively to expansion or contraction (Baker, Kuehnast, and Zandlo 1985). Because of topography, temperatures are sometimes cooler than the surrounding area at Buffalo Ridge, also called Coteau des Prairies, at the Alexandria Moraine, and in the highlands west of Lake Superior, and are sometimes warmer in the Red River valley.

Differences in temperature due to topography are also apparent in the differences in vegetation between north- and south-facing slopes. Southern exposures on high moraines and along major river valleys may have prairie vegetation, whereas northern exposures, which are cooler and moister, are likely to be forested. This microclimatic influence of topography on ecology can be observed on steep north- and south-facing slopes bordering the Minnesota and Mississippi Rivers.

Other microclimatic temperature differences occur because cool air is more dense than warm air and thus collects at the earth's surface and moves downslope because of gravity. This flowing cool air may collect in low areas, forming what are known as frost pockets.

Urbanization also influences temperature, mainly because of the alteration of surface conditions. Buildings, asphalt roads and parking lots, and other artificial surfaces absorb more solar radiation than vegetated areas, resulting in higher air temperatures. In winter, snow is removed from some surfaces, and what snow remains becomes dirty very quickly, thus reducing reflection. Minneapolis and St. Paul are sufficiently large to cause local climatic warming (Winkler, Skaggs, and Baker 1981).

The map showing normal summer temperatures (figure 2.3) illustrates many of the temperature characteristics that I have discussed. The warming effect of the Twin Cities is conspicuous, as are the topographic effects of Buffalo Ridge in the southwest and the Alexandria Moraine in the northwest. Lake Superior has a marked cooling effect along

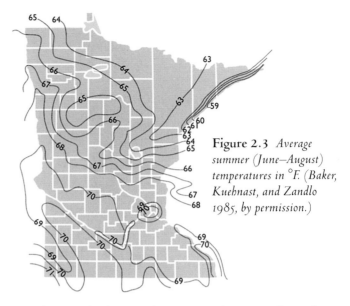

**Figure 2.3** *Average summer (June–August) temperatures in °F. (Baker, Kuehnast, and Zandlo 1985, by permission.)*

the North Shore and may contribute to a flow of cool air that extends to the southwest toward Mille Lacs Lake.

The average July temperature map for North America (figure 2.4) shows how the pattern within Minnesota fits into the continental pattern. Annual average temperatures in Minnesota range from 36°F in the north to 45°F in the south.

The length of the frost-free period, which is the growing season, is variable across the state from south to north. In spring in the southern half of Minnesota, frosts do not usually occur after mid-May. Farther north frosts may occur well into June. Usually the first frosts of fall occur in early September in the northeast but not until late September or early October in the Twin Cities and the southern part of the state. The length of the frost-free season (figure 2.5) is less than 100 days in some areas in the north and exceeds 160 days in the southeast. Heat generated by urbanization around the Twin Cities contributes to an earlier spring and later fall. Areas adjacent to Lake Superior also have a slightly longer growing season.

Extreme high or low temperatures interacting with availability of moisture undoubtedly influence the distribution limits of many species. This influence has not been documented, however, with data from field observations (Woodward 1987). Labo-

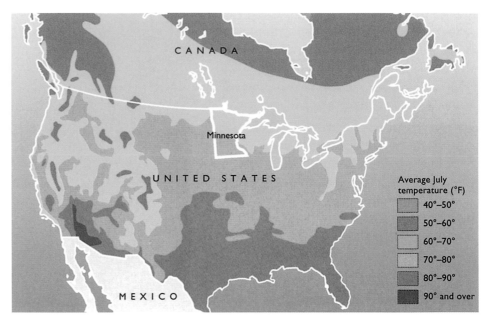

**Figure 2.4** *Average July temperatures in North America in °F. (Cohen 1973, by permission.)*

Average July temperature (°F)
- 40°–50°
- 50°–60°
- 60°–70°
- 70°–80°
- 80°–90°
- 90° and over

ratory experiments indicate that the northern limits for sugar maple, green ash, and American elm occur where average minimum temperatures reach -40°F, resulting in freezing and supercooling of buds and other tissues (Sakai and Weiser 1973). These data suggest that deciduous forests will not occur at latitudes where the minimum temperature is -40°F or colder (Woodward (1987). The actual northern limit of the deciduous forest, however, is somewhat to the south of these latitudes, probably because of the interactions between temperature and moisture.

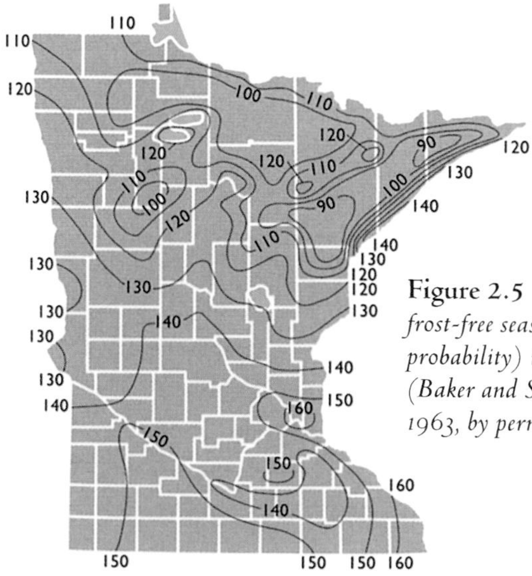

**Figure 2.5** *Length of frost-free season (50% probability) in days. (Baker and Strub 1963, by permission.)*

Animal distributions may also be limited by extreme temperatures, although specific data showing distributions controlled by temperature or other components of climate are not available. Northern bobwhite and opossum occur in Houston County and other southeastern counties but are rare in other parts of Minnesota (Hazard 1982; Janssen 1987). It is generally believed that their distribution is limited by low temperatures. The frostbitten ears and tails that I have seen on opossums are a strong indication that this mammal is not well adapted to cold.

Amphibian distributions may also be influenced by climate. Circumstantial evidence, such as that provided by the distribution of bullfrogs, suggests that temperature influences their rate of development. In the southern part of their range, bullfrogs metamorphose into adults in their second summer. In the northern part of their range, they metamorphose during their third summer. Such a long period of development, believed to be related to the cold environment, may restrict their northward distribution (Breckenridge 1944). The only breeding populations of bullfrogs in Minnesota are in southern Houston County (Vogt 1981), which is where the highest average annual temperature in the state occurs.

32

We do know that the climate warmed and cooled sufficiently in the past to significantly affect the distribution of ecological communities. The fossil remains of giant beavers, mammoths, and mastodons are frequently found in glacial till in many parts of Minnesota. Their large size probably helped them to survive the cold climate present at the end of the period of glaciation. Species that are now limited to tundra areas but which may have been present in Minnesota at one time include the reindeer, barren ground caribou, musk oxen, and lemmings.

Many of the mammals, and other organisms as well, presently living in Minnesota were already here at the time of the last glaciation. Presumably they moved south ahead of the ice and survived in sites with a more moderate climate. As the glaciers receded, they returned to Minnesota.

Some mammals became extinct at the end of the last glaciation. The reason for such massive extinction is not known, but several possible causes have been presented. One is the development of hunting skills by early humans and the use of fire and stone tools to kill animals. Another possibility is that the climate warmed and the vegetation growing in the new climate was not suitable as food.

There is no question that Minnesota has experienced warmer and cooler climates in the past, as evident in the changes in vegetation documented in the pollen record (see chapter 1). A more difficult question to answer is whether the present climate is becoming warmer or colder. We know that the average annual temperature varies from year to year. Ten-year temperature averages have therefore been plotted for eastern Minnesota in figure 2.6. The point for each year on the graph represents the average for a 10-year period ending in the year where the point is located. Using the 10-year average smooths the annual variations and highlights major trends during the 164 years during which data were recorded. A declining trend in the late 1800s was followed by a rising trend, which peaked in the 10-year period between 1930 and 1940. During this rise, the temperature increased by about 4°F. Further analysis of data through 1987 also indicates a warming trend (Skaggs and Baker 1989).

So is Minnesota's climate becoming warmer or colder? The best conclusion is that Minnesota's climate is dynamic and that the pattern illustrated in figure 2.6 does not allow us to confidently predict long-term changes. The decade of the 1980s, however, is one of the warmest decades recorded in

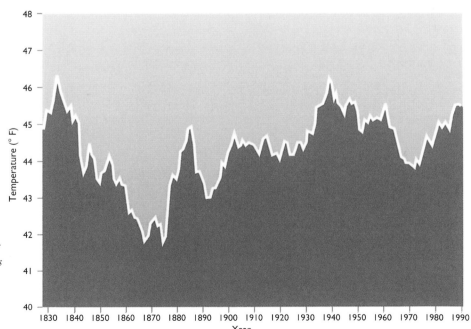

**Figure 2.6** *Ten-year average temperatures in eastern Minnesota, 1820–1990. The point for each year on the graph represents the average for a 10-year period ending in the year in which the point is located. (D. G. Baker, personal communication.)*

33

eastern Minnesota since 1820 (G. Spoden, State Climatology Office, Minnesota Department of Natural Resources, personal communication).

## Precipitation

Precipitation provides the soil moisture necessary for plant growth, recharges groundwater and springs, regulates flow in rivers, produces runoff that flows into wetlands and lakes, and cleanses the atmosphere of pollutants. These functions all occur as water moves through the hydrologic or water cycle (figure 2.7). To follow the movement of water through the hydrologic cycle, let your mind travel to a spot on a lake of your choice, where a bright sun warms the clear blue water. Imagine that you can see a microscopic diamond-

like molecule of water break from the tip of a wave, and, carried by a gentle breeze, sail upward into the atmosphere. This tiny molecule of water, buffeted by winds and temperature changes, bounces around in the upper atmosphere and eventually joins with other water molecules to create larger droplets. As more and more droplets form, the air becomes hazy, and clouds begin to develop. The clouds, pushed by global air currents, move over the landscape and when conditions are right, deposit their droplets of water as rain or snow.

Rain and snow meltwater generally first go into the soil to replenish soil moisture. When soil moisture is replenished, runoff begins. Runoff is the water available after the soil is saturated. On average, about 22% of the precipitation falling on Minnesota runs off into rivers, lakes, and wetlands

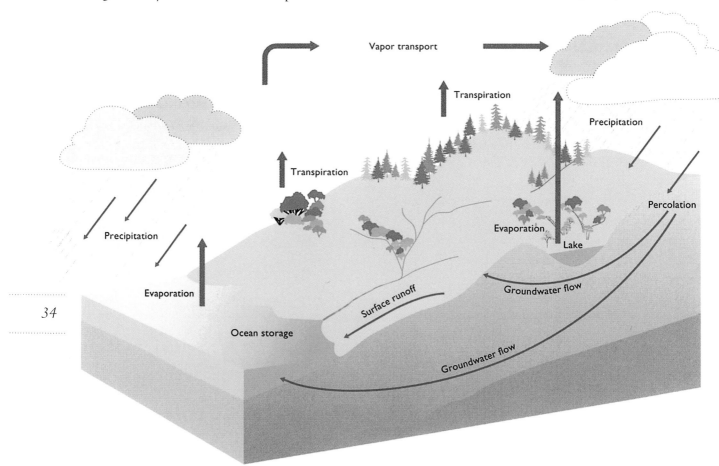

**Figure 2.7** *The hydrologic or water cycle shows the major pathways of water movement, precipitation, surface and underground flow, evaporation, transpiration, and vapor transport.*

**Figure 2.8** *Average annual precipitation in North America in inches. (Cohen 1973, by permission.)*

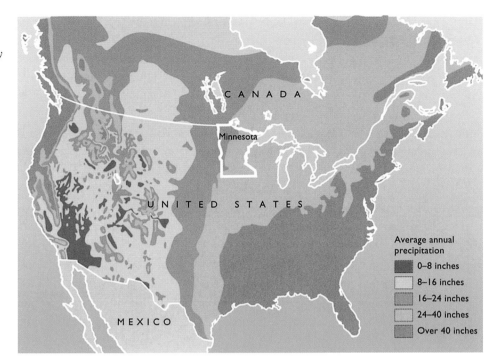

Average annual precipitation
- 0–8 inches
- 8–16 inches
- 16–24 inches
- 24–40 inches
- Over 40 inches

(Baker and Crookston 1980). Annual runoff in Minnesota varies from about 2 inches along the western border to 16 inches adjacent to Lake Superior (Borchert and Gustafson 1980).

Most of the precipitation is returned to the atmosphere by plants and by evaporation (figure 2.7). Plants take up water from the soil, incorporating some into plant tissue through photosynthesis and giving off the rest to the atmosphere through a process called transpiration. Water that evaporates from the earth's surface plus that lost by transpiration is referred to collectively as evapotranspiration. Evapotranspiration, which is influenced by temperature, vegetation, and other factors, is always occurring, even when the soil is dry and there is little or no runoff.

Our molecule of water from the lake enters a stream via runoff from the land surface or percolates through the soil and enters the groundwater. The molecule could remain in the groundwater for decades, eventually reaching a stream or perhaps a well. From the stream, the molecule may travel swiftly to the sea, or it may enter a drinking water system and end up in a glass of lemonade. Whatever the various detours taken by this molecule, it

will eventually return to a lake or an ocean, completing the cycle.

About 90% of the moisture that falls on Minnesota is carried from ocean sources by moist air masses. Most comes from the Gulf of Mexico, but a small portion originates in the Pacific Ocean. Local moisture sources, such as transpiration from vegetation and evaporation from lakes, are of lesser importance (Baker and Crookston 1980). The prevailing westerly winds push the moisture-laden Gulf air masses eastward. Annual precipitation in Minnesota varies from about 20 inches in the northwest to 32 inches in the southeast (figure 2.8; Baker and Kuehnast 1978).

Minnesota precipitation is highly variable in terms of both annual averages and daily extremes. The highest annual precipitation officially recorded was 53.52 inches in 1991 at St. Francis in Anoka County. This record was nearly broken in 1993, when the precipitation officially recorded at Fairmont in Martin County was 52.36 inches (U.S. Department of Commerce 1993).

Floods occur on the larger rivers about 1 or 2 years out of 10, mostly in April during snowmelt when the ground is still frozen, but sometimes in late spring and early summer. The most recent severe floods on the Mississippi and Minnesota Rivers occurred in 1965, 1969, and 1993.

The greatest 24-hour rainstorm of record was 10.8 inches on July 21–22, 1972, in Morrison County (Minnesota Department of Natural Resources 1988). Rainfalls of such magnitude may cause serious erosion and floods, and even loss of animal and human lives. Flash floods caused by heavy rainstorms are common in spring in southeastern Minnesota because of the hilly terrain and narrow valleys. Although these floods are usually of short duration, large amounts of topsoil may be carried to the valley floor, and roads and bridges may be destroyed.

The lowest annual precipitation was 7.8 inches at Angus in Polk County in the northwest in 1936, the time of the severe drought that affected all of central North America. During the mid-1930s, hot, dry summers killed much vegetation, small lakes and wetlands dried up, and dust storms often choked the landscape. Severe drought, based on the Palmer Drought Index (Palmer 1965), occurs once every eight to nine years except in the southwestern counties, where drought occurs once in every six years (Kuehnast 1972). The most severe droughts tend to last for several years.

Although total precipitation is important, the amount that falls during the growing season is perhaps more significant to natural communities and certainly to agriculture. In Minnesota approximately 65% to 75% of the annual precipitation falls from May through September (figure 2.9). Note the low growing-season rainfall in the extreme northeastern corner of the state. We might expect that proximity to Lake Superior would lead to high precipitation, but the opposite occurs. Lake Superior seems to stabilize the atmosphere in summer, reducing convection and cloud formation (G. Spoden, State Climatology Office, Minnesota Department of Natural Resources, personal communication).

Some of the growing season precipitation occurs during thunderstorms. Each year thunderstorms develop on about 45 days in southern Minnesota

and on about 30 days in northern Minnesota. Most occur from May through September, and some are accompanied by high winds, hail, heavy rains, and occasionally tornadoes. Temperature imbalances in the atmosphere force air movements that often lead to thunderstorms. Heat from the earth's surface warms parcels of air so that lighter warm air is beneath heavier layers of cool air. As the cool layers sink and the lighter layers rise, a tumbling motion is set into action. Cooling causes water vapor in the clouds to condense, thus releasing heat and creating energy that keeps the clouds growing. Giant anvil-shaped thunderheads (figure 2.10) are thus formed, which frequently lead to severe storms.

When the droplets of water in the clouds grow larger, they may fall as rain or they may be circulated upward and downward by the turbulence of the thunderstorm. If the upper layers of the thundercloud are cold enough, the droplets freeze into hailstones. As the hailstones are tossed up and down, they grow until they are tossed out of the updraft or become so heavy that they fall to the ground. Hailstorms occur an average of about four to five days annually in southwestern Minnesota and about twice per year in the north. Hailstorms occur most frequently in July and less frequently in June and August (Kuehnast 1972).

Lightning flashes characteristic of thunder-

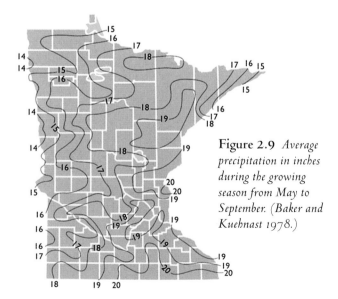

**Figure 2.9** *Average precipitation in inches during the growing season from May to September. (Baker and Kuehnast 1978.)*

**Figure 2.10** *Thunderhead clouds, with anvil-shaped tops, signal an approaching rainstorm. (Photo by Bob Firth, Firth PhotoBank, Minneapolis.)*

storms indicate the tremendous amount of energy in the clouds (figure 2.11). The bolts of energy are generated when strong updrafts and downdrafts separate the electrical charges. Sometimes the bolts hit trees or other tall objects on the ground, but often they jump from one part of the cloud to another without striking the ground. Superheating and expansion of air along the path of the lightning bolt create compression waves of sound that we call thunder. Fires ignited when lightning hits a tree or the ground may spread and burn large areas of forest or prairie. Fire plays an important role in

these ecological communities, as I will discuss in later chapters.

Precipitation during the nongrowing season is important because it adds moisture to the soil and contributes to groundwater. Freezing rain or sleet storms occur in early winter but are not common. They happen when rain falls on surfaces that are below freezing. Ice then coats vegetation (figure 2.1), roads, utility lines, and other surfaces. Sometimes the coating of ice is so thick that branches snap off trees, utility lines collapse, and food supplies are sealed from livestock and wild animals.

Snow and winter are almost synonymous to Minnesotans. One inch or more of snow covers the state on an average of about 110 days each year, ranging from 160 days in the northeast to 85 days in the southwest (Kuehnast, Baker, and Zandlo 1982). Annual snowfall averages about 60 inches along the North Shore and reaches 68 inches in the extreme northeast corner of the state. Annual snowfall decreases to about 40 inches in the south and to 36 inches along the western border (figure 2.12). Air makes up from 30% to 90% of new fallen snow, its density being determined by air temperature. Therefore, the ratio of new snow to water equivalent varies from about 7:1 to 15:1, meaning that a snowfall of from 7 to 15 inches is about equal to 1 inch of rain (Baker and Crookston 1980).

The greatest snowfall in a 24-hour period was 28.0 inches at Pigeon River in Cook County on April 4–5, 1933. The greatest annual snowfall, 147.5 inches, occurred at the same location during the winter of 1936–37. Duluth experienced the most snow in a single storm from October 31 to November 2, 1991, when 36.9 inches fell. When snow is combined with temperatures below 20°F and winds stronger than 35 miles per hour, the storm is defined by the National Weather Service as a blizzard. This combination affects Minnesota an average of two times each winter (Kuehnast 1972). Deaths and property damage attributable to blizzards are comparable in many ways to the effects of summer tornadoes.

Snow serves many useful functions. It acts as an insulation blanket, decreasing heat loss from the soil and preventing deep penetration of the frost layer. Mice and shrews use the space between the snow and the soil surface and are able to carry out their daily activities in this zone of relative warmth with little fear of predators and severe weather. Some species, such as ruffed grouse, actually burrow into loose, fluffy snow to roost, thus conserv-

**Figure 2.11** *Lightning slashes through a dark summer night. (Photo by Bob Firth, Firth PhotoBank, Minneapolis.)*

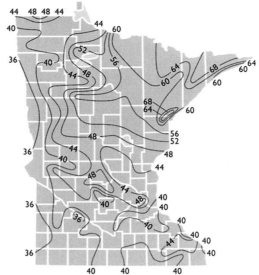

**Figure 2.12** *Average annual snowfall in inches from 1951 to 1980. Seven to 15 inches of snowfall equals 1 inch of rain. (Minnesota Department of Natural Resources, State Climatology Office, personal communication.)*

ing heat and avoiding predators (Huempfner and Tester 1988). A new-fallen snow reveals the activities of animals—the spot where a red fox caught a mouse, the trail of a gray squirrel searching for its buried acorns. The tracks and signs of animals are clearly visible to read and interpret.

## Clouds

Clouds are formed when air masses cool and the water vapor in the atmosphere condenses into tiny droplets or ice crystals. We can observe a small-scale version of this process when our warm, moist breath mixes with the cold winter air and creates a tiny cloud. In summer as thermals of air rise, cumulus clouds (figure 2.2) may form and sprinkle the sky like popcorn. Scattered cumulus clouds are indicators of a rising barometer and fair weather. If warm weather continues, the thermals increase in intensity, and the cumulus clouds are transformed quickly into anvil-topped cumulonimbus clouds with lightning, thunder, and rain. I have frequently observed these changes in clouds over the Alexandria Moraine in Mahnomen County. As the westerly winds push the air mass upward over the

39

moraine, the clouds change in form, and rain can sometimes be seen to the east of the moraine crest.

During winter when large layers of air are cooled without appreciable vertical movement, thin bands of stratus and cirrus clouds may stretch in a monotonous gray layer across the sky. As these bands meet and thicken, the sky darkens and snow may begin to fall. Under certain conditions, usually in late afternoon on a very cold day, bright spots called sun dogs, or parhelia, may be seen in the cirrus clouds, because of the bending of the sun's rays by ice crystals.

Cloud cover affects the amount of solar radiation that reaches the earth. Estimates of the average cloud cover in the Twin Cities area (figure 2.13) show that the clearest skies occur in July and August and that cloudy or overcast days are most likely to occur in November and December. Thus,

**Figure 2.13** *Average cloud cover and frequency of overcast days in the Minneapolis-St. Paul area. July and August have the clearest skies, and overcast days are most common in November and December. (Modified from Baker and Crookston 1980.)*

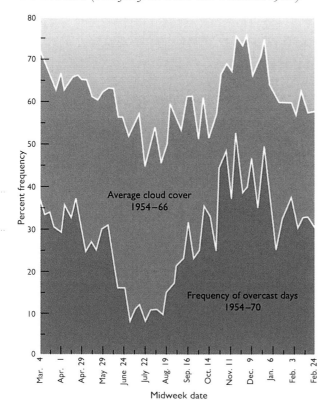

the sun warms the earth most on clear days in mid-summer and least on cloudy days in winter.

## Wind

The prevailing winds in Minnesota usually blow from the south or southwest from May through October. Winds that originate in the desert southwest are hot and dry. Winds that originate around a region of high atmospheric pressure called the Bermuda High in the Atlantic Ocean move clockwise carrying warm air masses over the Gulf of Mexico, where they pick up moisture. This Gulf air moves north across the central plains and arrives warm and moist. If the air collides with cool air from the north, violent mixing of the air masses may lead to thunderstorms and sometimes tornadoes (figure 2.14).

Tornadoes are caused by masses of horizontally rotating air with very low atmospheric pressure in the center. Wind speeds may exceed several hundred miles per hour. Tornadoes are characterized by a rotating funnel, which is visible only because of the moisture, dust, and debris carried by the wind. The life span of most tornadoes is only a few minutes, and most funnels do not touch the ground. Those that do usually affect areas several hundred yards wide and a few miles long.

Minnesota has an average of 17 tornadoes per year; 20% occur in May, 29% in June, and 24% in July. Most occur in the southern portion of the state and are associated with severe thunderstorms. According to historical records, the deadliest Minnesota tornado struck St. Cloud on April 14, 1886, killing 74 people (Kuehnast 1972).

From November through April the prevailing winds blow from the northwest, sweeping in cold air. The jet stream forms about seven miles above the earth at the boundary between the cold and warm air. A winter jet stream is like an 80-mile-wide ribbon of air moving at high speed, sometimes at over 200 miles per hour (Douglas 1990). As winter progresses, the jet stream dips south, causing turbulent swirls that become cells of low

**Figure 2.14** *(top) Summer storms in Minnesota have their origin in a region of high atmospheric pressure in the Atlantic Ocean called the Bermuda High. Winds move clockwise from the Bermuda High over the Gulf of Mexico, where they pick up moisture. This gulf air is heated as it moves northward across the central plains. If the warm, moist air collides with cool air from the north, violent mixing of the air masses often leads to thunderstorms and tornadoes. The rain that falls is water picked up from the Gulf of Mexico. (Modified by permission of the* Star Tribune, *Minneapolis-St. Paul.)*

**Figure 2.15** *(bottom) Winter storms in Minnesota are strongly influenced by the jet stream, which forms between the warm southern and cold northern air masses. When a cold northern air mass pushes the jet stream southward, turbulence creates areas of low pressure. These low pressure swirls carry warm, moist air in a northeasterly direction, and pull in more cold air from the north. Snow falls and strong winds blow until the storm moves off toward the northeast. (Modified by permission of the* Star Tribune, *Minneapolis-St. Paul.)*

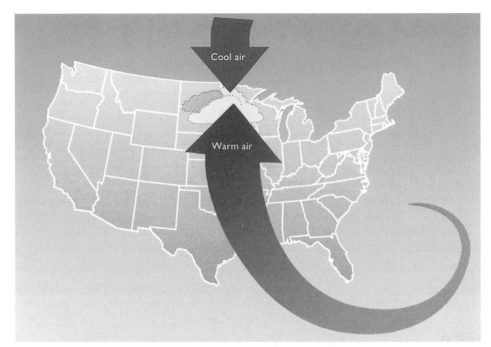

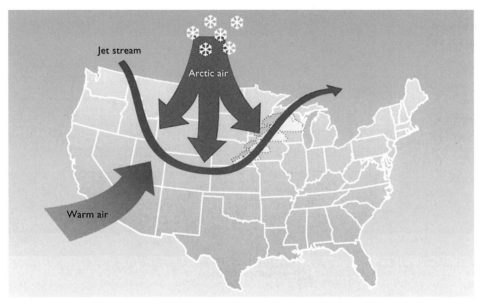

pressure. Winds shift and blow from the south as these cells follow the jet stream northeastward, growing in intensity until they draw in frigid arctic air and become winter storms (figure 2.15). When the storms subside or move off to the northeast, the winds usually shift back to the northwest.

Minnesota is indeed a theater of seasons, even if spring and fall seem to be all too short in most years. Our wide range of temperatures, rain and snow patterns, tornadoes, and blizzards can all be attributed to Minnesota's location near the center of the North American continent. Short of global change in the climate caused by human activities, our weather is likely to stay the same for a long time. Perhaps, with a twinkle in our eye, we simply need to embrace Governor Gorman's assertion that Minnesota's climate contributes greatness to the character of its people.

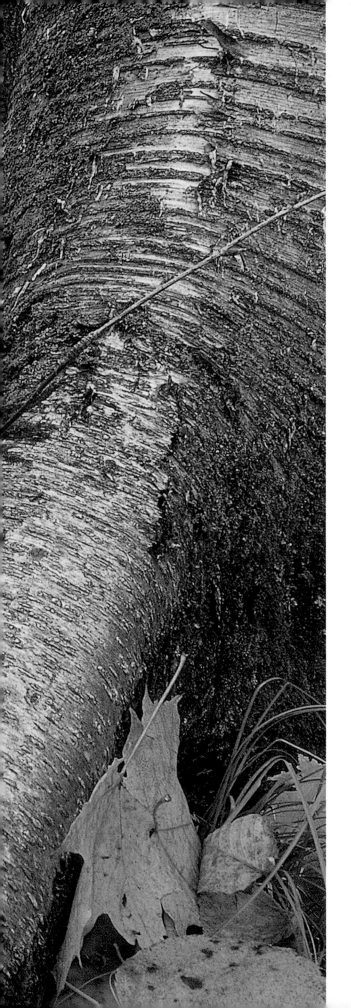

# PRINCIPLES
# OF ECOLOGY

The web of life that surrounds us is complex and interconnected (figure 3.1). Ecology is the study of the relationships and connections between the living things that constitute this web and their environment. In this chapter I will present some of the principles of ecology, which will provide a basis for understanding the structure of the many kinds of ecosystems in Minnesota and how they function.

To introduce the field of ecology, I will describe the relationships discovered by an ecologist between black cherry trees, tent caterpillars, and ants. Black cherry trees are common around the Twin Cities and in many deciduous forests in Minnesota. In early spring the buds open, producing leaves with tiny, glandlike nectaries along the edges. The nectaries on a single leaf last for only a few days, but the buds continue to produce new leaves for several weeks. Thus, the tree has nectaries for several weeks in the spring.

Thatching ants form colonies in open fields and in clearings in forests. The colony lives in a mound of dirt, grass, leaves, and twigs, and the ants prey on many kinds of insects, obtain nectar from plants and aphids, and also scavenge. They feed intensively from the nectaries on black cherry leaves for about three weeks in spring. David Tilman (1978) observed in a study conducted in Michigan that

**Figure 3.1** *Fungi, such as these mushrooms, help to decompose dead plant matter. In turn, mushrooms are often eaten by mice, insects, and other animals. (Photo by Richard Hamilton Smith.)*

other insects that approach a feeding thatching ant are quickly attacked and if captured, are carried back to the ant colony.

Tent caterpillars often cause severe damage to the leaves of black cherry trees. Eggs laid on the tree in the previous year hatch in spring, and the caterpillar larvae develop through five or six stages. Larvae in the tent caterpillar colony (figure 3.2) feed on the margins of leaves and are thus likely to come in contact with the nectar-feeding ants. Tilman watched the ants grasp young larvae in their jaws and carry them back to the ant colony. When he returned to the same tree one or two days later, all of the tent caterpillar larvae were missing. No mature tent caterpillar colonies were found in small black cherry trees located within 10 yards of an ant colony. It is likely that the thatching ants killed all of the tent caterpillar larvae.

Tilman's observations led him to suggest that the nectaries may result in less damage to the leaves by attracting ants that prey on tent caterpillars. This suggestion is supported by the findings that the nectaries are present when the tent caterpillar larvae are small, and that the nectaries are positioned on the margins of leaves, where the larvae feed.

This example illustrates how careful observations of the natural world can lead to an understanding of relationships at several levels of complexity. Similarily, we can examine the science of ecology at several levels. To know the ecology of an individual organism is to know its natural history. The description of the thatching ant illustrates this level. A second level of understanding is to know the ecology of a population, all of the organisms of a certain kind living in a specified area. All of the thatching ants in a colony represent a population. Some ecologists study the growth and regulation of populations. But the thatching ants do not live in isolation. They feed on insects and plants and interact with tent caterpillars and other organisms and are eaten by predators. They are part of a community, a third level of organization, which ecologists define as all the populations in a specified area. A fourth level consists of the community and its physical environment, the soil, air, nutrients, climate, and other nonliving components. This single, interacting unit is called an ecosystem.

When our observations enable us to determine how organisms are linked in food chains and food webs, we can begin to understand the structure and functioning of ecosystems. The cherry trees, the

44

**Figure 3.2** *A colony of eastern tent caterpillars. (Photo by David Tilman.)*

**Figure 3.3** *White pines were common over much of northern Minnesota in the early 1800s. (Photo by Richard Hamilton Smith.)*

ants, and the caterpillars are part of a food web. Sunlight, water, and nutrients are used by the cherry trees to produce nectar and leaves. The flow of energy and the transfer of nutrients continue when the tent caterpillar larvae feed on the leaves and the ants feed on the nectar and caterpillars. Competition may occur between ants and caterpillars for parts of the cherry trees. Predation occurs when ants kill and eat the larvae.

These and other aspects of the structure and functioning of ecosystems will be explained in this chapter. From this base we can build an understanding of the interactions of populations. Finally, we can examine the process of succession, by which communities and ecosystems change over time.

## Environment

Two terms are used to describe the place where an organism lives: environment and habitat. Environment encompasses everything, whether living or nonliving, that has an effect on an organism. Habitat generally refers to the place where we would go to find an organism and is analogous to an address. For example, the habitat of a white pine is a forest in northern Minnesota. The environment of a white pine, however, is much broader and includes physical factors, such as soil, water, nutrients, heat, and light, and biological factors.

The environment of a white pine growing in Itasca State Park in northwestern Minnesota (figure 3.3), for example, includes the following:

— The soil in which its roots are growing; all of the nutrients such as calcium, phosphorus, potassium, and nitrogen that are contained in this soil; the organisms in the soil such as worms and insects; and even the mice and chipmunks that have burrows in the soil.
— Water, both in the soil and as precipitation, and

all the elements and compounds that are brought to the white pine as part of the precipitation. These include both beneficial nutrients and harmful substances such as those associated with acid rain.

— Sunlight and warm temperatures, which are necessary for growth and reproduction.

— The other white pines and all of the other plants in the vicinity, which compete for the available light, water, and nutrients and for space in which to grow.

— All of the animals that might eat part of the white pine, which include white-tailed deer, porcupine, white-footed mouse, moose, spruce grouse, and the many species of insects that feed on the needles, bark, and wood of the white pine.

— Bacteria and fungi associated with the plant's roots, which help the plant to be more efficient in taking nutrients and water from the soil.

— Disease organisms, such as white pine blister rust, which may injure or kill the tree.

— Birds that nest in the tree and their droppings, which provide nutrients that are incorporated into the soil but which may also kill the stems and needles, as frequently happens in heron rookeries.

— The climate and weather, which may include high winds or tornadoes that can severely damage the white pine.

This list is not inclusive. I have not mentioned the birds and insects that prey on the insects eating the buds or needles, nor have I mentioned the para-

sites that might infect grazing or predatory insects. Furthermore, I have only briefly indicated how these environmental factors are interrelated. The more we look at the environment of a single white pine tree, the more we see the complexity of the natural world and its interrelationships. To illustrate, let us pick one component of the tree's environment, the white-tailed deer (figure 3.4), and describe its environment.

How does the environment of the deer compare with the environment of the white pine? Some factors, such as weather and topography, will be essentially the same for both, although the deer can moderate its local environment by sheltering under dense vegetation. Other factors, however, will be different. The environment of the white-tailed deer includes the following:

— The forest habitat of the deer, which includes white pine as well as many other species of trees, shrubs, and herbs. The forest provides both food and cover for the deer.
— Water from lakes, streams, and ponds.
— Other deer living in the same forest.
— Other kinds of animals, such as wolves, which are predators; ticks, which are external parasites; mosquitoes and flies, which in this case are pests; flukes, which are internal parasites; and snowshoe hares and moose, which are competitors for food.
— Rain, snow, heat, wind, and other weather factors that influence the physiological status of deer. For example, more body heat must be produced during winter than during summer.
— Hazards such as fire, which can destroy the habitat, and automobiles and other vehicles, which can kill or injure the deer.
— Humans, who during the hunting season are predators.

The environment of the deer seems to be slightly more complex than that of the white pine, although many similarities and overlaps are obvious. The deer does not have as direct or close relationship to sunlight or to the soil but is obviously dependent on these factors because they are essential for the production of cover and food. In fact, the bacteria in the rumen of the deer probably come from ingested soil.

We now see how describing the environment of an organism can show us many of the interrelationships in an ecosystem. Next we can examine the structure of the ecosystem and the interactions among the various parts to understand how it functions.

47

**Figure 3.4** *White-tailed deer are common throughout Minnesota, even in cities and in farmland, where they take advantage of small stands of trees and shrubs. (Photo © Jim Brandenburg.)*

## Ecosystem structure and function

Whereas environment refers to all things affecting a particular organism, structure refers to all the components or parts of an ecosystem. If we were to list all the parts of an ecosystem, we would include every plant, animal, bacterium, and fungus as well as all nonliving things. In addition, to really know the structure of an ecosystem, we would need to know how many or how much of each part is present. Function refers to the processes through which all of the parts of the ecosystem interact.

### Food webs

The structure of an ecosystem is in part determined by the feeding relationships between its organisms. Figure 3.5 shows several examples of food chains, or who eats whom. At the bottom of a food chain are green plants, or producers. Chlorophyll, the green pigment in leaves, enables the plants to use energy from the sun to drive the process of photosynthesis. In this process carbon dioxide and water are chemically combined to produce sugar molecules, the building blocks for new plant tissue. Plant tissue serves as food for over half of all the kinds of animals. Those animals that eat plant tissue are called primary consumers. Examples are meadow voles, muskrats, snails, worms, and a wide variety of insects and other invertebrates.

Secondary consumers are species that feed on primary consumers. Predators of all kinds—birds and mammals that prey on insects, mammals that eat other mammals and birds, and birds that prey on other birds and small mammals—are members of this category.

The entire complex diagram in figure 3.5 shows that many different species feed on a given kind of plant and that many different species may feed on the primary consumers. All these interrelated food chains are a food web.

Ecologists call the steps in a food chain trophic levels. The green plants, or producers, are the first trophic level. Animals that feed on the plants, col-lectively called grazers or herbivores, constitute the second trophic level. The third level includes those animals that prey on animals in the second trophic level. Mink preying upon meadow voles and musk-rats and northern harriers preying upon meadow voles are thus in the third trophic level. As figure 3.5 shows, however, some animals may occur in more than one trophic level. The northern harrier that feeds on a snipe that feeds on a grasshopper that feeds on a plant also belongs in the fourth trophic level.

Obviously, not all vegetation is eaten by grazers, and not all grazers are eaten by predators. The material that is not eaten by consumers is called detritus, and its fate is described by the detritus food web. Decomposition of detritus by processes such as tearing, shredding, mixing, ingestion, and elimination results in the fragmentation and ulti-mate breakdown of plant and animal tissues into inorganic compounds and elements. Detritivores such as protozoa, nematodes, insects, earthworms, and snails feed on dead plants and animals. They reduce the particle size of the detritus, making it more suitable for bacteria and fungi, which com-plete the decomposition process. Decomposers act as grazers, carnivores, or both, depending on the kind of foods they consume.

### Energy flow

We can now use our understanding of food webs and trophic levels to follow the flow of energy through an ecosystem (figure 3.6). The source of all energy is the sun. Through photosynthesis plants convert to plant tissue only a small percent-age of the total amount of energy falling on a given area. The remainder of the solar energy is dissi-pated as heat. The energy contained in plant tissue is available for consumption by the primary con-sumers of the second trophic level. Grazing animals consume leaves and stems of nonwoody plants, commonly referred to as herbs. Browsing animals, like deer, consume leaves, twigs, stems, and bark of trees and shrubs. Some fish and small invertebrates consume algae. Some of the energy in this plant

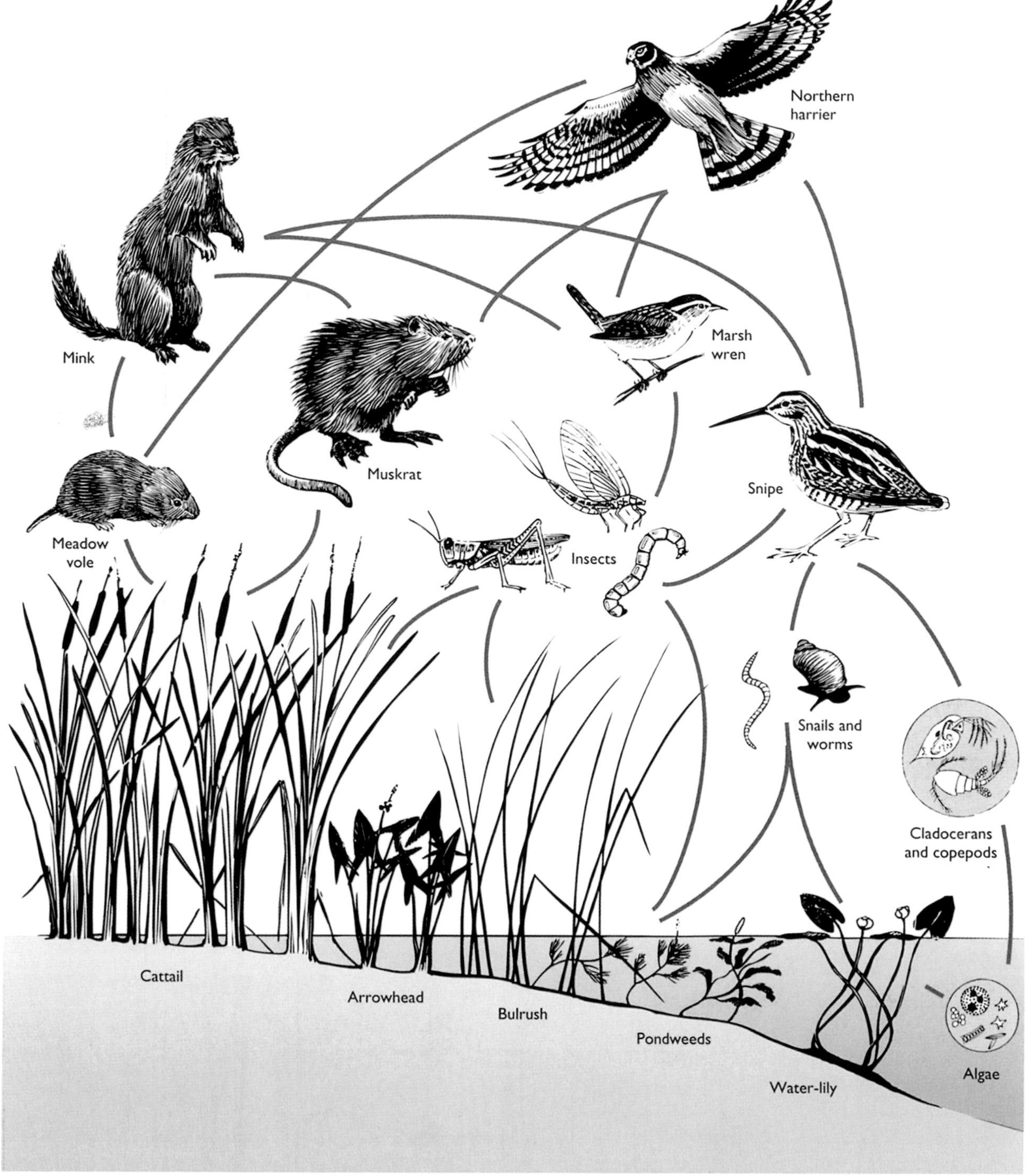

**Figure 3.5** *One of many food webs occurring in Minnesota ecosystems. Each pathway is a food chain, and some kinds of animals may function in several trophic levels. Organisms within circles are microscopic in size. Energy and nutrients move through these food chains.*

**Figure 3.6** *Ecology is about the complex and interconnected web that surrounds us. Plants and animals, living and dead, combined with soil, water, and other nonliving components make up the web. Canada lynx hunt hares, mice, and other prey amid the fallen logs of this northern Minnesota forest. (Photo by Bob Firth, Firth PhotoBank, Minneapolis.)*

**Figure 3.7** *(right) Snow-shoe hares are characteristic of the northern coniferous forest. Their fur is brown in summer and white in winter. (Photo courtesy of Cedar Creek Natural History Area.)*
**Figure 3.8** *(below right) Canada lynx are effective predators on showshoe hares. (Photo © Jim Brandenburg.)*

material is converted into animal tissue. The remainder of the energy in the plant tissue is passed through the animal in undigested form, is used by the animal in the process of respiration, or is lost as heat. The energy in the animal tissues of those species in the second trophic level is then available for predators in the third trophic level. Similarly, the third trophic level provides energy to the fourth, and the fourth to the fifth.

At the fifth trophic level the energy received is only a small portion of the energy present at the beginning of the food chain. This major finding that only about 10% of the available energy in one trophic level is actually transferred to the next higher level was the result of the best known ecological study ever conducted in Minnesota. Raymond Lindeman (1942), a graduate student at the University of Minnesota, investigated the transfer of energy from one group of organisms to the next, that is, from one trophic level to the next, at Cedar Bog Lake in Anoka County. From these observations, Lindeman developed the concept of trophic levels and quantified the efficiency with which energy was transferred from one level to another. Energy conversions were complex in Cedar Bog Lake because of the many species of algae present. Lindeman found that energy flow was related not

only to biomass, the weight of the organisms present at a given instant, but also to the rate of production of different kinds of organisms. For example, algae reproduce in hours or days whereas fish reproduce only once per year. Some algae were eaten by grazers, but most were not eaten at all, going directly to the sediment at death and becom-

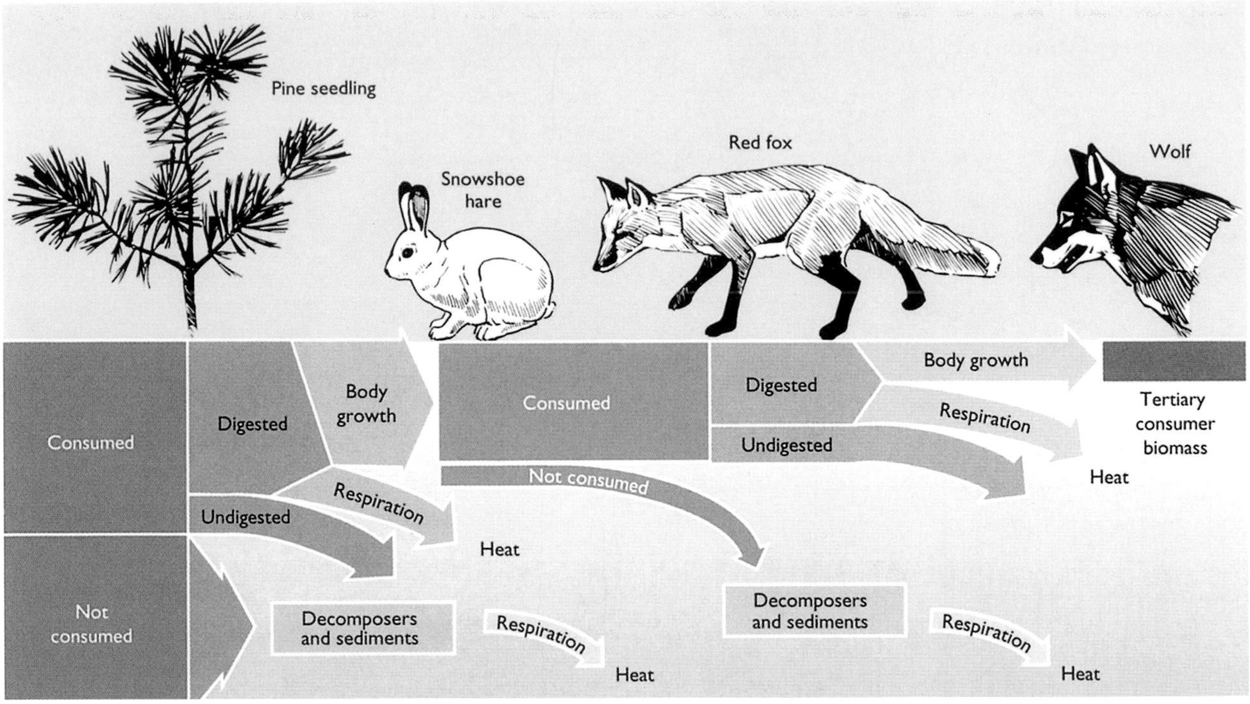

**Figure 3.9** *Energy becomes less and less available to successive consumers in a food chain. Because of this decrease in available energy, typical food chains consist of only four or five links. (Modified from Cunningham and Saigo 1990, by permission.)*

ing part of the detritus food web. Lindeman's findings in Minnesota laid the groundwork for future research throughout the world on trophic dynamics, energy flow, and the functioning of ecosystems.

To see how the transfer of energy works in an actual food chain, let us begin with the 3.5 ounces of white pine seedlings consumed in one meal by a snowshoe hare (figure 3.7). This meal can be converted to calories by multiplying the 3.5 ounces by 117, the number of calories in 1 ounce of seedlings, which equals 410 calories of plant material. The hare converts approximately 10% of the 410 calories to animal tissue, thus increasing the caloric value of its body tissues by approximately 41 calories. The hare is then captured and eaten by a fox. The fox consumes the 41 calories—and the rest of the hare—and converts them into approximately 4 calories of new fox tissue. This fox is subsequently captured by a larger predator, such as a lynx (figure 3.8) or timber wolf. The larger predator takes in the 4 calories and produces approximately 0.4 calorie of new lynx or wolf tissue. Thus, by the fourth

trophic level, the 410 calories of plant material have been reduced to 0.4 calorie (figure 3.9). Because the efficiency of energy transfer from one level to another is so low, food chains typically do not exceed four or five steps. The amount of energy available in the fourth or fifth trophic levels is too small to support additional trophic levels. This step-by-step reduction in available energy explains why predators are usually rare in comparison to their prey, and why grazing animals are so common.

What has happened to the 410 calories that we started with in the meal eaten by the hare? As just described, some of this energy has been transferred through the food chain. Some of it has been used in the process of respiration, that is, metabolism, which keeps the organisms alive, and is eventually lost as heat (figure 3.9). Some of the calories are passed through the digestive tracts of the organisms in undigested form and are eventually broken down or decomposed by the action of bacteria and fungi. These bacteria and fungi also carry on metabolism, and the energy from the plant material is eventually lost as heat through their respiration. In fact, all of the energy entering an ecological system will eventually be lost in the form of heat (figures 3.9 and 3.10). At the same time, however, energy from the sun is being added to the system by plants through photosynthesis whenever growing conditions are suitable. Thus, in every ecosystem there is a never-ending, one-way flow of energy from the sun through plants and then through consumers or decomposers.

### Nutrient cycling

Nutrients, essential to all plants and animals, move through an ecological system in much the same way as energy, with one major difference. Energy flows one way through an ecosystem and is eventually lost, whereas each nutrient is recycled or used over and over again by different plants and animals (figure 3.10). The nutrients important to any ecosystem are inorganic chemicals, both elements and compounds. A total of 23 inorganic substances are

**Figure 3.10** *Energy (red arrows) flows and nutrients (green arrows) cycle through the major components of every ecosystem. All ecosystems have abiotic components, producers, and decomposers, and most also have consumers in the form of herbivores and carnivores. (Modified from Miller 1992, by permission.)*

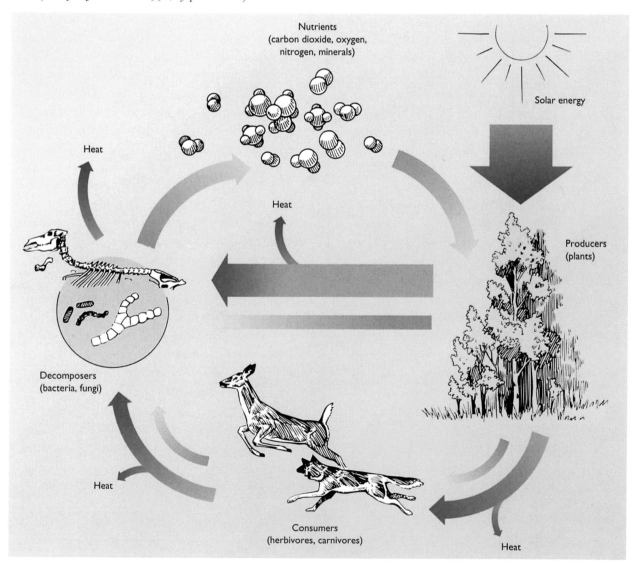

considered essential nutrients for plant growth. Hydrogen, oxygen, and carbon, in the forms of water and carbon dioxide, are used by plants in the process of photosynthesis. Nitrogen, phosphorus, and potassium (the common elements in a typical fertilizer) are nutrients that become incorporated into plant tissues. Small amounts of other elements such as iron, boron, and copper may also be required by plants.

Animals require water, oxygen for respiration, and nutrients in the form of plant or animal tissue made up of carbohydrates, proteins, fats, and oils. Vitamins and minerals must also be provided by the food that is eaten. Vitamins are complex organic molecules that help to regulate normal body functions. Minerals are inorganic elements, such as calcium and iron, that are required for growth.

From where do plants obtain nutrients? Most come from the soil, but some come from the air and from rain and snow. The gradual breakdown of bedrock or other parent material is a primary source of elements for the soil. In addition, elements are added to the soil from decomposition of plant and animal remains. Thus, the calcium present in the soil may have come from the breakdown of limestone bedrock deposited on the ocean floor millions of years ago or from the decomposition of sugar maple leaves during the past year or two. Runoff carries calcium from the land into streams, lakes, and wetlands. Carbon dioxide is obtained from the air and water, and a variety of chemical substances are obtained from precipitation.

To see how nutrients move through an ecosystem, we will follow an atom of calcium in an ecosystem containing white pine and white-tailed deer. Calcium is present in most soils in the form of calcium carbonate, commonly called limestone. Erosion breaks down the limestone into smaller and smaller particles, which eventually dissolve in the water present in the soil. The roots of the white pine are then able to take up the calcium, which becomes incorporated into the tissue making up the tree. Let us assume that the atom of calcium that we are following is incorporated into new

needles in a young tree. A white-tailed deer nibbles on these young, tender needles and takes in the atom of calcium. Digestion and assimilation within the body of the deer then incorporate the calcium into muscle or bone. If the deer is eventually killed by a timber wolf, the atom of calcium moves into the wolf and may be incorporated again as muscle or bone. If the deer dies of some other cause, or if the wolf dies, the bones and flesh will decay or decompose, and over a long period of time, the atom of calcium will be returned to the soil and be available to another plant.

This movement of nutrients from plants to grazing animals is straightforward. Most plants, however, are not eaten by grazing animals, and most animals are not eaten by predators. Thus, most nutrients follow a different path through the ecosystem. Annual plants die, leaves drop from deciduous trees in the fall, and trees are killed by disease or blow down in a windstorm. This plant material is broken into smaller and smaller particles by such animals as woodpeckers, beetles, ants, and other invertebrates. These small particles are eventually consumed by bacteria and fungi in the process of decomposition or rotting (figure 3.11). Some plant parts, such as sugar maple leaves, decompose rapidly because they are rich in nutrients used by bacteria and fungi. In contrast, other plant parts, such as pine needles and oak logs, decompose slowly because they contain resins and other compounds that resist decay.

Nutrients can also be lost from a particular ecosystem only to turn up in another ecosystem. For example, calcium dissolved from limestone can be carried by the water in a stream or river and move hundreds or even thousands of miles to a new location. Migrating birds and insects also carry nutrients from one ecosystem to another. New supplies of nutrients are made available through precipitation and by the continual breakdown of rocks.

Nutrient cycles in natural ecosystems, those relatively undisturbed by human activities, are more or less in balance. In the absence of disturbance, a wetland, forest, or prairie community on a given

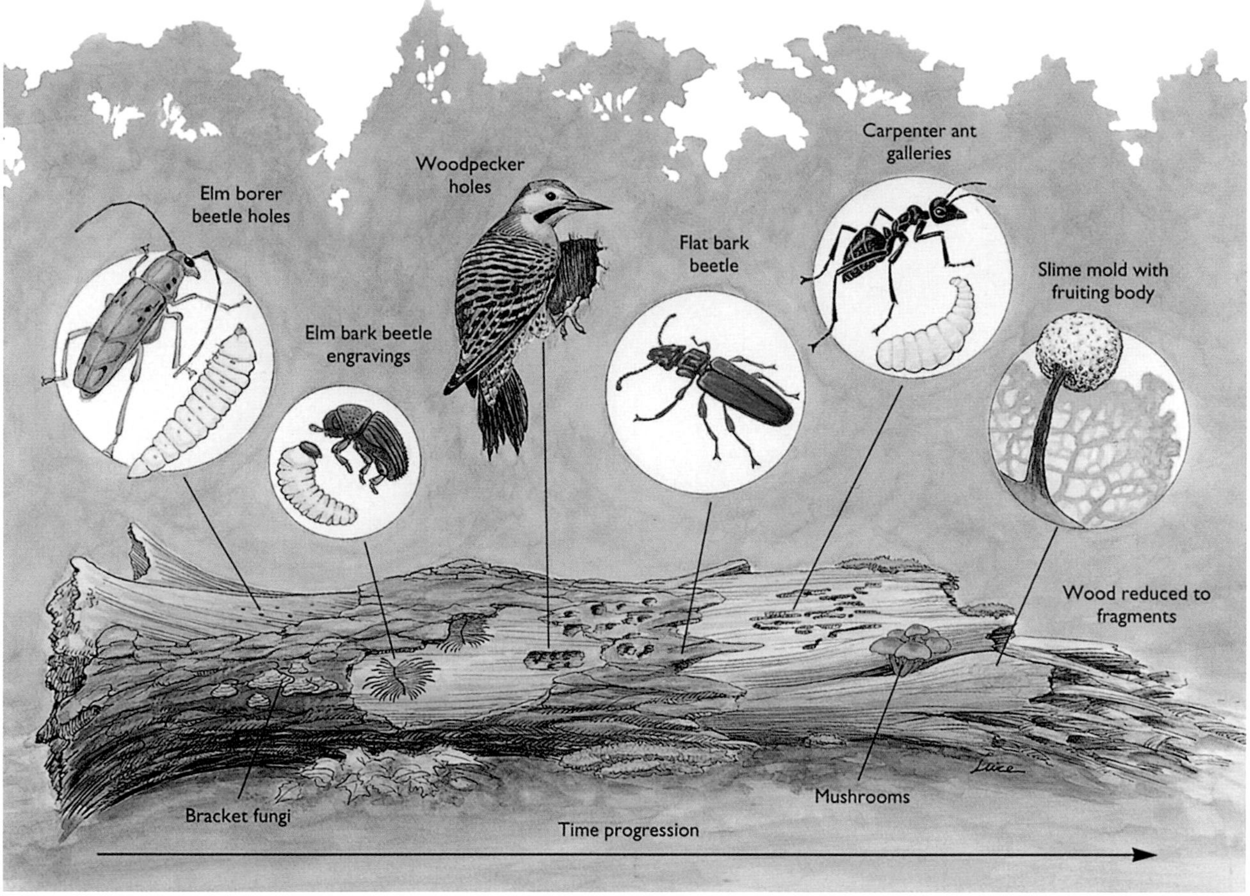

**Figure 3.11** *Many kinds of organisms break up and consume dead plants and animals. Some insects on this old tree trunk feed directly on the wood, woodpeckers feed on the insects, and fungi and bacteria break down complex molecules into simpler chemicals. These nutrients are returned to the soil, where they can be again taken up by plant roots. (Illustration by Don Luce; modified from Miller 1992, by permission.)*

site will continue to survive for hundreds and even thousands of years. The nutrients taken from the soil each year by the growing plants are replenished naturally by decomposition, breakdown of parent material, and precipitation.

The nitrogen cycle is especially significant in ecosystems because nitrogen, an essential component of proteins, is one of the most important nutrients for plant growth. Although the atmosphere is approximately 78% nitrogen, the cycling of this element and its uptake by plants is complex (figure 3.12). Plants cannot take nitrogen directly from the atmosphere and combine it with other elements to form proteins. Molecular nitrogen must first be broken down and then combined with

such elements as oxygen and hydrogen before it can be used by plants. Bacteria, industrial processes, automobile engines, and even volcanoes change nitrogen in the air to a form that can be taken up directly by plants. Of these factors, bacterial transformation of atmospheric nitrogen is by far the most important.

In recent years, however, human production of nitrogen fertilizers from petroleum products has become almost as large a source of input of nitrogen to ecosystems. When farmers harvest corn or wheat from a field, they remove many of the nutrients that would have been recycled in a natural ecosystem. To continue harvesting year after year, farmers must replace the nutrients with artificial

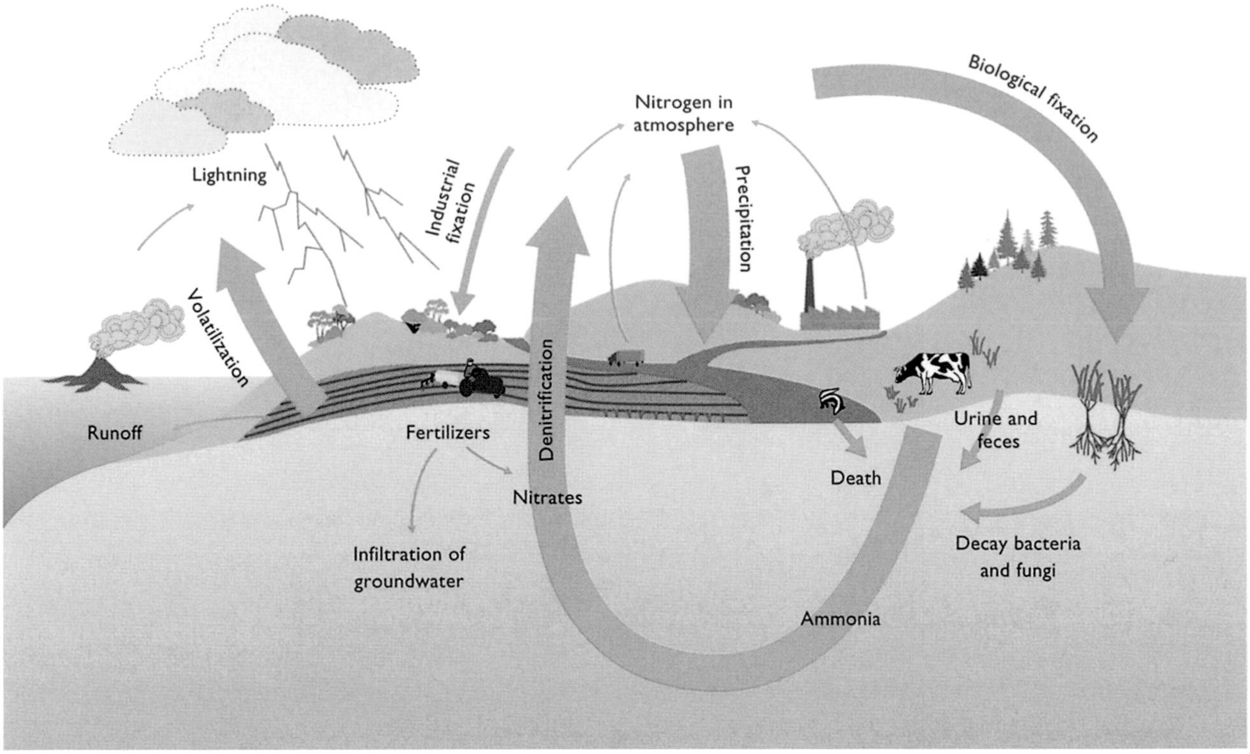

**Figure 3.12** *The atmosphere is the major source of nitrogen, an important element essential to all organisms. Biological fixation is an important source of nitrogen for many plants. Nitrogen-fixing bacteria living in soil and in the roots of some plants convert atmospheric nitrogen and oxygen into a form suitable for plant use. Commercial fertilizers are included in the arrow showing industrial fixation. Nitrogen compounds useful to plants are also found in precipitation, where nitrogen is captured from the atmosphere after being formed by lightning, volcanoes, and industrial and vehicular exhaust gases. Animals obtain nitrogen by eating plants. Nitrogen is returned to the atmosphere through decomposition of dead organic matter, urine, and feces and through the volatilization of gaseous forms of nitrogen, thus completing the cycle. The width of the arrows gives an indication of the amounts of nitrogen moving through the various pathways. (Modified from Cunningham and Saigo 1990, by permission; D. Biesboer, personal communication.)*

fertilizers (figure 3.13), animal manure, or composted plant material.

### Productivity

Energy from the sun and nutrients from many sources interact in the production of new plant tissue. Ecologists refer to this plant tissue as net primary productivity. All of the new tissue produced during a single growing season—tissue in stems, trunks, branches, leaves, roots, buds, flowers, and seeds—should be measured accurately to represent net primary productivity. But determining how much tissue is added to a living tree or to a complex, finely divided root system is very difficult. Most measurements, therefore, reflect only that part of the production that can be easily collected and weighed. When comparing the productivity of different ecosystems, it is essential to take into account exactly what has been measured.

Productivity is usually expressed in terms of the biomass, the weight of the tissue, added each year for some specific area. For example, the measurement of annual primary productivity can be expressed in pounds per acre per year for both terrestrial and aquatic habitats. This type of measurement is familiar to farmers, who express their annual crop yields in bushels or tons per acre, and to loggers, who express their harvest in cords or board feet per acre. The productivity of animals can be measured in a similar way. Fish farmers express their production in pounds per acre of pond per year.

The productivity of agricultural or forest crops is relatively easy to measure, but the values represent only the harvest. Cornstalks, straw, sugar beet tops, tree leaves, bark, branches, and root systems are not included. Thus, these values do not represent the total productivity of the species or the community.

Additional complications arise when we try to measure the productivity of a natural ecosystem. In a prairie, for example, some plants, such as pasque flower and prairie smoke, complete their growth in the spring, dry up, and begin to decompose by early summer. Others, such as gentians and warm-season grasses, do not finish growing until late summer or until the first frost of autumn. To obtain an accurate measurement of productivity, the biomass of these prairie species should be

measured several times during the entire growing season, but repeated measurements are generally difficult and impractical to make. As a result, the prairie vegetation is harvested at a selected date, usually in late summer, when the biomass present is believed to be at its maximum. In most studies only the aboveground biomass is measured, and productivity is expressed as aboveground biomass per acre.

Measurement of total productivity in forests containing both coniferous and deciduous trees is even more difficult. Most coniferous trees begin to grow in early spring and continue to grow until they freeze. By this date, however, the deciduous trees have dropped their leaves. Measurements should be made of each species at the optimum time. But measurement of all the tissue, including

**Figure 3.13** *Chemical fertilizers are applied each year to many fields to maintain the fertility needed to produce large harvests of corn, soybeans, and other crops. (Photo by Bob Firth, Firth PhotoBank, Minneapolis.)*

leaves or needles, on a tree can only be carried out by careful harvesting of the entire tree. This is expensive and time consuming, and as a result, is rarely done. Instead, an estimate of a tree's biomass is often made based on the diameter of the trunk, the height of the tree, and other parameters. Belowground roots are seldom included, but an estimate of belowground productivity can be made by multiplying aboveground values by 0.2 (Bray and Dudkiewicz 1963).

In spite of the difficulties of measuring productivity, ecologists often use the net productivity of plants as a measure of the condition of an ecosystem by comparing it with the productivity of similar or different ecosystems. Net productivity is influenced by nutrient and moisture availability, weather conditions, and other factors. Thus, productivity is one measure of how well an ecosystem is functioning. In addition, changes in net primary productivity can provide a measure of the effects of natural disturbances, such as drought or fire, or of experimental manipulations, such as fertilizing or shading. In the following chapters, I discuss the productivity of many communities in Minnesota and compare the communities with each other and with world averages.

In all ecosystems productivity is limited by some component of the system. Organisms require certain things to survive and grow. For example, white pine, and all other plants, require organic forms of nitrogen. If nitrogen is in short supply, the pine cannot grow, regardless of how much water, phosphorus, and other requirements are available. Thus, its productivity will be reduced until more nitrogen becomes available. But then something else will become limiting. In wetlands and lakes, phosphorus is often the most important factor limiting productivity. In the case of terrestrial animals, food, water, or cover, rather than a single element, is likely to be limiting. The limiting factor for any organism may change over time and may be different in different localities. For a pine tree, water is limiting during a drought, but in wetter years phosphorus or calcium may be limiting.

58

## Populations and communities

The ecosystems that we observe as we drive along the highway or hike along a trail are made up of many kinds of interacting organisms. We have explored several aspects of the structure and function of ecosystems. Now we will examine the dynamics of the populations of organisms that are involved in these interactions.

In most communities a few species will be very abundant, some will be common, and many will be rare. Where did all these species come from? Why are some abundant and some rare? Questions like these are studied by ecologists interested in population dynamics. A brief introduction to this aspect of ecology provides background for understanding the structure of the communities that exist in Minnesota.

### Population dynamics

All individuals of a species living in a prescribed area are called a population. If the area is a lake, its borders are easily determined. Boundaries around terrestrial habitats such as a prairie or forest, however, are often poorly defined. Thus, determining the prescribed area for a population may be difficult. The number of organisms in a population is often expressed in terms of commonly used land measures. For example, deer populations may be expressed as deer per square mile.

Within a population, births (germination in the case of seeds) and deaths occur, and individuals (in the case of animals) may move into or out of the prescribed area. When a population first becomes established in an area, or when it is recovering from a low level, it increases rather slowly, as did the population of partridges in figure 3.14. As numbers build, the rate of increase rises sharply, in what is called exponential growth, and in time reaches the maximum, or biotic, potential of the species. But this growth pattern of the population cannot continue forever. At some level, food, space, or some other environmental factor or combination of factors becomes limiting, and the number of

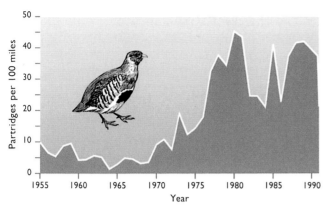

Partridges per 100 miles — 50, 40, 30, 20, 10, 0

1955 1960 1965 1970 1975 1980 1985 1990

Year

**Figure 3.14** *This graph illustrates growth and fluctuations in the population of gray, or Hungarian, partridges in the agricultural region of Minnesota, based on birds observed per 100 miles of roadside counts. The population remained relatively stable from 1955 to about 1970. Then it increased rapidly until about 1980. This type of increase is characteristic of populations that have recently become established in an area. From 1980 to 1991, the population fluctuated from highs of 41 to 45 to lows of 20 to 25 birds per 100 miles of roadside counts. One could predict from these data that the carrying capacity of the habitat is between 20 and 45, and that the population will probably fluctuate within this range in the future if the habitat remains unchanged. (Data from Dexter 1993.)*

individuals born declines or the number dying increases until the population size stabilizes or fluctuates above and below an average (figure 3.14). The number of individuals that can be supported is referred to as the carrying capacity of the prescribed area. If the limiting factor is modified by weather, human manipulation, or some other cause, the population will respond by increasing or decreasing until the same or a different limiting factor leads to stabilization.

For most plants and animals population growth is also dependent on the number of individuals of each sex and age. Human populations provide a good example of this dependency. Most young are produced by the middle age classes—very young and very old individuals cannot reproduce at all. Mortality also varies, being highest in the very young and very old age classes. Thus, the proportion of individuals in the reproductive age classes predicts whether the population will increase or decrease. Human populations with a large proportion of children will grow at a rapid rate when the children reach reproductive age. Populations with

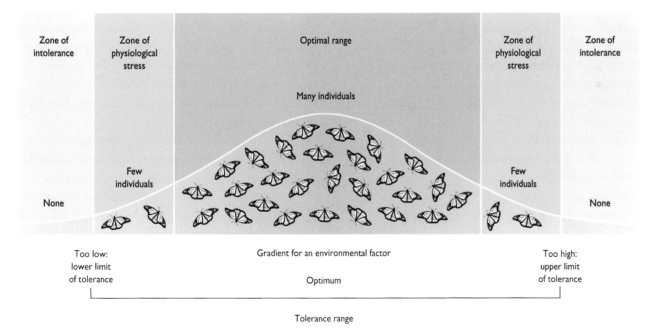

Zone of intolerance | Zone of physiological stress | Optimal range | Zone of physiological stress | Zone of intolerance

Many individuals

Few individuals | Few individuals

None | None

Too low: lower limit of tolerance | Gradient for an environmental factor | Too high: upper limit of tolerance

Optimum

Tolerance range

**Figure 3.15** *The ranges of tolerance for each environmental factor, such as temperature, shade, and availability of each nutrient, are unique for each species. In many cases, tolerance levels determine the spatial distribution and abundance of a species. (Modified from Cunningham and Saigo 1990, by permission.)*

more individuals in older age classes are likely to remain stable or perhaps decrease.

Some organisms, such as algae and bacteria, have the potential to reproduce rapidly when environmental conditions are favorable. Their life span, however, is very short, sometimes only minutes or hours. Thus, numbers of such organisms may be very high at certain times and very low at other times. In contrast, other populations of organisms, such as white pine and white-tailed deer, increase more slowly, but individuals may live for many years. Thus, the numbers of individuals in the populations do not fluctuate over such a wide range.

In a specific community at a specific time, conditions will be optimum for some species, which will be abundant, and these same conditions will be marginal for other species, which will be rare. To what degree conditions are favorable for a species depends on the species' status in the community, called its niche. The niche of a species is defined by all aspects of its environment and by its genetic makeup, which determines its range of tolerance or adaptability (figure 3.15). If a species can live under a wide range of conditions, such as heat or cold, or wet or dry, it is said to have a broad niche and to be highly adaptable. These species are better able to survive changes in environmental conditions than species with narrow ranges of tolerance. High adaptability is a great advantage when organisms must face unpredictable natural changes, such as prolonged drought or human-induced stresses such as acid rain. This combination of environmental conditions and the genetic makeup of each species determines the distribution and abundance of organisms in a community.

### Species interactions

Competition within and between populations is a vital mechanism in determining the structure and function of all ecosystems. Competition occurs among individuals of the same species as well as among individuals of different species. Plants compete with one another for the available nutrients, moisture, light, and space. Similarly, animals compete for the available food and cover necessary for their survival. For example, both sheep and deer eat shrubs and other woody vegetation, as well as grasses and other plants. The sheep in a pasture compete with each other for the available food. They also compete with the deer that jump the fence to feed in the pasture.

Extensive study of the role of competition between populations has led to the idea that two species cannot occupy the same niche. In other words, the two species cannot coexist if their requirements are too similar. One species will always do better and will outcompete and eventually eliminate the other. This idea, known as the competitive exclusion principle, was first summarized by G. Hardin (1960).

Competition between different kinds of plants and animals living in the same area is less common than one might expect because organisms have many different ways of obtaining needed resources. Competition between different species of animals is decreased because they use different types of cover or food or feed at different times of the day. Different kinds of plants may use moisture from different levels in the soil, and some plants can use the low levels of light under the forest canopy whereas others need full sun. The result of the use of these different strategies is that many species of plants and animals can live together in a community.

Other interactions among populations occur in all ecosystems. Some species are mutually beneficial to each other. Good examples of mutualism are the relationships between plants and the insects that pollinate their flowers while feeding on nectar. Another example is the relationship between rhizobium bacteria and certain species of plants, mostly in the legume family. These bacteria, which live in nodules on the roots of the plants, take nitrogen molecules from the atmosphere and convert them first to ammonia and then to nitrates. Because plants are not able to use molecular nitrogen, they depend in part on this process of nitrogen fixation for their supply of nitrogen.

**Figure 3.16** *Predation is common and natural. The banded garden spider will eat the grasshopper entangled in its web. In turn, the spider must be cautious, because it may become prey to a bird or other predator. (Photo by Barney Oldfield.)*

In many situations, interactions between species have been occurring for thousands of years. Each species may have adaptations related to the interaction that allow the interaction to continue. For example, certain plant species have chemical defenses that prevent grazing by most animals, but the digestive systems of insects that are adapted to grazing on those plants have evolved ways of detoxifying the chemicals. The term coevolution is used by population biologists to describe the process by which such close relationships develop in ecosystems.

Predation, parasitism, and grazing are other kinds of interactions in populations. The impact of one animal killing and eating another (figure 3.16) is simple enough from the standpoint of population dynamics. The prey population is reduced by one, and the predator gains energy and nutrients. In some cases, predators may limit or control prey populations, as the thatching ants control the tent caterpillars on the black cherry trees. In other cases, populations of prey and predator may fluctuate independently or in synchrony. The numerical relationships between predators and prey are not well understood in most ecosystems in part because of the difficulty of determining the amount of predation that occurs.

In the case of parasitism, the parasite lives in or on its host. Usually the host does not die from the effects of the parasite, and no change occurs in the size of the host population. A wood tick that attaches itself to a deer to obtain a meal of blood may have little effect on the deer. Hundreds of ticks attached to a deer, however, may weaken the deer by causing a significant loss of blood, and the deer may then become ill and suffer an early death.

Grazing is the term used to describe consumption of vegetation. Usually grazing does not kill the plant being eaten. Death does occur, however, if the entire plant is eaten. In some instances grazing may even stimulate growth of new plant tissue, just as mowing stimulates the growth of grass in a lawn.

## Succession

In the foregoing discussion we looked at the structure and function of ecosystems and communities as if they were stable over time. All communities, however, undergo changes over long periods of time—years or even centuries. Ecologists call this process of change succession.

The mechanisms causing succession are complex. Many of the changes in communities are brought about primarily by the plants and animals

61

**Figure 3.17** *Sand dunes change in shape as wind moves the grains of sand. Plants that grow on the dunes are often buried by the moving sand. Eventually, pioneer plants become established and anchor the sand, stabilizing the dune. (Photo by Bob Firth, Firth PhotoBank, Minneapolis.)*

themselves. Plants and animals can change their environment in a number of ways, as we will see as we follow the process of succession on a sand dune (figure 3.17).

Succession begins when a new deposit of sand is laid down by wind or water. Wind-carried reproductive spores of lichens fall on the sand and develop as small spots of gray green crusts on the surface of the sand. Seeds of many kinds of plants

are carried to the sand dune by wind, water, and animals. Those species adapted to living in the hot, dry, infertile sand grow, and their roots begin to stabilize the dune. The water-holding ability of the sand is increased by the presence of the roots and by the accumulation of organic materials from the decay of the lichens, plant stems, leaves, and roots. Certain insects and other kinds of animals live in the habitat created by these colonizing plants.

These first plants and animals to occupy the sand dune are called pioneer species.

As time goes on, the amount of organic matter and moisture in the sand increases. Because of the increased available moisture, other plant species are able to grow in this habitat, and a more dense vegetation develops (figure 3.18). But some of the pioneer plant species are not able to obtain the light and nutrients they require in this dense vegetation, and consequently they disappear from the community. Similarly, some of those animals that were able to grow and reproduce in the sparse pioneer vegetation are unable to survive in the changed

environment of this second stage in succession. They also disappear from the community, and other kinds of animals find homes in the denser habitat.

These same processes of succession continue as time passes, and the kinds of plants and animals in this locality change. Sometimes succession proceeds quickly, and sometimes it takes hundreds of years. How quickly succession proceeds depends on the type of habitat, on the climatic conditions, and in many cases on the history of land use, for example, whether the area has been logged or cultivated. Eventually a community develops in which

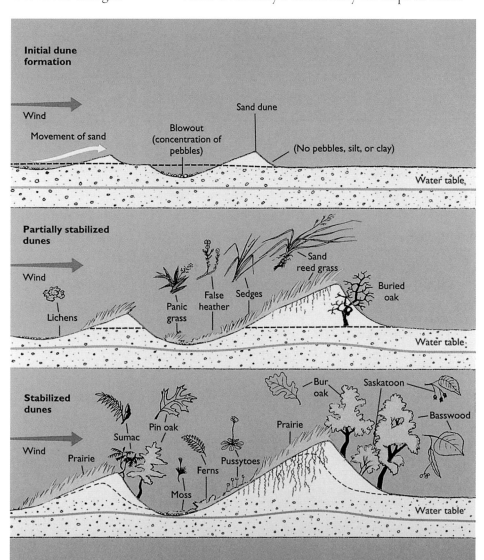

**Figure 3.18** *The process of succession, or change in a community over time, is illustrated in this diagrammatic representation of vegetation on a sand dune. (Illustration by Don Luce.)*

63

**Table 3.1** *Comparison of early- and late-successional communities*

| Characteristic | Early-successional community | Late-successional community |
|---|---|---|
| Size of plants | small | large |
| Number of species | low | high |
| Change in species composition | rapid | slow |
| Conditions for seedling establishment | excellent | poor |
| Seed production | high | low |
| Rate of seed dispersal | high | low |
| Seed dispersal distance | large | small |
| Trophic structure | mostly producers | mixture of producers and consumers |
| Food web | simple | complex |
| Transpiration rate | high | low |
| Net primary production | high | low |
| Resource acquisition rate | high | low |
| Root-to-shoot ratio | low | high |
| Efficiency of nutrient cycling | low | high |
| Efficiency of energy use | low | high |
| Growth rate | rapid | slow |
| Tolerance to stress | intolerant | tolerant |

*Sources:* Modified from Grime (1979); Huston and Smith (1987); and Odum (1969).

the plants and animals are able to reproduce and replace themselves, thus perpetuating the existing community. Characteristics of both the structure and function of these late-successional communities differ from those of early-successional communities (table 3.1). Some of these differences, such as the size of plants, are obvious. Some, however, are difficult to observe, and their existence has not been well documented.

In the first half of the twentieth century, community succession was studied intensively as an important part of the newly developing science of ecology. Observations made by founding ecologists, especially F. E. Clements (1916) and H. C. Cowles (1899), indicated that the process of succession,

that is, the direction of change, was ordered and predictable, leading from the pioneer stage through intermediate stages to a steady-state, or climax, community. Development of the community was known to be gradual and was believed to be due to biological reactions or responses. The pioneer plants and animals altered the environment— shade, soil organic matter, available moisture, and so on—making it unsuitable for their own reproduction but favorable for a new group of species. This process was thought to repeat itself until the climax was reached. The climax in any given area was believed to be determined primarily by climate. Some ecologists felt that a mysterious force might be guiding the successional process, and they viewed the community as a highly integrated superorganism (Clements 1916).

This perspective on succession and the climax stage persisted for several decades. Recent research, however, has provided a new understanding of succession (Connell and Slayter 1977; Glenn-Lewin, Peet, and Veblen 1992). Today we recognize that the end point of the successional sequence in a specific locality is quite unpredictable. Extrinsic disturbances, those originating outside the community, such as fires, flood, drought, and windstorms, also influence succession. The variation in these disturbances is evident from the examples in table 3.2. Some disturbances occur frequently and some rarely, and some influence small areas and some large areas. At any given place in a community, changes may occur at any time because of a disturbance, for example, a tree falling. This opening may be invaded by pioneer species, or seedlings of the original trees may be present and may grow quickly, preventing establishment of pioneer species. Replacement of individual trees depends on many factors, including the presence of seeds or seedlings and adaptations of the new trees to the conditions of light, moisture, and nutrients. In turn, these conditions may be influenced by short- or long-term changes in the climate resulting in periods of drought, excess moisture, or extreme heat or cold. The unpredictable nature of these events means that the future status of a given community is also unpredictable.

Thus, a community may actually be a mosaic made up of small tracts in different stages of succession. If we were to walk through a northern coniferous forest, for example, covering hundreds or thousands of acres, we might pass through climax stands of mature white spruce and balsam fir with seedlings and saplings of these species in the understory, dense pioneer stands of young aspen and paper birch growing on sites that had been logged or where wild fires occurred 10 or 20 years ago, and stands of 50- to 60-year-old aspen and paper birch with young spruce and fir in the understory. This mosaic occurs even within stands. When one or several trees blow down in windstorms, the openings created receive full sunlight compared to the shade in the surrounding forest. Pioneer species, such as aspen and paper birch, may fill the sunny openings very quickly, creating small patches of pioneer successional stages within midsuccessional or late-successional stands.

Long-term changes, such as a change in the climate, can disrupt the process of succession. For example, marked changes in the ozone layer or the carbon dioxide content of the atmosphere may cause changes in the climate that could affect the three major biomes—the deciduous forest, the northern coniferous forest, and the tallgrass prairie—that occur in Minnesota today. The communities making up these biomes, whose distribution is influenced primarily by differences in climate, developed when the climate changed from cold and wet following glaciation to warmer and drier during the past few thousand years. If Minnesota's climate becomes even warmer and drier, as some climate change models predict, the location of the biomes will probably change: for example, the deciduous forests will likely expand their range to the north and east, replacing part of the northern coniferous forest.

The remaining chapters of this book discuss these three biomes and the three kinds of aquatic ecosystems—wetlands, lakes, and streams and rivers—which occur in each of the biomes. I discuss conditions in these biomes as they existed before European settlement and as they are today. In each chapter, the structure of the various communities is described and the functioning of the ecosystems is evaluated in terms of such concepts as productivity, energy flow, nutrient cycling, species interactions, and succession. Thus, the principles of ecology presented in this chapter are applied to places and times in Minnesota.

**Table 3.2** *Events affecting ecosystems in Minnesota*

| Event frequency (years) | Event | Area (acres) | Geomorphic response | Vegetation response |
|---|---|---|---|---|
| 0.1 | daily temperature change | 0.001 | frost action | seedling disturbance |
| | precipitation | 10–100 | soil erosion | seedling disturbance |
| 1 | seasonal temperature change | 10,000–100,000 | | seedling change |
| 10 | fire | 1–100 | soil erosion | destruction, succession |
| | pocket gopher activity | 0.0001 | soil surface disturbance | seedling establishment |
| 10–100 | abandonment of farmland | 0.5–40 | soil surface disturbance | succession |
| | extreme storms (rain, ice, wind) | 0.2–100 | flooding, erosion | destruction, succession |
| | drought | 1,000 | erosion by wind | physiological response, death |
| | early or late frost | 100 | | physiological response, death |
| | acid rain | 1,000-10,000 | leaching | physiological response |
| | climate change | continental | | physiological response |
| 100–10,000 | climate change | continental | | succession |
| | glaciation | continental | landform | destruction, succession |

## DECIDUOUS FOREST  ④

Sap buckets on the trees, signs announcing the availability of fresh maple syrup along the roads, and patches of last fall's leaves showing through the melting snow are sure indicators that winter is over in the deciduous forest. Where the stately sugar maples have escaped the logger's saw and the farmer's grub hoe, collecting the spring run of sap to process into sugar and syrup is an annual rite of spring (figure 4.2).

Soon the spring flowers put on a grand show, before the leaves of the trees and shrubs have shaded out the sunlight. During summer in the deciduous forest, the tree canopy, which includes the trunk, branches, and leaves of the tops of the tallest trees, is dense, creating deep shade on the forest floor. A subcanopy of smaller trees and saplings occurs beneath the canopy layer. Maple seedlings may be abundant, and many kinds of shrubs and small nonwoody plants, called herbs, are present in the ground layer in spite of the low light level.

As daylight becomes shorter and shorter in autumn, many deciduous trees and shrubs produce a hormone that weakens the cells at the point of leaf attachment and stimulates formation of a protective layer of tissue. The cells in this abscission layer block off the circulation of water and nutrients between the roots and the leaves. As photosynthesis gradually comes to a stop, chlorophyll,

67

Figure 4.1 *Bracken ferns sometimes seem to cover the forest floor in deciduous forests. (Photo by Bob Firth, Firth Photo-Bank, Minneapolis.)*

**Figure 4.2** *Containers attached to mature sugar maple trees collect sap in early spring. About 40 gallons of sap are needed to make 1 gallon of maple syrup. (Photo by Jim LaVigne.)*

from which the leaves get their green color, breaks down into colorless compounds revealing yellow, gold, and red pigments, which have always been present in the leaves (figure 4.3).

With their circulatory system cut off, the leaves fall. (The word *deciduous* comes from the Latin word meaning to fall off.) The trees and shrubs are now ready for winter. The abscission layer and scales covering the buds that will open in the spring protect the plants from rapid loss of water during winter. Resistance to freezing is accomplished by a process called hardening, which involves cell dormancy and an increase in the content of sugar and other solids in the cell fluids. As the contents of the cells become more concentrated, their freezing point decreases, much as the freezing point of water gets lower when salt is added. The freezing of water between the cells is not fatal. Hardening enables plants to survive very cold temperatures until winter is once again over.

The first colonists on the Atlantic coast and the first hardy pioneers that pushed westward found endless expanses of dense deciduous forest broken occasionally by rivers, lakes, and wetlands and by openings caused by fires and windstorms. Before European settlement, this biome covered a vast area in North America, extending from the upper St. Lawrence River valley westward through the Great Lakes region and southward to Georgia on the east and Texas on the west (figure 1.6).

Before the rapid expansion of farming in the late 1800s, deciduous forests occurred in a broad region in southeastern and south-central Minnesota and in a narrow band extending northwestward between the coniferous forest to the northeast and the prairie to the west. The forest's range was probably limited by low temperatures in the northeast and by high temperatures, low precipitation, and fire in the south and west. Patches of deciduous forest also occurred in the north-central counties and in the Lake Superior highlands, areas with high precipitation and somewhat warmer temperatures than the rest of northern Minnesota.

The earliest map showing the forest and prairie areas of Minnesota (figure 4.4) was published as part of the *Flora of Minnesota* by Warren Upham (1884). He listed the species of plants that he and earlier botanists had found in the state and gave distribution limits for certain plants. Upham's map and the map of pre-European settlement vegetation (figure 1.25) both show that the deciduous forest was not continuous. Although large stands of deciduous forest occurred in some places, the vegetation in the region was a mosaic of forests, brushlands, oak woodlands, prairies, and savannas. Only a small fraction of these presettlement deciduous forests remain today—most of the forests have been cleared for farms and urban development.

## Deciduous forest ecology

The deciduous forest region in Minnesota exists under the range of climatic conditions that extend

**Figure 4.3** *(right) Fall colors enhance the beauty of Minnesota's deciduous forests. (Photo by Bob Firth, Firth PhotoBank, Minneapolis.)*

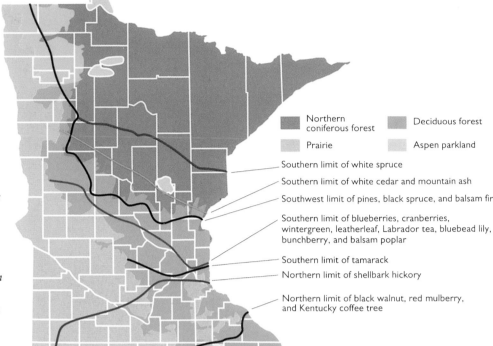

**Figure 4.4** *One of the earliest maps of vegetation in Minnesota, redrawn from the original by Warren Upham (1884), shows the borders between northern coniferous forest, deciduous forest, and prairie in 1884. Distribution limits of trees and shrubs characteristic of the two types of forests are also shown.*

Northern coniferous forest
Deciduous forest
Prairie
Aspen parkland

Southern limit of white spruce
Southern limit of white cedar and mountain ash
Southwest limit of pines, black spruce, and balsam fir
Southern limit of blueberries, cranberries, wintergreen, leatherleaf, Labrador tea, bluebead lily, bunchberry, and balsam poplar
Southern limit of tamarack
Northern limit of shellbark hickory
Northern limit of black walnut, red mulberry, and Kentucky coffee tree

from the northwest to the southeast corners of the state. Average annual temperatures range from 37°F in the northwest to 45°F in the southeast. Average annual precipitation ranges from 20 inches in the northwest to over 32 inches in the southeast. The frost-free season exhibits great variation, ranging from only about 100 frost-free days per year in the northwest to about 160 in the southeast (figure 2.5). Many deciduous trees and shrubs are tolerant of broad ranges of moisture and temperatures, which enables them to live under these differences in climatic conditions.

Today in Minnesota we find seven types of deciduous forest: maple-basswood, aspen, aspen-birch, paper birch, oak, northern hardwood, and lowland hardwood (Aaseng et al. 1993). Each type has its own characteristics with respect to soil moisture, the frequency and kind of disturbance, and its successional status. Many of these forests appear to have developed from deciduous brush-land and savanna in areas where fire was suppressed following European settlement. These different types of forest, interspersed with areas of prairie,

brushland, savanna, and numerous wetlands, provide a wide variety of habitats within the deciduous forest biome.

## Maple-basswood

The late-successional, or climax, deciduous forest, which developed in areas protected from frequent fires, is often referred to as maple-basswood forest in recognition of the two dominant trees (Braun 1950; Daubenmire 1936). Much of our knowledge of the maple-basswood forest in Minnesota comes from studies of a large tract known as the Big Woods, which once covered over 2 million acres west and south of the Twin Cities (figures 1.25 and 4.4). Nearly all of the Big Woods has been destroyed by clearing for agriculture and other uses. In Carver County, which was totally covered by the Big Woods, only one stand of 74 acres remains (Kurt Rusterholtz, Minnesota Department of Natural Resources, Natural Heritage Program, personal communication).

Early studies of several remnant stands by R. F. Daubenmire (1936) suggested that the Big Woods was a relatively old, homogeneous forest with sugar maple and basswood as the dominant trees. In fact, this research probably formed the basis for the use of the term maple-basswood to describe this forest in Minnesota. Daubenmire's findings were accepted for many years. Examination of pollen diagrams of sediment cores taken from lakes in the Big Woods in the 1960s and 1970s, however, suggested that the historical sequence reported by Daubenmire might not be correct. Eric Grimm, a graduate student at the University of Minnesota working with Edward Cushing, decided to investigate this matter further. His objectives were to determine how forest invaded the prairie in Minnesota during the past 4,000 to 6,000 years and to examine the details of succession and the causes of the expansion of the Big Woods.

Grimm used data from pollen records to study the invasion of forest into the prairie, and data from the Public Land Survey records of the mid-1800s to determine more accurately the character and history of the Big Woods. By examining the original survey notebooks, which listed the tree species that the surveyors marked to witness section corners, he determined that the characteristic trees of the Big Woods in the mid-1800s were American elm, comprising 27%; basswood, comprising 14%; sugar maple, comprising 12%; and ironwood, comprising 7%.

He was surprised, however, to discover from the pollen records that these species had only recently invaded the area and that the first trees to invade the prairie were oaks, aspen, and willow (Grimm 1983, 1984). Furthermore, he found that these species occupied the area of the Big Woods for thousands of years in spite of the presence of fire. Large bur oaks are resistant to fires, and aspen and willow sprout quickly after burning. These species persisted until the 1600s or 1700s. Then, probably because of a climate change that produced cool, wet conditions, the oak and aspen were replaced by elm, basswood, ironwood, and sugar maple.

The frequency of fires strongly influenced which tree species were present in different parts of the Big Woods. At any given location, the probability of fire was influenced by many factors: climate and weather patterns; soil and slope; the presence of firebreaks in the form of lakes and rivers; the size, pattern, and fuel content of the vegetation; and the presence of American Indians (Grimm 1984). The role of American Indians in starting fires had been thought to be minor compared to lightning. But a review of 247 accounts of prairie fires in central North America in which the source of ignition was known revealed that only one was caused by lightning and that 246 were set by American Indians (Moore 1972). Fires that started on the prairie spread into the Big Woods when and where conditions permitted. American Indians set fires also within the forest to make travel easier and to improve food conditions for themselves and for game animals. Grimm felt that fires were a more important influence on vegetation than the physical factors of climate, soils, and topography. He concluded that the historical factors of when and where fires occurred determined the overall pattern of vegetation in the Big Woods at the time of the land survey.

In maple-basswood forests today, the canopy layer is composed of maple and basswood together with elm, red oak, ironwood, ash, butternut, and bitternut hickory (Rogers 1981). It is sugar maple, however, that seems to determine much of the character of the forest. Its seedlings often blanket the forest floor (figure 4.5), its bursting buds provide the first bright color in the forest in early spring, and its leaves create the deepest shade in summer and the most brilliant displays of red and orange in the fall.

In addition to these visual attributes, sugar maple makes a significant contribution to the functioning of the forest through its role in nutrient cycling. Each spring prolific leaf production brings nutrient elements from deep in the soil into the forest canopy, where they are held in the leaf tissue. In sugar maple, as well as in basswood, slippery elm, and black cherry, these nutrients are not moved into the trunk in the fall. Rather they are returned to the soil through the leaves, which drop in autumn and decay or are consumed by insects, earthworms, millipedes, and other invertebrates. Because the leaves decompose so rapidly and return their high levels of calcium, phosphorus, and magnesium to the soil, sugar maple is sometimes referred to as a nutrient pump. The effect of this pumping is a rapid enrichment of the forest soil. Given enough time, the sugar maples and other species ultimately produce fertile soils called Alfisols. As nutrient levels increase in these soils, which have good internal drainage and moderate water-holding capacity, conditions become more

**Figure 4.5** *Sugar maple seedlings often form a continuous carpet on the floor of a late-successional forest. (Photo by Richard Hamilton Smith.)*

and more favorable to the continued growth and reproduction of the late-successional maple-basswood community.

Sugar maple plays an important role in the maple-basswood forest also because of its high reproductive rate, its great tolerance for shade, and its long life. In 1953, John Curtis (1959) studied sugar maple reproduction in southern Wisconsin by following the progress of 5.3 million seeds found on 1 acre. Twenty-eight percent (1.5 million) of the seeds germinated the following spring. Nearly 95% of these seedlings died during the summer, leaving 80,420 seedlings. Of these, 57% survived the first winter. By the spring of 1956, 14,320 seedlings remained. In a typical sugar maple forest more of these seedlings, or saplings as they are called as they grow older, die each year. The number of mature trees, those over 4 inches in diameter, is usually about 124 per acre. Thus, maple reproduction is adequate to replace the old trees as they die. Furthermore, these seedlings can survive in the dense shade, and once the canopy opens up and more light becomes available, they grow very rapidly. Individual trees grow to as much as 40 inches in diameter and may live to be 400 years old (Curtis 1959).

Early European settlers divided stands of large sugar maples into family territories, called sugar bushes. Many of these sugar bushes still exist and are used for the production of maple syrup. Sap is collected in March and April, when the days are warm and the nights are cold. Several V-shaped cuts are made in the trunk, and a wooden or plastic spigot is inserted at the base of each cut. American Indians collected the sap in birch bark bowls, but today plastic pails and bags are usually hung from the spigots. A single tree may yield 5 gallons of sap each day, but about 40 gallons of sap are needed to produce 1 gallon of maple syrup.

Basswood (figure 4.6) is the second most important tree in many late-successional forests, although many of the basswood trees probably began growing in an intermediate stage of succession and survived because of the great longevity of

**Figure 4.6** *Basswood trees, with their spherical seeds, are important members of mid- and late-successional forests. (Photo by Jim LaVigne.)*

their root systems. Basswood seedlings require high levels of light and therefore have difficulty becoming established in dense shade. Basswood can reproduce in heavily shaded sites, however, by sending up sprouts from the base of the trunk. If the sprouts survive and the old tree dies, a ring of uniform-sized trees occurs, about which legends of fairies and mystical events have been fashioned.

Shrub cover is sometimes sparse in deciduous forests, primarily because of the low light intensity. From measurements of light that I made in several stands dominated by basswood, sugar maple, and elm in Itasca State Park, I determined that the trees shaded out over 95% of the available sunlight in midsummer. Leatherwood, bush honeysuckle, and mountain maple are typical shrubs in northern Minnesota, whereas red-berried elder, choke cherry, pagoda dogwood, and prickly ash predominate in southern Minnesota forests.

Herbs are also somewhat sparse because of the dense shade. Early in spring, however, before the leaves of the trees and shrubs have shaded out the sunlight, trillium (figure 4.7), bellwort, blood-root, violets, Dutchman's breeches, and many other species create a beautiful mosaic of bright colors

on the forest floor. These flowers fade quickly as the plants rapidly produce seeds and fruits and store energy for next year's growth before the light necessary for photosynthesis is shaded out by the leaves above. Later in the summer, shade-tolerant herbs, such as zig-zag goldenrod and enchanter's nightshade, display their delicate flowers, and ostrich and lady ferns produce spores. In some stands, especially those having wet, loamy soil, wood nettles may cover large areas (Rogers 1981).

Ginseng, a small perennial herb with red berries, grows in many deciduous forests. Extracts from ginseng have long been valued in China and Korea, both as medicine and as an aphrodisiac. This latter quality is derived from the root, which is often forked and therefore bears a resemblance to a human figure. Some people believed that ginseng had powers to cure every ailment and to restore virility, and paid the highest price for roots that most closely resembled the human torso (Pfann-muller and Coffin 1989). Many homesteaders supplemented their livelihood by digging ginseng roots in the forests along the Minnesota River, and depots that purchased ginseng were set up in Nicollet and Carver Counties. Dried roots sold for

73

**Figure 4.7** *Big white trillium, and many other flowering plants in deciduous forests, bloom early in spring before the leaves of trees and shrubs shade out the sunlight. (Photo by Jim LaVigne.)*

$8 per pound in 1915. In recent years the price has fluctuated between $225 and $300 per pound. About 2,000 pounds of ginseng were dug annually in the 1970s in Minnesota (Henderson 1980). It has been said that ginseng from Minnesota's forests helped pay for many farms and enabled many settlers to survive during years of heavy grasshopper infestations (Jones 1962). Because of public pressure to maintain this source of income, ginseng is still being harvested in Minnesota, in spite of the disappearance of local populations and the reduction of its range.

74

The Minnesota trout lily, a plant endemic, or unique, to Minnesota, lives in mature maple-basswood forests and adjoining floodplains. This plant grows only in Rice and Goodhue Counties and nowhere else on earth (Coffin and Pfannmuller 1988). Because it is so rare, it was classified as an endangered species by the federal government in 1986.

Where did this plant come from, and why is it so rare? Botanists believe that the Minnesota trout lily evolved from the more common white trout lily after the glaciers receded from Minnesota. Minnesota trout lilies are rare because they almost never produce seeds but grow annually from an underground bulb. Reproduction occurs when the underground stem produces a runner that bears a new bulb. The only likely way for Minnesota trout lilies to spread to new sites is for the bulbs to be transported by floodwaters or perhaps by animals.

The Minnesota Department of Natural Resources (DNR) Natural Heritage Program monitors all of the known populations of this plant. In Nerstrand Woods State Park a boardwalk was specially constructed to allow visitors to observe and photograph Minnesota trout lilies.

*Aspen forests*

Aspen forests are most common along the western edge of the deciduous forest biome primarily on moist soils. The canopy is dominated by trembling aspen (figure 4.8). Paper birch, balsam poplar, bur oak, pin oak, green ash, and basswood may also be present in the canopy and may be common as seedlings or saplings. American hazelnut often forms a dense shrub layer. Characteristic herbs include wild sarsaparilla, Canada mayflower, false melic grass, mountain rice grass, and Pennsylvania sedge.

Aspen readily invades disturbed areas on moist soils. Its abundant seeds are very light and seem to be blown into every opening and clearing. They root quickly, and the young seedlings grow rapidly, shading out new aspen seedlings and many other species that cannot tolerate low light levels. Aspen also replace themselves by sprouting from their roots when the adult tree is cut or burned. Root sprouts produce very large leaves, sometimes 6 or more inches wide. These large leaves capture much sunlight for photosynthesis, and the sprouts quickly develop into saplings.

The aspen forest that develops is an early-successional community. Land survey records provide evidence that relatively pure stands of aspen historically occurred on level terrain and may have been maintained by fire and windthrow. If protected from fire and other disturbances, aspen forests will succeed to midsuccessional forests composed of oaks and the other species present with the aspen (Aaseng et al. 1993).

### Aspen-birch forest

Aspen-birch forest, an early-successional community, is widely distributed in northern Minnesota in areas that have been disturbed by logging and fire. The canopy contains a mixture of trembling aspen, paper birch, and large-toothed aspen; birch is more common on hillsides, and aspen is more common on flat terrain (Aaseng et al. 1993). Beaked hazelnut, mountain maple, and juneberry are common shrubs. Bracken ferns (figure 4.1) sometimes create dense ground cover. In the absence of disturbances, these forests may succeed to one of several forest types because the understory layers often contain both deciduous and coniferous trees.

### Paper birch forest

Paper birch forests are also early-successional communities, usually originating after a fire. Paper birch dominates the canopy, and mountain maple and beaked hazelnut are common tall shrubs. Mid- and late-successional tree seedlings and saplings, such as sugar maple, basswood, yellow birch, and balsam fir, are common in the understory, indicating a successional trend toward northern hardwood forest or spruce-fir forest. The herb layer is similar to that in aspen-birch forest; bluebead lily, shining clubmoss, and other mosses are especially common (Aaseng et al. 1993).

### Oak forest

Oak forests occur along the southern and western edges of the biome and in some locations within other deciduous forests. Oaks make up more than 30% of the canopy; aspen, paper birch, and other deciduous trees are also present. Some of these forests on dry, sandy soils originated from oak savanna. Prevention of fires following European settlement resulted in a shift from savannas to dense forests (Aaseng et al. 1993; Grimm 1983).

On moister sites with lower fire frequency, forests of red, white, and bur oak are midsuccessional stages that may progress toward a late-successional maple-basswood forest.

### Northern hardwood forest

The northern hardwood forest is distinguished by the presence of conifers, including white pine, balsam fir, white spruce, and white cedar (Aaseng et al. 1993). The canopy is dominated by sugar maple, basswood, and yellow birch; red oak occurs on drier sites, and black ash and American elm occur on wetter sites. Common shrubs include fly honeysuckle, leatherwood, mountain maple, and beaked hazelnut. These forests, found mainly along the North Shore of Lake Superior (Flaccus and Ohmann 1964), develop on loamy sites protected from fire and are a late-successional stage.

### Lowland hardwood forest

Lowland hardwood forests occur throughout Minnesota on sites where the soil is periodically saturated. These forests are dominated by American elm and black ash. Slippery elm, rock elm, basswood, bur oak, hackberry, yellow birch, green ash, aspen, balsam poplar, and paper birch may also be present. Fire is rare in these forests, and windthrow and flooding occur occasionally. They are considered late-successional communities.

A somewhat different type of wet forest occurs on seasonally flooded soils on the floodplains of many rivers. The DNR classification system (Aaseng et al. 1993) identifies these forests by a separate category called floodplain forest, even though mature stands contain many of the same species as lowland hardwood forests. Annual cycles of river flooding often leave characteristic windrows of debris on the ground and ice scars on the trees in floodplain forests. Black willow and cottonwood are common pioneer species on newly formed

75

**Figure 4.8** *(overleaf) Aspen forests are common along the western edge of the deciduous forest biome. (Photo by Blacklock Nature Photography.)*

**Figure 4.9** *The border between the deciduous forest biome and the tallgrass prairie biome was irregular and dynamic during pre-settlement time. Fingers of forest extended into the prairie along streams, which served as firebreaks. Fingers of prairie extended into the forest when repeated fires killed the trees. Today, the impact of agriculture and control of fires has stabilized much of the prairie-forest border. (Photo by Bob Firth, Firth PhotoBank, Minneapolis.)*

floodplains and sandbars. American elm, black and green ash, silver maple, and swamp white oak occur in mature forests.

### The prairie-forest border

Before European settlement, the border between the prairie and forest (figure 4.9) was dynamic, changing over time depending on the frequency and severity of fires and on weather patterns (Wheeler et al. 1992). Several kinds of communities occurred along the border. In some locations brush prairie occurred. M. L. Heinselman (1974) described brush prairie as a mosaic of low shrub thickets, patches of small trees, and prairie. Brush

prairies burned frequently but were quickly regenerated because of the rapid growth of willows, dogwoods, hazelnuts, cherries, and aspen.

Savannas, communities of scattered trees with prairie beneath (figure 4.10), also occurred along the prairie-forest border. The Reverend H. M. Nichols described a savanna near Chanhassen in western Hennepin County in a letter published in 1855 in the *Minnesota Republican* (Nichols 1939, 143):

The face of the country here differs from almost every other portion of the Territory. It is not a prairie, neither is it timber, nor yet is it openings, such as we call openings in other places. There is plenty of large heavy timber, maple, ash, oak, bass, &c., but the trees stand alone, as if a part of what had once been a heavy forest, while the rest had been taken off, without leaving a vestige behind.

Sometimes not more than a dozen or twenty of these trees will be found on an acre; while again, on an acre or two they stand like a forest. They shade the ground but little, being tall with small tops, like forest trees, as indeed they are, with but little of the forest. Thus a man has the strong soil of a timber farm, without the labor of clearing off the timber.

Here are also some splendid meadows, where the grass grows higher than a man's head. Take it all in all a more desirable place for residence could not well be found in Minnesota. It is only some 3 or 4 miles from the Minnesota River, and about the same distance from Minnetonka Lake.

The tall grass referred to was probably big bluestem or Indian grass. These species are both common on most of the native tallgrass prairie remnants in Minnesota.

A community such as that described by Rev. Nichols exists today at the Cedar Creek Natural History Area in Anoka County. Cedar Creek lies in the transition zone between the tallgrass prairie biome to the west and the deciduous forest biome to the east. Before settlement, the vegetation in this region was probably a mosaic of small patches of oak savanna scattered throughout larger expanses of oak woodland and brushland. Pollen studies in east-central Minnesota indicate that wildfires caused by lightning strikes or humans occurred every few years (Grimm 1984). This high frequency of fires is believed to have been responsible for preventing the spread of the deciduous forest westward into the brushland and savanna. Prairie plants are adapted to frequent fires, and the thick bark and sprouting capability of the oaks allow them to survive fires.

The first white settlers arrived in the vicinity of Cedar Creek in the 1850s. Selective logging provided timbers and lumber for homes and other buildings. By the early 1900s many farms existed in the area. Most fires were controlled to prevent damage to buildings, crops, livestock, and the surrounding woodlands. Without fires, the savanna

**Figure 4.10** *Savanna, prairie with scattered trees, occurred in some places along the forest-prairie border. At Cedar Creek Natural History Area in Anoka County, pasque flowers, Indian grass, and big and little bluestem occur with bur and pin oak (Photo courtesy of Minnesota Department of Natural Resources.)*

79

**Figure 4.11** *Prescribed burning has been carried out at Cedar Creek Natural History Area since 1964 to restore oak forest to savanna. Fire breaks are disked, and the fire is started with drip torches. (Photo by author.)*

developed through the process of succession into oak forest with a dense understory of shrubs (Wovcha, Delaney, and Nordquist 1995). In some localities the change was very rapid—openings became filled by saplings and shrubs within a decade.

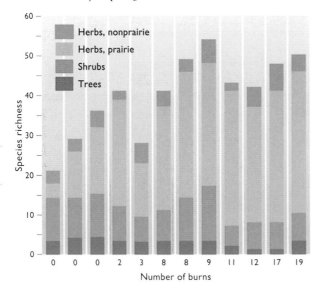

**Figure 4.12** *The number of prescribed burns affects the number of species of trees, shrubs, and herbs on experimental plots at Cedar Creek Natural History Area. Herbs are divided into prairie and nonprairie categories. (Tester 1989, by permission.)*

In 1964 an experiment using prescribed burning (figure 4.11) was begun to restore the oak forest at Cedar Creek to savanna (Irving 1970; Tester 1989). Burning every year or every other year in spring has resulted in a decrease in trees and shrubs (White 1983) and a striking increase in prairie species, which grow under and between the scattered oaks (figure 4.12). Today portions of the experimental area probably closely resemble presettlement savanna.

The suppression of fire also allowed the invasion of the prairie by other species. Before settlement, aspen, hazelnut, and willow as well as oak grew in areas bordering the Big Woods. In places where settlers came and controlled fires, these species began to encroach on the prairie. N. H. Winchell (1884, 278–79) recorded this process in Fillmore County in the late 1800s:

The area covered by native timber is steadily increasing. A large proportion of the county is covered with bushes which are composed of hazel, aspen, oak (two sorts) and, where these are wanting, a species of low willow which seems to come up first after the prairie fires are stopped. After the willow, hazel and oak and aspen gradually come in, and in time convert the original prairie to a bushy or timbered region. Over some large tracts in the county this process is going on. There are thousands of acres of young native timber, not exceeding five or six inches in diameter, due to this gradual change since the suppression of the prairie fires.

The same process, with the same species, also occurred along the prairie-forest border in northwestern Minnesota (Ewing 1924) and is still occurring on the remaining prairies (Tester and Marshall 1961, 1962). Invading trees are primarily trembling aspen, which thrive in the absence of fires. Aspen seeds are very light and are distributed widely by wind. Seedlings require high light for growth and are frequently found on bare pocket gopher mounds. Aspen also reproduce by suckering from roots, which produces dome-shaped clones in which every tree develops as a sucker from the roots of the first seedling (Buell and Buell 1959). Thus, all trees in a clone are connected and are genetically identical. Clones are common in the prairie-forest border in northwestern Minnesota and Manitoba. This community of scattered aspen clones interspersed over the prairie is referred to as aspen parkland (Bird 1961). Balsam poplar, bur oak, and paper birch are sometimes associated with the aspen, and on some dry sites bur oak groves can be found.

The prairie has long contained corridors of deciduous forest along the rivers and streams that extend westward into the tallgrass prairie region and far beyond the western border of Minnesota. Along these waterways grow long, narrow stands of trees, called floodplain or gallery forests. The rivers and streams provide moisture for tree growth, and frequent floods deposit layers of silt over the floodplain, providing a rich supply of nutrients for tree growth. Common trees are cottonwood, willow, hackberry, box elder, American and slippery elm, silver maple, and ash (Noble 1979). Floodplain trees are tolerant of flooding, damage from floating ice and debris, and heavy siltation. These stands provide habitat and travel corridors for many kinds of forest animals, enabling them to survive far beyond their usual distributional range.

*Ecosystem function*

Each year in deciduous forests new leaves are produced by the trees and shrubs, trunks and branches increase in size, and many herbs produce entire new plants. In addition, old roots increase in size, and new roots are produced. Most of this biomass is in the tree trunks and branches; less than 1% is represented by the herbs. Comparison of world averages of the annual productivity of deciduous forests to the productivity of communities in the other two biomes shows that the deciduous forest has the highest biomass and productivity, followed by northern coniferous forest and tallgrass prairie. These estimates included the dry weight of all living trees, shrubs, and herbs, plus all dead plant material aboveground.

Because these average values integrate a variety of soil, topographic, and nutrient conditions, the differences in the values probably reflect major climatic differences between the biomes. Maximum biomass and productivity on land is found in locations with high temperatures and high precipitation; thus, the highest productivity in the world is found in tropical rain forests. In Minnesota, the temperature gradient increases from northeast to southwest. The precipitation gradient increases from northwest to southeast. Because the deciduous forest is more productive than the other biomes in Minnesota, we can reasonably assume that the optimum conditions of temperature and moisture for plant growth in Minnesota occur where this forest occurs, in a northwest to southeast strip across the state (figure 1.25). To the northeast of this strip, precipitation is adequate for deciduous forest, but the temperature is cooler; to the southwest, the temperature is warmer, but precipitation is lower.

The variations in biomass and productivity shown for the different stands in table 4.1 are, to a

**Table 4.1** *Comparison of approximate biomass and productivity of selected forest, prairie, and agricultural communities*

| Community | County | Biomass (pounds per acre) | Productivity (pounds per acre per year) | Source of data |
|---|---|---|---|---|
| *Deciduous forest* | | | | |
| Red oak | Anoka | 230,000 | 7,500 | Ovington et al. 1963 |
| Aspen | Hubbard | 205,000 | 9,000 | Bray and Dudkiewicz 1963 |
| Mixed swamp forest | Anoka | 90,000 | 6,500 | Reiners 1972 |
| *Northern coniferous forest* | | | | |
| Black spruce | Itasca | 90,000 | 3,000 | Grigal et al. 1985 |
| Jack pine | Itasca | 110,000 | | Ohmann and Grigal 1985 |
| Red pine | Itasca | 240,000 | | Ohmann and Grigal 1985 |
| White cedar | Itasca | 135,000 | | Ohmann and Grigal 1985 |
| 10 Minnesota and Wisconsin stands | | | 8,000 | Crow 1978 |
| *Tallgrass prairie* | | | | |
| Sand prairie | Anoka | 400 | 1,000 | Ovington et al. 1963 |
| Mesic prairie | Polk | 4,000 | | Smeins and Olsen 1970 |
| Mesic prairie in Red River valley, North Dakota | | | 4,000 | Hadley 1970 |
| *Agricultural land* | | | | |
| Alfalfa hay (aboveground only, wet weight) | | | 6,000 | Rock et al. 1993 |
| Sugar beets (belowground only, wet weight) | | | 37,000 | Rock et al. 1993 |

*Note:* Except where noted, all values are aboveground dry weight.

great extent, related to temperature and moisture. Within a biome, however, differences in productivity are often related to the availability of nutrients. For example, the high production of sugar beets on agricultural land in the tallgrass prairie biome is due mainly to the addition of fertilizer.

What happens to all the biomass produced each year in the deciduous forest, the old leaves, the dead plants, and the twigs, bark, branches, and tree trunks that fall to the ground? Grazing animals, including insects, eat only a small amount, usually less than 10%, of the annual production of plant tissue. These thousands of pounds of litter, plus carcasses of insects, birds, mammals, and other organisms, if allowed to accumulate, might cause profound changes in the functioning of the forest ecosystem. This dead biomass, however, constitutes the basis of the detritus food web, described in chapter 3. The breakdown of the forest litter is caused by both physical and biological forces. Wind, rain, ice, and frost may physically break leaves and twigs and cause entire trees to fall.

Earthworms, bacteria, and fungi plus protozoans, insects, and other invertebrates are the primary decomposers, however (figures 3.1 and 3.11). Even complex litter components such as lignin and cellulose can be broken down by the digestive enzymes of fungi.

In mature or climax forests, the rate of accumulation of litter is about equal to the rate at which the litter is broken down or decomposed. Thus the community remains in a somewhat stable state with regard to litter (Ovington et al. 1963), just as it remains stable in terms of species composition. The end products of decomposition are the chemical elements and compounds used by the plants as they grow. These are returned to the forest soil to be used again (figure 3.10). With sugar maple acting as a pump, some nutrients cycle rapidly, adding to the productive capacity of deciduous forests.

Movement or cycling of specific nutrients has been studied in several deciduous forests, although none of these are in Minnesota. In an undisturbed New Hampshire forest, input of sulfate, nitrate, and hydrogen ions was found to be primarily from rain and snow. As the water moved through the forest, its composition changed. Measurements of the chemistry of the water in streams flowing out of the forest showed that the forest ecosystem lost calcium, magnesium, sodium, and potassium and gained nitrogen and phosphorus (Likens et al. 1977). The first four elements were replaced by weathering of the bedrock. Gains in nitrogen and phosphorus came from precipitation and gaseous input. In general, nutrient cycling in these deciduous forests appeared to be similar from year to year, and losses and gains of specific nutrients were quite predictable. The system appeared to be stable when no disturbances caused disruption of energy flow or nutrient cycling.

## Animals and community interactions

I will not discuss or even mention all of the kinds of animals that occur in each biome. Rather I will discuss in some detail the most important and the most conspicuous species and treat species of lesser importance in groups based on taxonomic relationships or on similarities in their role in the functioning of the ecosystem. Lists of vertebrates with information on their distribution in Minnesota are presented in appendixes E, F, and G. Readers are encouraged to consult the references associated with each list to obtain detailed information about the distribution and ecology of individual species.

### Mammals

Evan Hazard concludes in his book *The Mammals of Minnesota* (1982) that there is more human concern for white-tailed deer than for any other wild mammal in Minnesota. Deer are of interest as a source of food for people and wolves, as objects of aesthetic beauty, as tourist attractions, as a source of danger on highways, as pests to farmers, foresters, and orchard operators, and perhaps primarily as the state's most important big-game species.

Today white-tailed deer (figure 3.4) are present in every county in Minnesota. This was not always

the case, however. Deer are usually most abundant in the pioneer and early stages of forest succession, where the shrubs and small trees provide excellent browse and cover. Before European settlement, deer were common in the deciduous forests and along rivers and streams in southern Minnesota, where food and cover were plentiful, but "up to about the year 1860 deer were very rare in the heavy evergreen forests of the northeastern counties from about Chisago and Morrison Counties northward" (Surber and Roberts 1932, 17).

As settlers moved into Minnesota, deciduous forest was cleared to make fields, and deer were taken to feed families and to sell to markets. Market hunting was at its peak in Minnesota in the 1870s, and venison could be purchased in St. Paul butcher shops for 8 to 10 cents per pound. E. B. Swanson (1940) reported that in December 1872, 6 tons of venison were shipped from Litchfield in Meeker County, near the western edge of the Big Woods, to Boston, Massachusetts. In 1876, 7,409 venison saddles and carcasses plus 2 tons of venison hams were shipped to St. Paul.

Uncontrolled hunting and the reduction in habitat due to land clearing and intense farming reduced the number of deer in the southern counties so that by the late 1800s they had all but disappeared (Petraborg and Burcalow 1965). In the early 1900s, however, logging and burning in the central and northern counties stimulated the growth of pioneer shrubs and trees, and the deer population increased markedly there. More recently, deer have increased in agricultural and urban areas and are even considered to be pests in some places. Minnesota's deer herd is thriving today, and the population is large enough to allow for an annual hunting season.

Deer are primarily browsers, which means that they eat mostly leaves, twigs, and buds of trees and shrubs. They also eat a wide variety of other foods, such as grasses and herbs, acorns, mushrooms, aquatic plants, and orchard and agricultural crops. Deer are important residents of the deciduous forest because their foraging can have a great impact

83

on the community. Deer sometimes remove nearly all the plants of a certain species from a community. Sumac, red osier dogwood, and mountain maple are sometimes browsed so severely that the plants die.

In winter their food is limited to the browse available above the snow. When the snow reaches about 18 inches in depth, deer move to a yarding area where they may spend the remainder of the winter. Deer yards are often located where winter conditions are somewhat moderated such as on south-facing hillsides warmed by the winter sun. White cedar swamps, where snow may not drift and where radiation of body heat to the cold night sky is reduced, are also often used for yarding. Some of the deer fitted with radio transmitter collars at Cedar Creek Natural History Area moved as far as 20 miles from their summer range to their winter yard (Rongstad and Tester 1969). The deer showed a marked preference for the white cedar stands, which occur in lowlands within the oak forests (figure 4.13). Once in the yard, they feed on whatever they can reach. If the snow persists into spring, all the browse may be consumed. Then, many deer, especially fawns, die of starvation, and some pregnant does resorb their fetuses.

Consequences of a deer population exceeding the carrying capacity of the habitat may be severe. If numbers become too large, the food supply will be exhausted, and starvation is likely. Predation is one of the ways in which deer populations are controlled. In the north, timber wolves and humans are the main predators. In most deciduous forests of Minnesota, however, humans alone are the predators, and in parks, refuges, and urban areas deer often have no predators. In these locations, deer may be very abundant and often cause damage to reforestation efforts, gardens, and yards.

Other grazers in deciduous forests are cottontail rabbits and woodchucks, sometimes called groundhogs or marmots. Cottontails, which prefer edges of the forest and somewhat open brushy areas, are found throughout the deciduous forest and in the prairie along streams and in other areas where

trees and brush occur. The rabbits are known for their high rate of reproduction, sometimes having four litters of four to eight young in a single year.

Like cottontails, woodchucks prefer edges and clearings, where they dig extensive burrow systems, usually with several entrances. They eat green vegetation primarily and put on a heavy layer of fat during summer and fall. Before winter sets in, they move into their dens and begin hibernation. While they sleep their body temperature drops to within a few degrees of the surrounding temperature, and their metabolic rate may be reduced by half. Woodchucks and other hibernating mammals alternate periods of torpor, similar to sleep, with periods of arousal. During the 10% of the time they spend in arousal, they use about 90% of their stored fat. When the hibernator is in torpor, the amount of

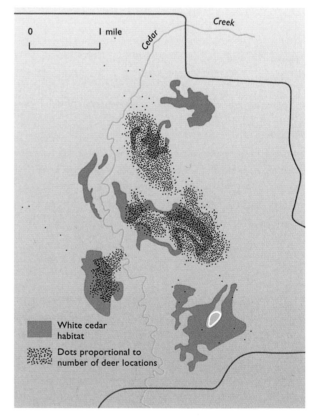

Figure 4.13 *Seven radio-tracked white-tailed deer showed a marked preference for white cedar stands at Cedar Creek Natural History Area in late winter. Some deer moved as far as 20 miles from their summer range to winter in these cedar swamps. (Rongstad and Tester 1969, by permission.)*

stored fat needed for energy to maintain body temperature is greatly reduced. Energy is conserved because less heat is lost to the air and ground when the animal's body is nearly as cool as its surroundings. Other hibernators in the deciduous forest include several species of bats, ground squirrels, chipmunks, and jumping mice.

Some mammals such as raccoons, skunks, and bears are dormant during winter but are not true hibernators. They are called lethargic carnivores. They increase body fat in the fall, but their body temperature drops only a few degrees, and their metabolic rate remains quite high during their winter sleep. On warm winter days some animals may waken and travel in search of food. In very long, cold winters, these lethargic carnivores may deplete their fat reserves and die in their dens (Mech, Barnes, and Tester 1968).

A great variety of other kinds of mammals live in deciduous forests. Gray and fox squirrels commonly den in hollow trees or build leaf nests in branches. The squirrels are primarily grazers, feeding on nuts, fruits, berries, and other plant parts, but they will take insects, birds, eggs, and other animal food when they can. Raccoons (figure 4.14) feed on a wide range of plant and animal foods and often can be observed hunting along a lakeshore or stream. Both squirrels and raccoons are hunted for their meat. In frontier times, raccoon fur was used extensively for coats and caps. For many years raccoons have been extending their range northward and westward and are now found throughout Minnesota.

Opossums, which are poorly adapted to cold weather, are common in southern Minnesota forests but are rare north of the Twin Cities (Hazard 1982). Their distinguishing characteristic is the marsupial pouch in which the female carries her young. Gestation is only 12 or 13 days, and the tiny young climb into the pouch on their own and attach themselves to a nipple, where they remain for about two months. After leaving the pouch, young opossums may ride on their mother's back by clinging to her fur.

**Figure 4.14** *Raccoons are common in deciduous forests but also live in cities and in farmland. (Photo by Bob Firth, Firth PhotoBank, Minneapolis.)*

Striped skunks and gray and red foxes are common in Minnesota's deciduous forests, but only the gray fox is found exclusively in this habitat. All three species are omnivores and eat a wide range of foods, including rodents, birds, eggs, insects, frogs, seeds, and fruits. All are nocturnal, or active mainly at night, and are therefore not commonly seen. Gray foxes have the ability to climb trees and use this ability both to obtain food and to escape from predators.

Six species of bats live in both the deciduous and the coniferous forest biomes, and one, the eastern pipistrelle, lives only in the deciduous woods of the southeast. They feed almost exclusively on insects. Some species of bats are migratory, leaving Minnesota in the fall and returning in the spring. Others hibernate in caves, mines, rock crevices, and buildings. Proper temperature and humidity are requirements for overwintering sites, and water, insects, and a place to rear young are requirements

**Figure 4.15** *White-footed mice (on rock) and short-tailed shrews are common in many forests. The mice eat mostly seeds and nuts, where-as the shrews are carnivores, feeding on insects, worms, and sometimes on mice and other small mammals. (Photo by author.)*

for breeding. Some females gather in maternity colonies in warm, sheltered locations, but others bear their young in isolation. Most species have low reproductive rates, usually bearing only one or two young per year, but are long-lived. Little brown bats may live 20 years or more (Hall, Cloutier, and Griffin 1957).

Bats are active at night, capturing insects while flying. Their echolocation system enables them to find insects in the air. Capture is made when the bat uses its wing to sweep the insect near enough so that it can grasp it with its teeth. During the day, bats roost in trees, under bark, in buildings, in rock crevices, and in woodpiles.

Many people have a strong fear of bats. Contrary to popular opinion, bats generally avoid people and are not attracted to hair. They often have an erratic flight path when they are chasing insects, however, and may swoop down close to the ground to gain speed. A nearby person may think that the bat is attacking, but in fact the bat is careful to avoid any stationary object. Bats can be beneficial to humans because of the large number of insects they eat. In addition, research on bats has provided information on birth control, artificial insemination, vaccine production, survival in cold temperatures, drug testing, and navigation (Galli 1991).

Bats can cause problems for people, but the problems are neither common nor very serious. Accumulations of bat droppings in an attic or building may contain spores of histoplasmosis, a common fungus in soil and dust. Breathing these spores may cause a cough or mild flu, but the possibility of contracting histoplasmosis is remote (Galli 1991). Bats are known to carry rabies, but in a typical year only about six infected bats are reported in Minnesota, and no outbreaks of rabies in bats have been observed.

On average, about 12 species of mice, chipmunks, and shrews are common in deciduous forests. The white-footed mouse (figure 4.15) is likely to be most abundant, followed by the red-backed vole and the masked shrew. Mice and chipmunks may be either herbivores or omnivores, but their most common food is vegetation. Shrews (figure 4.15) are meat-eaters, feeding on insects, worms, snails, and other invertebrates and sometimes on other mammals. In turn, small mammals are eaten by numerous predators, including weasels, foxes, hawks, and owls.

One of the most secretive and interesting mammals in the deciduous forest is the southern flying squirrel. These squirrels are only active at night and are seldom seen. Furthermore, their high-pitched and almost birdlike calls may not be within the range of hearing of some people. Flying squirrels live in dens and leaf nests and eat a wide variety of plant and animal foods, similar to gray and fox squirrels. These squirrels cannot truly fly, but they do have a flat tail and folds of skin on each side that stretch from the front to the hind leg, giving them a flat gliding surface. They glide to cover distances quickly and to escape from predators. Flying squirrels have been known to visit bird feeders, so perhaps a periodic check at night will be rewarded with a look at one of Minnesota's seldom-seen mammals.

## Birds

Because of the deciduous forest's structural diversity, it provides a large variety of habitats for birds. Areas with a large variety of habitats can support a large number of species; thus, many species of birds breed in the deciduous forest (appendix F). Characteristic breeding species in upland sites include the broad-winged hawk, barred owl, black-billed cuckoo, least flycatcher, eastern wood-pewee, yellow-bellied sapsucker, hairy woodpecker, pileated woodpecker, white-breasted nuthatch, rose-breasted grosbeak, red-eyed vireo, veery, and ovenbird. On moister sites along water where cottonwood, silver maple, American elm, and green ash are the common trees, wood ducks, red-shouldered hawks, eastern screech-owls, red-bellied woodpeckers, tufted titmice, yellow-throated vireos, northern orioles, wood thrushes, and scarlet tanagers are likely to breed. In more open, savannalike stands with several species of oak as the common trees, the breeding birds include the red-tailed hawk, great horned owl, red-headed woodpecker, gray catbird, house wren, brown thrasher, and indigo bunting.

We can further understand how the deciduous forest supports a large number of species of birds by looking at how they gather food. The variety of habitats is provided by the patchiness of the forests, which is caused by disturbance and by the layers, or strata, in the vegetation. Many birds search for their food and locate their nests in a

specific stratum, such as the herb layer, shrub layer, or the canopy layer. More bird species are therefore likely to be present in deciduous forests, which may have four or even more layers, than, for example, in northern coniferous forests, which usually have only two or three.

Thus, a useful way of looking at birds in the deciduous forest is to group species with similar feeding methods into foraging guilds. B. R. Noon, V. P. Bingham, and J. P. Noon (1979) summarized the results of 70 censuses of breeding birds in northern deciduous forests by guilds and found that the following species were most likely to be recorded in a census:

| | |
|---|---|
| Foliage gleaners (obtain food from leaves and stems) | red-eyed vireo scarlet tanager cardinal |
| Ground foragers | ovenbird wood thrush brown-headed cowbird |
| Salliers (fly from a perch to catch flying insects) | eastern wood-pewee great crested flycatcher Acadian flycatcher |
| Bark gleaners | white-breasted nuthatch tufted titmouse black-capped chickadee |
| Bark drillers | hairy woodpecker downy woodpecker red-bellied woodpecker |
| Hover gleaners (glean from bark and foliage while hovering) | blackburnian warbler parula warbler golden-crowned kinglet |
| Hawkers (catch insects in flight) | whip-poor-will chimney swift tree swallow |
| Predators | barred owl red-tailed hawk broad-winged hawk |

One of the few species of birds that affect their community in a major way is the great blue heron, a colonial nester. Many pairs, sometimes more than a thousand, gather in April at the same location, called a rookery, and build large stick nests near the tops of trees. The nests are often only a few feet apart, and a single tree may hold many nests. Young herons remain in the nest until July or August, when they are nearly full grown. The parents catch fish, crayfish, frogs, and other aquatic animals, which they partially digest before regurgitating the food into the beaks of their young. When the young fledge and leave the nest, the colony remains empty until the following spring.

A rookery may be used annually for many years. The Cold Spring heron colony in Stearns County has been known since the late 1880s. Records have been kept since 1954, when the colony consisted of 386 nests in 102 trees. A peak population of 1,584 nests occurred in 1974 (Searle and Heitlinger 1980). The Nature Conservancy has established a 62-acre preserve to protect the colony from disturbance.

How do great blue herons affect the community in which they live? The answer is that the birds may actually kill many trees within the rookery with their excrement and regurgitations, which may kill leaves and twigs on contact and increase the acid and nutrient content of the soil in which the trees grow. Stinging nettles, which flourish in rich, acid soils, are often thick beneath a rookery. Unfortunately, when a tree dies, the birds may abandon their nests and move to live trees nearby, thereby leading to the death of still more trees.

In terms of size, wild turkeys are a dominant species in the southeastern forests. They have probably always been present, but their numbers were low until the 1960s. Introductions of turkeys from other states by the DNR beginning about 1960 coincided with several years of favorable weather, resulting in a sharp increase in the population. Establishment of food patches of corn in strategic locations by the DNR and private groups enhanced survival over winter. By 1978 the population numbered in the thousands, and since that date, hunting of gobblers has been allowed.

Another important game bird, the ruffed grouse, is common year-round in deciduous forests as well as in the northern coniferous forest. The courtship drumming of male grouse during the spring breeding season is often heard in early morning and evening hours. More is said about ruffed grouse in chapter 5.

**Figure 4.16** *Passenger pigeons are now extinct. This photograph was taken of the exhibit at the James Ford Bell Museum of Natural History, University of Minnesota. Large flocks, probably of millions of birds, fed on acorns and nuts in Minnesota's deciduous forests. (Photo by Jim LaVigne.)*

Woodcock are found in many of the same locations as ruffed grouse. They breed throughout the deciduous and northern coniferous forest biomes in open habitats with shrubs, brush, or small trees as cover. Their favorite food is earthworms, which they seek out in rich, dark soils, often under alder thickets. Because of their remarkable cryptic coloration, they are rarely observed until they flush just before they are about to be stepped on. Woodcock migrate south and east from Minnesota, and many winter in the states bordering the Gulf of Mexico.

The passenger pigeon, although extinct, deserves to be mentioned. Passenger pigeons (figure 4.16) were reported to be the most abundant birds in North America, and for that matter in the entire world, about 300 years ago. They traveled in flocks of millions, which reportedly darkened the sky. By the late 1800s only a few wild passenger pigeons remained, and in 1914 the last one died in the zoo in Cincinnati.

Passenger pigeons were strongly associated with deciduous forests because the birds' main foods were the acorns and nuts produced by forest species such as oak, chestnut, and hazelnut. Flocks would travel over the forest, searching for areas with large acorn or nut crops, and then would settle in nesting colonies containing millions of pairs. When the food was exhausted, the colony would search for a new location. This search usually happened each year, because the incredibly large numbers of birds would consume the entire food supply in a single nesting season. The largest nesting colony ever recorded covered 850 square miles and contained about 135 million adults (Schorger 1955). Although the cause of extinction of this species is not known for certain, it is generally believed to be a combination of netting, shooting, and removal of nestlings by market hunters.

### Amphibians and reptiles

More species of amphibians and reptiles live in southeastern Minnesota than in any other part of the state, perhaps because of the warmer climate. Thirty-seven of the 48 kinds of amphibians and reptiles present in Minnesota (appendix G) occur in the southeast, and some, such as the bullfrog, are at the extreme northern limit of their range. Most are found in the deciduous forest, in open or brushy clearings on the edge of the woods, or on the steep, rocky bluffs along the Mississippi River and its tributaries. Most amphibians and reptiles are shy and secretive, and many live in inaccessible places, making them difficult to observe and study. As a result, we know little about the abundance of these animals in Minnesota. Some indication of their abundance can be obtained from public records, which reveal that 6,965 painted turtles were harvested and sold in the state in 1978 (Minnesota Department of Natural Resources, Section of Fisheries, unpublished records) and that bounties of $1 each were paid for 3,000 rattlesnakes in Houston County in 1982 (Oldfield and Moriarty 1994).

We can infer that amphibians and reptiles are just as important as birds or small mammals in terms of their role in the trophic structure of ecosystems from data collected in a hardwood forest in New Hampshire. T. M. Burton and G. E. Likens (1975) found that the biomass or weight of salamanders alone was twice the biomass of all birds combined and equal to that of small mammals. The annual productivity of salamanders was about five times that of birds.

One of the more abundant amphibians in the deciduous forest is the gray treefrog. Its most remarkable characteristic is that the color of its skin changes to match the surroundings. This change is controlled by the pigment cells in the skin, called chromatophores, which expand or contract to cause variations in color. A color change from green to gray or brown can occur within a few minutes. Adhesive discs on the toes of gray treefrogs (figure 4.17) enable them to climb nearly any surface, and they can be found high in a tree or even climbing up a window of a house. They prefer breeding in undisturbed ponds with a scattering of logs, sticks, and brush.

Since treefrogs and other amphibians have moist skins, lay their eggs in water, and complete

89

their development in water, they must live in habitats with adequate moisture. Their preference for moist conditions is apparent during rainstorms, when many frogs, toads, and salamanders, moving beyond their usual habitat because of the rain, are observed on country roads.

Reptiles are adapted to existence on land throughout their life. Their bodies are covered with scales or plates, and their eggs are laid on land or develop within the body until hatched. Reptiles are often seen basking in the sun and can be observed in the driest habitats in Minnesota.

Fox and gopher snakes, which may reach 5 feet or more in length, are predators of treefrogs, as well as other amphibians, small mammals, birds, and eggs. They often crawl into burrows of pocket gophers, where they will eat both adults and young. Fox snakes seem to prefer moist habitat in river valleys and near wetlands. Gopher snakes are more

**Figure 4.17** *Gray treefrogs use the adhesive discs on their toes to climb trees and other objects. They breed in forest ponds and spend winters buried in the leaf litter on the forest floor. (Photo courtesy of Artis Orjala.)*

likely to be found in the rocks on steep slopes and in dry, savanna-type habitats, such as occur on the Anoka Sand Plain. Both species lay eggs, which hatch in about two months. Newly hatched young are about 10 to 12 inches long.

Garter snakes are probably the most common reptiles in the deciduous forest. They prey on amphibians, earthworms, and other invertebrates. Although they prefer moist, shady forest habitats, they often wander into gardens, lawns, and roads, where many are killed by traffic. Garter snakes give birth to live young; broods as large as 22 have been reported (Dunlap and Lang 1990). The ratio of males to females in the broods varies with body size. Larger females produce more male young.

Like many other kinds of snakes, garter snakes spend the winter in a place called a hibernaculum, located in such places as an old ant mound, an abandoned burrow system, or a crevice among the rocks. Hundreds of snakes of several species may winter in one hibernaculum.

Minnesota's only poisonous snakes live in the southeastern counties. Timber rattlesnakes (figure 4.18), which are relatively common, and massasaugas, or swamp rattlesnakes, which are rare, live in forests and open fields adjacent to streams in summer. When fall arrives, timber rattlesnakes move to the steep, rocky bluffs and outcrops near their hibernacula, where many individuals overwinter together. Massasaugas winter individually, sometimes in crayfish burrows (Vogt 1981) in streamside forests. Rattlesnakes are not vicious and will usually try to escape when disturbed. Furthermore, they frequently sound a warning rattle before striking. If surprised, however, they will strike without rattling. Hikers and rock climbers would do well to remember this and use caution when moving about in rattlesnake habitat in southeastern Minnesota.

Rattlesnakes prey on a wide variety of animals, including mice, gophers, ground squirrels, rabbits, and birds. Although they are formidable predators, rattlesnakes also have their enemies. Foxes, coyotes, and a variety of hawks and owls prey on all kinds of snakes.

**Figure 4.18** *A timber rattlesnake basks in the sun on a rock overlooking the Mississippi River valley. This poisonous snake has rattles, vertical eye pupils, and facial pits below its eyes. Its rattle is held vertically when crawling or when alarmed. (Photo by Barney Oldfield.)*

How do these animals that cannot regulate their body temperature and that do not appear to have any type of insulating body cover, such as fur or feathers, survive Minnesota's winters? There are several answers. Turtles and some frogs, for example, spend winter in the mud at the bottom of ponds, lakes, or streams, where the temperature is above freezing. Some snakes, salamanders, and toads spend winter burrowed in the soil below the frost line (Breckenridge and Tester 1961). A few amphibians have evolved special mechanisms that allow them to withstand below-freezing temperatures.

W. D. Schmid (1982) has studied three frogs in Minnesota forests—gray treefrogs, spring peepers, and wood frogs—that can survive freezing. These frogs spend the winter in or under the leaf litter on the forest floor, where temperatures are often below freezing. Schmid found that the water content of these amphibians is about 80% of their body weight. Some of this water is inside the cells, and some is free in the body. During winter the water in the body may freeze, giving the impression that the frog is frozen solid. The water inside the frog's cells, however, contains sufficient glycerol, which acts like antifreeze, to prevent the cells' contents from freezing. In spring the temperature gradually warms, and the frozen frogs thaw. The glycerol is metabolized, and none is present in the body fluids in summer. This mode of overwintering is an excellent example of the special adaptations that allow animals to survive in Minnesota's climate.

### Insects and other invertebrates

Insects abound in the deciduous forest, as they do in every community in Minnesota. Each commu-

**Figure 4.19** *This inch-worm, a moth larva, is so well camouflaged that it appears to be a twig of the basswood tree. (Photo by Don Rubbelke.)*

nity, and even each species of tree or shrub, has a characteristic association of insects (figure 4.19). My discussions of insects in this and the following chapters are based on ecological rather than taxonomic aspects.

Deciduous forest trees are affected by insects in a variety of ways. The most conspicuous impact is probably caused by defoliators such as webworms, sawflies, caterpillars, cankerworms, and leaf miners, which eat leaves and needles. Defoliation rarely kills trees but does reduce growth and may weaken trees, making them susceptible to other insect pests or to disease.

Scale insects, aphids, and spittlebugs suck juices from leaves, twigs, or roots. If abundant, these sucking insects may weaken or even kill trees. Boring insects, including species of moths, wasps, and beetles, make holes in wood. Usually they infest old or damaged trees, but they can also be found in dead trees, logs, and pulp wood (figure 3.11). Bark beetles bore through the bark and lay eggs in tunnels or special galleries between the bark and the wood. Feeding larvae may girdle the trunk with tunnels, preventing movement of nutrients and water. This may cause wilting and death. Some insects feed on buds and shoots, leading to stunting or malformation of branches. Other species,

the root weevils, feed on roots of seedlings and young trees, especially conifers. They also may cause stunting or death.

One insect, the forest tent caterpillar, can have a significant impact on a deciduous forest. Entire communities may go through nearly complete defoliation by forest tent caterpillars. At about 10 to 20-year intervals their numbers peak, and extensive areas in the forest seem to be covered by their silken cocoons. These cocoons are spun by the mature larvae (figure 3.2), which then form a pupa inside the cocoon. In about 10 days the adult moths emerge, mate, and deposit egg masses around twigs. The eggs overwinter, protected by a tough casing that prevents them from drying out. In spring, the eggs hatch, and the larvae, or caterpillars, feed on leaves.

Outbreaks of tent caterpillars in Minnesota typically last about three years. When high populations result in complete defoliation of a tree, large numbers of larvae move over the ground searching for suitable sites for spinning cocoons. Such movements have given the name armyworms to tent caterpillar larvae.

In a severe outbreak in May 1951, tent caterpillars consumed nearly every leaf of all the trees and shrubs in an aspen-birch forest in northwestern Minnesota (Fashingbauer, Hodson, and Marshall 1957). Starving larvae literally covered the forest floor, and roads were slippery with the large numbers of caterpillars. Many died, but the survivors spun webs, formed cocoons, and later emerged as adult moths ready to lay eggs. The forest slowly recovered and by late June took on the appearance of early spring as the aspen and dogwood put out new buds and leaves.

Humans, of course, consider mosquitoes to be one of the worst insect pests in the forest. But mosquitoes serve as food for predators and do not seem to bother many kinds of animals. Mosquitoes breed in nearly every puddle and pond. Their aquatic larvae live at the water surface and are eaten by minnows, dragonfly nymphs, ducks, and many other predators. Adult mosquitoes are eaten by frogs, toads, bats, damselflies, flycatchers, and other birds.

An important beneficial activity of forest insects is pollination, the carrying of pollen from a male flower to a female flower. Sweet smelling flowers, such as basswood, attract bees and other nectar-loving insects. Other plants, such as Jack-in-the-pulpit, smell like decaying meat. This smell attracts flies, which serve inadvertently as pollinators. The hood of the Jack-in-the-pulpit flower has vertical strips of almost clear tissue that look like openings to a fly. As the fly buzzes about trying to escape through these windows, it contacts the reproductive parts of the flower and transfers the pollen.

The relationships between grazing, or herbivorous, insects and the plants they eat involve many complex interactions through which the plant and the insect become adapted to existing together. For example, a plant may develop a chemical compound in its leaves that may discourage feeding by a certain species of insect. In time, the insect may develop tolerance to the chemical compound and is then able to eat the leaves. Such cycles of action and response in which organisms adapt to a changing environment probably explain much of the diversity of insects and plants, and, in fact, of many organisms that we know today.

Another group of invertebrates, the ticks, are also common in deciduous forests. Ticks are not insects but are relatives of mites and spiders. In spring ticks climb onto plant leaves and stems, where they wait patiently for a bird or mammal, including a human, to pass by. They seize their host with their forelegs and climb aboard for a meal of blood. A barbed mouth part, the proboscis (figure 4.20), inserted into the prey anchors so securely that pulling on the tick is likely to separate the body from the head, which remains in the wound. After engorging, the tick releases its proboscis and drops to the ground, where it molts and again searches for a meal. When mature, the female deposits her eggs in the soil.

Two kinds of ticks, the wood tick and the tiny deer tick, live in Minnesota. The wood tick is not known to be dangerous to humans. The tiny deer tick, however, carries Lyme disease, caused by a spirochete, or bacterial pathogen (Johnson 1985). This disease can be serious in humans and is the subject of much research.

Deer ticks are active from April to October and attach themselves to deer, chipmunks, white-footed mice, and almost any other mammal, including humans. About two dozen species of migratory birds are also known to carry the larval stage, thus providing potential for rapid spread of the disease (Anderson et al. 1986).

The tick requires two years and three hosts to complete its life cycle (Habicht, Beck, and Benach 1987). Soon after hatching from the egg, the larva attaches to its prey, often a white-footed mouse, for about two days. The larva then drops to the ground, where it remains for about one year. After winter dormancy, the nymph, as it is now called, again climbs up on a twig, stem, or blade of grass and waits in ambush. It attaches to another prey for three or four days. The following spring or summer the tick, now an adult, attaches to a deer or other prey for its final meal. Mating occurs while the tick is on this host, but we do not know how the male and female find each other. The male tick dies soon after mating, and the female drops

93

Figure 4.20 *The head of this deer tick was magnified with a scanning electron microscope. The white line in the lower right corner of the photograph is 0.002 inch long. Adult ticks are about the size of the head of a pin. (Photo courtesy of Gilbert Ahlstrand, Minnesota Agricultural Experiment Station, University of Minnesota.)*

off the host and overwinters. No one has yet discovered where the eggs are laid, but they hatch the following spring, and the cycle begins again.

The range of the deer tick in Minnesota is spreading, and therefore more cases of Lyme disease can be expected in the future. Early symptoms of Lyme disease may include a ring-shaped rash at the site of the bite, fever, headache, and fatigue. About a month later problems may occur in nerves and muscles, possibly leading to encephalitis and heart pain. Months later, if the disease is not diagnosed and treated, changes in the body's immune system may lead to severe arthritis and paralysis. If the disease is diagnosed quickly, it can be treated successfully with antibiotics (Habicht, Beck, and Benach 1987).

### Forest pathogens

Another type of community interaction concerns diseases that affect important tree species in the deciduous forest. Oak wilt and Dutch elm disease, both caused by fungi, have killed many trees and have changed the character of many of our deciduous forests. Oak wilt is caused by the fungus *Ceratocystis fagacearum*, which grows in the vessels that carry water and nutrients from the roots to the leaves. Oak wilt has probably been present in deciduous forests for centuries, but it was not until 1940 that the causal agent was identified.

All varieties of oak tested are susceptible to oak wilt; the red oaks are the most vulnerable, and the white oak group is the most resistant (French and Stienstra 1975). Infected trees attempt to protect themselves from the fungus by producing tyloses and gums, substances that block the vessels. This blockage causes the leaves to wilt. Leaves of infected trees turn brown, yellow, or tan beginning at the tip and outer edges. The last part of the leaf to turn color is the base near the central vein. The entire crowns of infected red oaks wilt within a few weeks after the first symptoms appear. Once a tree is infected with oak wilt fungus, there is no way to save it.

Oak wilt spreads from tree to tree in two ways. Roots of adjacent oaks often form natural grafts, thus allowing interchange of water that can carry fungal spores. Groups of such infected trees, called infection centers, are common in southeastern Minnesota. Oak wilt fungus remains alive for some time in the roots and trunks of dead trees. Here spore masses are produced on fungal mats underneath the bark. Pressure pads produced in the mat develop sufficient pressure to split open the bark, thus allowing dissemination of the infectious spores. The mats have a sweet, fermenting odor that attracts insects such as sap-feeding picnic beetles (Nitidulidae). Spores cling to the beetles as they crawl over the mat. If the beetle flies to another tree to feed on sap from a fresh wound, the spores may drop into the wound and infect the healthy tree. The adult beetles breed in galleries

94

or tunnels beneath the bark of infected oaks and may carry infective spores to healthy oaks. Prompt removal of dead and dying trees and breaking of the root grafts between infected and healthy trees by use of a trencher or vibrating plow or by application of chemical barriers are the recommended ways to control the spread of the disease (Rexrode and Brown 1983).

Dutch elm disease is also caused by a fungus, *Ceratocystis ulmi,* which belongs to the same genus as the fungus responsible for oak wilt. Thus, the two diseases have many similar characteristics. Dutch elm disease was first found in the United States in 1930. It was introduced from Europe, probably in logs containing the fungus. The first cases were found in Minnesota in 1961, when one infected tree was found in Ramsey County and seven were found in Wright County. By 1973 the disease had appeared in 61 Minnesota counties (French, Stienstra, and Noetzel 1974).

Symptoms are a rapid wilting of leaves, which turn dull green to yellow, and a brown discoloration in the outer sapwood of infected branches. These symptoms, like those of oak wilt, are caused by the accumulation of tyloses, gums, and fungi, which plug the tree's vessels, thus preventing uptake of water. The fungus is spread by both the native elm bark beetle and the introduced European bark beetle, which overwinter in burrows or under the bark of dead elms. Fungal spores are carried to healthy trees by the adult beetles in the spring. Root grafts between elms also provide pathways for the spread of the fungus. The control methods of sanitation and disruption of root grafts described for oak wilt are also recommended for Dutch elm disease. Insecticides may provide some additional protection by killing the beetles, and further research may discover effective fungicides that can be injected into healthy trees.

Unfortunately, in some towns the loss of elms and oaks has been severe, and the stately trees that shaded the streets are gone. The loss of even one favorite shade tree is discouraging. In the forest, however, when individual trees and small groups of

trees die, they are replaced by other trees, perhaps of a different kind, by natural succession. The general character of the deciduous forest is maintained by these natural processes, which have been going on for thousands of years.

## Present status of the deciduous forest

The acreage of Minnesota's deciduous forest that remains today is difficult to estimate. In the 12 million acres comprising the 28 counties identified as the Central Hardwood Region, 2.3 million acres of forest remained in 1990 compared to 2.7 million acres in 1962 (Hahn and Smith 1987; Jakes 1980; Leatherberry 1991). Much of this 15% loss represents forests converted to agriculture and urban development. As a result, the remaining tracts become smaller and more isolated. This process of fragmentation results in islands of deciduous forest surrounded by cropland or buildings and streets, as illustrated in the changes that have occurred in Rice County (figure 4.21). These islands may be too small for animal species with large home ranges. These species disappear, and species that thrive on habitat edges become more common. An ecological theory referred to as island biogeography tells us that the number of species in a patch of forest is related to the patch size, that the rate of immigration of new species into the patch depends on the distance from similar habitats, and that the rate of extinction of species from the patch is a function of patch size. Thus, as human activities shrink the deciduous forest, we can expect to see fewer kinds of plants and animals, that is, lower species diversity. The remaining forest patches will become less and less like the forests that existed before European settlement.

Of the 287 species of animals and plants listed as endangered, threatened, or of special concern in Minnesota, 51 occur in deciduous forests (Pfannmuller and Coffin 1989). Five of the 17 listed mammals, 3 of the 27 birds, 11 of the 17 amphibians and reptiles, and 32 of the 191 plants occur in this community. Identification, protection, and

95

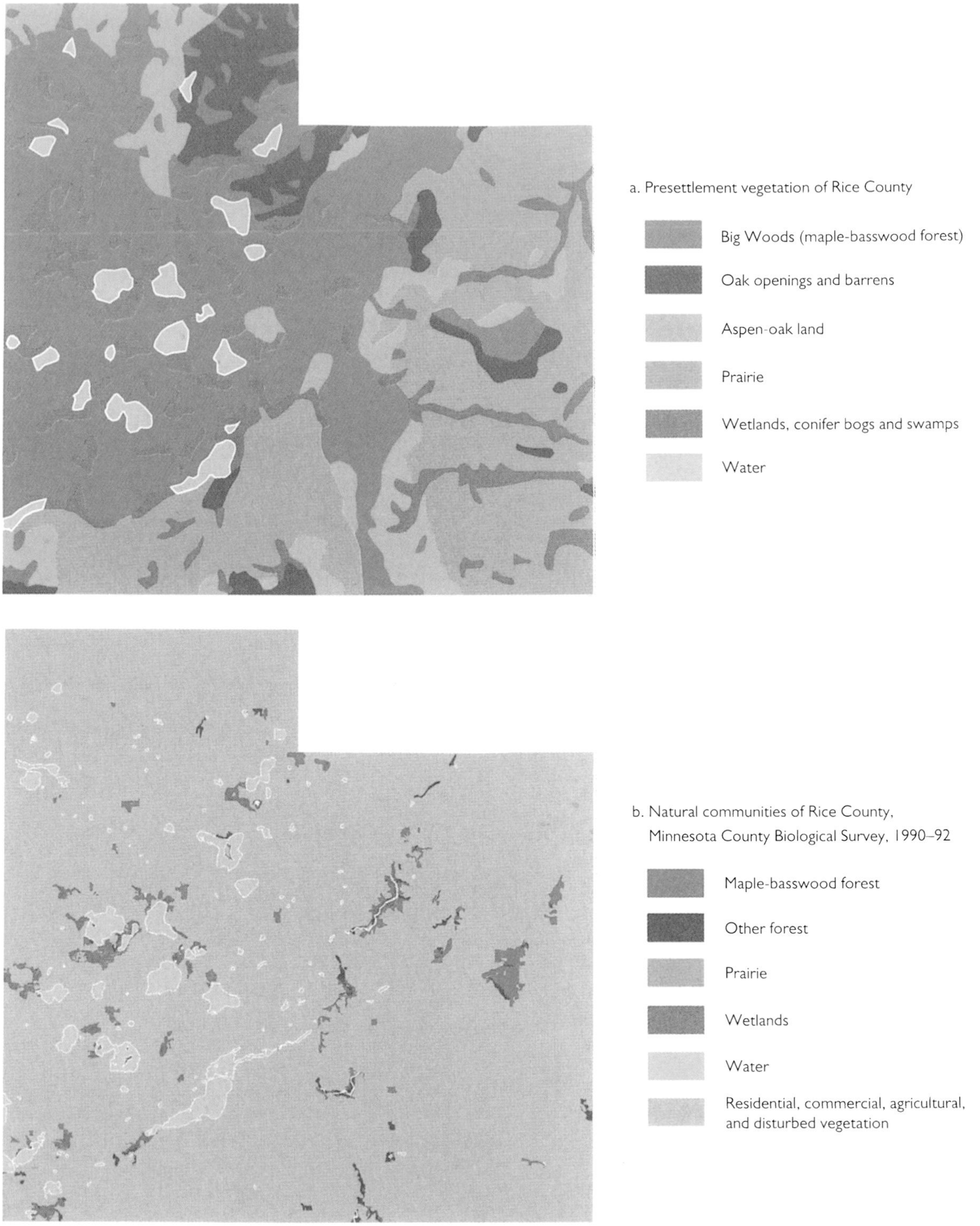

a. Presettlement vegetation of Rice County

Big Woods (maple-basswood forest)

Oak openings and barrens

Aspen-oak land

Prairie

Wetlands, conifer bogs and swamps

Water

b. Natural communities of Rice County,
Minnesota County Biological Survey, 1990–92

Maple-basswood forest

Other forest

Prairie

Wetlands

Water

Residential, commercial, agricultural,
and disturbed vegetation

**Figure 4.21** (a) *Before European settlement, the Big Woods covered a large area in east-central Minnesota, including much of Rice County. This map was prepared from records made by government surveyors conducting the General Land Office survey in the mid-1800s.* (b) *Today, only small remnants of the Big Woods remain. Nearly all of the natural communities in the county have been replaced by land development for agriculture, cities, and roads. (Adapted from Dunevitz and Epp 1995, by permission.)*

management of Minnesota's endangered species presents a challenge to both government and private agencies as well as to individual citizens. Uncommon plants and animals enhance our world and are part of our natural heritage. We owe it to ourselves not to destroy parts of our environment but rather to maintain and enhance the natural world in which we live.

Minnesotans are fortunate that a number of excellent deciduous forest stands have been preserved by government and private agencies. Remnants of the deciduous forest are in parks, Scientific and Natural Areas and other lands owned by the DNR or The Nature Conservancy, and on private lands. The Richard J. Dorer Memorial Hardwood State Forest in the southeastern corner of the state offers rustic and even majestic vistas, excellent fishing and hunting, and beauty and solitude to match the northern forests. The Dorer Forest, established in 1963 and named in honor of one of Minnesota's outstanding conservationists, extends from the Twin Cities to Iowa, following the west bank of the Mississippi River and extending away from the river along major tributaries. The official boundary of the Dorer Forest encompasses about 1 million acres, but only about a fourth of that is forested. Present plans call for the state to purchase about 83,000 acres to provide recreational and aesthetic benefits. About 46,000 acres were acquired from 1963 through 1994 (John Hellquist, Minnesota Department of Natural Resources, Division of Forestry, personal communication). A secondary objective is to restore deciduous forest on cutover or burned areas and on abandoned farmland. No attempt has been made to create a wilderness area in the Dorer Forest. Rather, it has supported multiple uses such as water conservation, control of erosion, timber harvests, fishing, hunting, and other forms of public recreation.

Although considerable controversy exists regarding the purchase of more and more land for the Dorer Forest, much progress has been made in controlling erosion, improving trout streams, and reforesting tracts owned by the state. A large popu-lation of wild turkeys roams the Dorer Forest, and white-tailed deer, raccoons, and ruffed grouse are common.

Other tracts of deciduous forest, remnants of the Big Woods, exist in parks, in natural areas along streams and rivers cutting through a sea of corn and soybeans, and on steep, hilly sites not suited for farming. Some of these forests have not been subjected to major disturbances for over 150 years. In Minnesota maple-basswood forests of this age are classified as old growth. Such stands may have existed for hundreds of years, shaped by fires and natural succession. Wolsfeld Woods in Hennepin County and Nerstrand Woods in Rice County, both designated as Scientific and Natural Areas by the DNR, are excellent examples. However, both of these areas have been disturbed to some extent by grazing and selective logging.

Another old-growth forest occurs in the Wilderness Scientific and Natural Area in Itasca State Park. No roads cut through this forest, and one must walk about half a mile to reach the old-growth stands. The giant trees create a cathedral-like feeling that is enhanced by the solitude and silence. Some of the trees appear to be hundreds of years old, some are dead, and some have fallen and are decaying. Often I have sat against the base of a gigantic oak, a landmark in the forest, and thought about how this forest has functioned for centuries. All of the principles of ecology can be observed and studied here.

Some people feel that old-growth forests should be harvested so that the trees will not be "wasted." Others feel that these forests are valuable as sanctuaries and living laboratories and that they should be preserved as examples of rapidly disappearing ecosystems. In recent years both Minnesota and the federal government have shown increased interest in old-growth forests and are developing plans for identifying and managing the few remaining stands. Considering that it was the deciduous forests of the northern hemisphere that nurtured much of modern civilization, let us hope that we in Minnesota can live in harmony with its remnants.

97

# NORTHERN CONIFEROUS FOREST

**5**

The smell of wood smoke, the sounds of waves smashing on the rocks, and the wind whistling through the pine trees mean the north woods to many Minnesotans. Our north woods, characteristic of most of the northeastern third of the state (figure 1.25), are part of the northern coniferous forest, which stretches across North America in a broad band covering parts of the northern states and Canada and extending into Alaska (figure 1.6). Throughout the northern coniferous forest, stands of red, white, and jack pine are interspersed with communities dominated by aspen, birch, spruce, and fir. Major peatlands occur north and west of Upper and Lower Red Lakes and in St. Louis, Carlton, and Aitkin Counties. This mixture of habitats and the many lakes and streams make the region diverse and thus subject to a wide range of recreational uses, such as fishing, hunting, and canoeing, and to economic uses, such as logging and mining.

The area comprising the northern coniferous forest in Minnesota was completely glaciated. Rock outcrops and glacial moraines provide a varied and scenic landscape in northeastern Minnesota (figure 1.15). As a result of the glaciation nearly all of the

99

**Figure 5.1** *Red pines grow tall and straight and are highly prized by loggers. Forests of huge red and white pines occurred in many locations in northern Minnesota in the early 1800s. These forests were dependent on wild fires, which removed the shrubs and hardwood trees and created excellent growing conditions for the fire-resistant pines. (Photo by Bob Firth, Firth PhotoBank, Minneapolis.)*

**Figure 5.2** *The gray and rust-colored lichens on rocks lining the shore of Lake Superior are a combination of a colorless fungus and an alga containing chlorophyll and, in some cases, other pigments. They become established on bare rock and thus are pioneers in succession. Freezing and thawing of water held on the rock surface by the lichens eventually creates tiny cracks in the rocks. Seeds may germinate here, and roots will penetrate the cracks, where water and nutrients collect. (Photo by Bob Firth, Firth PhotoBank, Minneapolis.)*

soil has developed on glacial deposits and lake beds, or on rock surfaces scraped bare by the ice. The weathering that initiated soil formation began with rain falling on the bare till and rock. Carbonic acid in the rain leached dissolved minerals out of the surface horizons. The soil was further developed by the growth and decomposition of vegetation, a process that continues indefinitely. As the needles, leaves, and twigs drop from the trees in the northern coniferous forest, a layer of litter develops on the ground. When this litter decays into finely divided particles, it is referred to as humus.

In many parts of northeastern Minnesota, bare rock surfaces are exposed, and little soil has developed. Lichens, a symbiotic combination of two organisms, a colorless fungus and an alga containing chlorophyll, are common on many exposed rocks (figure 5.2). The fungus attaches to the rock, absorbs water, and protects the enclosed algal cells. Each of the many species of lichens consists of a different kind of fungus and its associated alga. Patches range in size from a few hundredths of an inch to many inches and range in color from light gray to bright orange to brownish black.

**Table 5.1** *Comparison of ecological characteristics of deciduous forest and northern coniferous forest in Minnesota*

| Ecological characteristic | Deciduous forest | Coniferous forest |
| --- | --- | --- |
| Number of species | many | few |
| Soil | relatively rich brown | poorly developed |
| Litter | neutral | acidic |
| Strata | three to four | two to three |
| Ground cover | herbs | low and dwarf shrubs |
| Mosses and lichens | sparse | abundant |
| Light intensity | high in winter and spring, low in summer | low to moderate through the year |

*Source:* Modified from Kornas (1972).

Often lichens are the first organisms to appear in an area and are thus pioneers in succession. Eventually bits of rock and soil collect around the lichens and provide an environment for other plants. In these areas, seedlings begin to grow and manage to survive by sending their roots into cracks in the rocks, where nutrients and water accumulate. Eventually soils will form over the rock surface, but the time required for this process is far longer than the 10,000 years since the last glaciers left Minnesota.

The physical characteristics of the northern coniferous forest can be characterized simply as cool and moist. Average annual temperatures range from about 36° to 41°F, and the frost-free season may be less than 100 days (figure 2.5). Average annual precipitation ranges from about 21 to 28 inches going from west to east. Snowfall may exceed 68 inches per year, and on average the ground is covered with 1 inch or more of snow for at least 125 days per year (Kuehnast 1972).

Minnesota's coniferous trees are well adapted to living in this cold and often dry environment. Because conifers retain their needles throughout the year, they begin photosynthesis as soon as water is available in the spring and continue until early winter. Furthermore, neither energy nor time is wasted in the production of a complete new set of leaves each year. More mineral nutrients are thus retained rather than recycled each fall, as in deciduous trees, enabling conifers to grow well on coarse soils with low nutrient content. Conifers' needlelike leaves (figure 5.3) have thick protective layers and sunken

**Figure 5.3** *Flat needles help to identify balsam fir. (Photo by Jim LaVigne.)*

air passages that minimize moisture loss. In addition, the needles expose little surface area to the elements. These adaptations allow the trees to survive the extreme moisture stress caused by the cold, dry conditions of midwinter, when most water is in the form of ice.

Although some deciduous trees, such as aspen and paper birch, thrive in the northern coniferous forest, the evergreen conifers are the most visible difference between the coniferous forests and the deciduous forests. Table 5.1 lists other important differences. The ground litter in coniferous forests is more acidic, the soil is not as rich in nutrients as in deciduous forests, and light on the coniferous forest floor is low to moderate throughout the year. These environmental conditions result in the presence of fewer plant species and fewer layers, or strata, in the coniferous forest.

## Northern coniferous forest ecology

The northern coniferous forest, like every ecosystem, is dynamic. Before European settlement, the forests of northeastern Minnesota were a mosaic of types determined by site conditions such as soil and moisture and by the occurrence of fires. As settlers began to log the forests and as fires were suppressed, the forests began to change in response to the disturbance and the absence of fire. Today the northern coniferous forest consists of the same forest types, but the abundance and distribution of the types reflect changes in the conditions in the region brought about by human activities.

### Balsam fir–white spruce forest

On upland sites with a moderate amount of moisture, especially those protected from frequent fire,

**Figure 5.4** *Paper birch grows in most of the successional stages of the northern coniferous forest. (Photo by Jim LaVigne.)*

**Figure 5.5** *Bunchberry is a common herb in northern coniferous forests. It has white flowers in late spring, which develop into red berries in late summer. These bunchberries appear to have been nipped by an early frost. (Photo by Bob Firth, Firth PhotoBank, Minneapolis.)*

the forest is composed primarily of balsam fir (figure 5.3) with some white spruce and paper birch (figure 5.4). On some sites these species occur in combination with black spruce (Aaseng et al. 1993). Balsam fir and white spruce are shade-tolerant, late-successional species.

In a detailed study of upland forest communities in the Boundary Waters Canoe Area Wilderness (BWCAW), D. F. Grigal and L. F. Ohmann (1975) examined 68 logged stands and 106 stands undisturbed by logging. They concluded that early-successional aspen, birch, and maple communities as well as jack pine and black spruce communities tended to change over time to communities dominated by balsam fir, birch, and white spruce.

Beneath the spruce and fir light intensity is generally low, and thus shrub cover is sparse. Typical species are beaked hazelnut and mountain maple. The herb layer is also sparsely developed, most

commonly with bunchberry (figure 5.5), big-leaved aster, twin flower, red baneberry, wild sarsaparilla, bluebead lily, and Canada mayflower. The moss layer, however, may be well developed in some stands, and several species of feathermoss may be abundant.

Gaps created when a tree falls or is blown down by the wind are especially common in mature fir-spruce forests. These gaps provide an environment for light-loving species and for species that occupy edges between habitats. The yearly rate of gap formation is about 1% of the area in northern coniferous forests in Minnesota (Heinselman 1973). Filling of these gaps by saplings present in the understory or by early-successional species leads to patchiness in the forest, and the degree of patchiness is a reflection of the frequency of disturbance.

On dry, upland sites with low nutrient content, where balsam fir, white spruce, and birch will likely be the climax species, one finds pioneer forests dominated by jack pine, or by mixed jack pine and black spruce (Ohmann and Grigal 1979). On moist, rich soils, aspen is the dominant pioneer species. Note that some of these pioneer species are the same as in the deciduous forest. Pioneer species often have small, light seeds that are easily carried by the wind. Their rapid and widespread seed dispersal and their ability to grow rapidly under high light conditions account for their wide distribution.

### Black spruce–feathermoss forest

In low, wet areas black spruce is likely to be the dominant tree. Northward into Canada, black spruce takes on a more important role and often grows in extensive stands covering thousands of acres on the upland. This community is characteristic of the boreal forest biome, which occurs across northern North America. Shrubs and herbaceous plants in the black spruce forest are similar to those in the upland forests dominated by white spruce and balsam fir, except that mosses are usually more common and in many areas cover the ground surface like a soft carpet.

A small plant parasite that plays a surprisingly important role in the black spruce forest is the eastern dwarf mistletoe (*Arceuthobium pusillum*). This plant is an important cause of tree mortality in the state and is especially severe on black spruce. The parasite has a root system that absorbs nutrients from within the host and a reproductive system of aerial shoots that rise from the branches of the host tree. The shoots are usually less than an inch in length and are covered with scalelike leaves. Five to 10 years after being infected, trees respond by growing branches at the site of the infection, resulting in bushy masses of branches and twigs called witch's brooms (figure 5.6). These growths persist and may reach 10 feet in diameter. Eventually the infected tree weakens and dies. Dwarf mistletoe spreads by explosively expelling its sticky seeds in fall. The seeds may land on a nearby healthy tree or may be dispersed long distances by birds and mammals, thus initiating new infections. Selective harvest and clear-cutting appear to be effective controls (Ostry and Nicholls 1979); fire may have controlled dwarf mistletoe in presettlement forests.

Another unique plant that gives us insight into the complex interrelationships of the forest ecosystem is Indian pipe (figure 5.7), which grows in coniferous forests and is often thought to be a parasite on tree roots. However, it obtains nutrients by sharing the mycorrhizal fungi that grow on the roots of most conifers (Björkman 1960). Filaments from the fungi are attracted by a chemical secreted by the Indian pipe plants. The fungi penetrate the Indian pipe roots, forming a

**Figure 5.6** *Eastern dwarf mistletoe, sometimes called witch's broom, is a common parasite on black spruce. Trees respond to the parasite by growing branches at the site of the infection, producing bushy masses of branches and twigs. Infected trees usually weaken and die. (Photo courtesy of Jon Ross.)*

**Figure 5.7** *Indian pipe, a plant without chlorophyll, is often thought to be a parasite on tree roots, but it obtains most of its nutrients by sharing the mycorrhizal fungi that grow on the roots of most conifers. (Photo by Jim LaVigne.)*

bridge with the conifer roots. Indian pipes obtain at least some of their food from the fungi, which obtain some of their nutrients from the tree.

### White cedar forest

In a few locations nearly pure stands of white cedar occur. These forests, some well over 200 years old, are classified as old-growth. Kurt Rusterholz (1990), a forest ecologist with the Minnesota Department of Natural Resources (DNR), believes that the stands originated after catastrophic fires over 200 years ago and have experienced no significant fires since establishment. The stands contain many fallen cedar logs but few cedar seedlings or saplings. White cedar will undoubtedly dominate these stands for many years, but the absence of

cedar reproduction makes predictions about the future forest very difficult. In some stands small cedar seedlings can be found growing on decaying logs. Even in these stands, however, seedlings more than a few inches tall are completely absent. Only within and near the BWCAW, where deer populations are lower, are small cedars fairly common. Heavy browsing by deer is likely responsible for the lack of white cedar seedlings in some stands, especially in areas where deer concentrate during winter.

### White and red pine forests

Before European settlement, magnificent forests of red pine (figure 5.1) and white pine (figure 3.3) occurred in many locations in northern Minnesota. Logging eliminated most of these forests, but about 25,000 acres of these old-growth forests still exist in Minnesota today, mostly in the BWCAW (Rusterholz 1990). About 1,500 acres of 200- to 250-year-old red and white pine forests remain in Itasca State Park, which was established in 1891, just in time to prevent the loggers from removing the last few acres of pine from this part of Minnesota.

The presettlement pine forests were a mosaic of red pine stands on dry sites prone to frequent and intense fires and white pine stands on moister sites that burned less intensely (Aaseng et al. 1993; Coffin and Pfannmuller 1988). Early scientists investigating forest communities and succession did not recognize that fire was a critical ecological factor in the maintenance of the pine forests. Beginning in the 1950s, ecologists (Ahlgren and Ahlgren 1960; Buell and Niering 1957; Spurr 1954; Swain 1973) began to develop an understanding of the importance of fire in succession in northern Minnesota. These scientists established that fires started by lightning were frequent and that they played a significant role in the structure of plant communities. Later, M. L. Heinselman (1973, 1981), working in the BWCAW, and S. S. Frissell (1973), working in Itasca State Park, explained the mechanism by which fire originated and maintained the forests of red and white pines before logging.

**Figure 5.8** (*right*) *Large scars at the bases of old red pines were caused by fires. An accumulation of needles at the base of the tree probably burned slowly on the lee side of the trunk, injuring the growth layer under the heavy bark. The injury healed, but repeated fires sometimes injured the same tree. (Photo by Jim LaVigne.)*

**Figure 5.9** (*below*) *This cross section of a red pine trunk shows growth rings and at least one fire scar. A count of the growth rings indicates that this red pine was at least 274 years old when it blew down in a windstorm. The location of the fire scar tells us that the fire occurred when the tree was about 50 years old. (Photo by author.)*

Much of the history of the role of fire was worked out by examining fire scars in trees (figures 5.8 and 5.9). These scars provide evidence of each fire and make it possible to determine precisely the year when the fire occurred. Red pine is particularly resistant to decay and has proved to be the most useful tree for dating fires (Frissell 1973). In addition to fire scars on trees, bits of charred wood and ash deposited in the sediments on the bottom of lakes and ponds have also been useful in determining fire history (Clark 1993). Such evidence combined with analysis of pollen grains found in the sediments has allowed scientists to reconstruct the history of forests. This process is similar to the reconstruction of the vegetation history of Minnesota since glaciation described in chapter 1.

Frissell (1973) reported that during the 272 years from 1650 to 1922 a fire occurred somewhere in Itasca State Park on average once every 8.8 years. At a given location, the average frequency of fires was about every 22 years. He also found that the

date of origin of most of the stands of red and white pine in Itasca State Park corresponded closely with a year in which a major fire burned on that site.

Comparison of the dates of major fires in Itasca State Park and the BWCAW suggests that conditions favorable to intense burns, probably dense dry fuel, low precipitation, and high winds, occurred simultaneously in both areas in 1727, 1759, and 1864. Heinselman (1973) reported that a fire occurred somewhere in the BWCAW at least once every 8 years from about 1600 to the present. But he also found that the interval between burns on a given area was about 80 years, nearly four times longer than at Itasca.

But how are the pines able to survive fires? The forces of lightning are awesome. Every tree is a potential lightning rod able to discharge strikes so powerful that the entire tree may be shattered. Numerous strikes may occur in a few moments during an electrical storm, and scattered fires from these strikes can result in large forest fires. Few acts of nature are more frightening than a raging fire sweeping through a forest (figure 5.10). Sparks fly and dense smoke obscures the sun. But, to answer the question, all forest fires are not the same. If a site is to be suitable for pine reproduction, either a ground or crown fire must be intense enough to expose the mineral soil by burning the accumulation of litter. The fire must also remove the shrub and herb layer so that light intensity at the ground

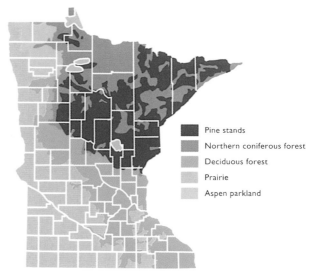

Pine stands

Northern coniferous forest

Deciduous forest

Prairie

Aspen parkland

**Figure 5.11** *Pine forests covered large areas in northern Minnesota before logging. (Data from Heinselman 1974.)*

**Figure 5.10** *Forest fires often move along the ground, burning dead leaves and branches, but sometimes the flames shoot up a tree trunk. The most severe fires are those in which the crowns of conifers ignite and flames jump rapidly from tree to tree. (Photo by author.)*

level will be high. Because such fires destroy many of the seeds of red and white pine, seeds must be available from nearby trees shortly after the fire and before the establishment of a dense cover of shrubs and herbs. Survival of the seedlings requires favorable light, moisture, and temperature conditions during the period immediately following germination. If another fire occurs before the seedlings have had a chance to develop to the point where they can withstand its heat, the process must begin again.

Once the bark of the young trees has developed its ability to withstand heat, the red and white pines are relatively safe from surface fires. Of course, they will die if a crown fire sweeps through the forest, as sometimes happens during hot, dry, and windy periods. As the pines grow taller and more majestic each season, shrubs and deciduous trees such as aspen, birch, and oak begin growing and form a deciduous layer in the forest. These species, however, are very sensitive to heat, and most individuals will be killed by the next fire that moves through the area. The pines, on the other hand, which are now even more resistant to the fire, continue to grow. This process continued for hundreds of years and resulted in the development of the red and white pine forests over large parts of northern Minnesota (figure 5.11).

107

A basic understanding of the importance of fire in the establishment and maintenance of forest ecosystems throughout North America evolved from the work of Heinselman and others. In Minnesota this understanding made clear how control of fires by humans has contributed along with other activities such as logging to significant changes in the northern coniferous forest. Before logging and settlement in the late 1800s, most fires probably burned until they reached a natural firebreak, such as a lake or river, or were put out by rain or wet fuel conditions. Beginning in the early 1890s, an effective program of fire prevention and control was established by the DNR's Division of Forestry and the U.S. Forest Service (figure 5.12).

**Figure 5.12** *Modern fire fighting involves high-tech equipment, as well as old-fashioned pick and shovel work on the ground. (Photo by Bob Firth, Firth PhotoBank, Minneapolis.)*

In Itasca State Park, for example, fires have been controlled relentlessly during the past 75 years. Suppression of fires has allowed the late-successional species to develop a dense understory. On some sites balsam fir, white spruce, and paper birch are growing beneath the pines, and on other sites sugar maple and basswood dominate the understory. Competition for light, moisture, and nutrients by the understory vegetation, together with browsing by deer, is severely limiting pine reproduction (Hansen, Kurmis, and Ness 1974). Few pine saplings or seedlings can be found. As the old-growth pines become overmature and weaken because of disease and heart-rot, they become more susceptible to windthrow. Once they are gone, the process of succession will lead to late-successional forests of spruce, fir, and birch or of sugar maple and basswood. Zealous protection from fires has not benefited the pine forests.

Today managers of some of Minnesota's parks and forests are urgently searching for ways to create natural pine forests similar to those that existed hundreds of years ago. Many red and white pines in existing forests are several hundred years old, and blowdowns are common when high winds occur. Fires appear to be needed in these stands to reduce competition from the understory species, and in some localities, natural fires will be allowed to burn. In addition, managers are planning to use prescribed burning to mimic the natural disturbances needed to set back succession and restore the pine forests.

### Jack pine forests

Jack pine forests also characteristically occur following fire on sites that are dry and poor in nutrients. Jack pine is a shade-intolerant pioneer species that is especially well adapted to growing after a hot fire. The seed cones of some jack pines may remain attached to the tree for many years with the scales tightly closed by resins. If the temperature exceeds 116°F during a fire, the resin softens and allows the scales to open and discharge seeds (Fowells 1965). These seeds are heat resistant and are not damaged when they fall on a bed of ashes and coals. Most germination occurs in the first few years. The trees grow so rapidly that they are able to produce seeds themselves in about five years.

Much sunlight passes through the sparse foliage of the jack pines and reaches the forest floor. As a result, shrubs and herbs adapted to growing in a dry environment are present. Bush honeysuckle, beaked hazelnut, mountain maple, and blueberry are common shrubs. Blueberries grow about a foot tall, have oval, pointed leaves, and produce tiny pink flowers in June and delicious berries in July and August. The plants grow best under open conditions where abundant sunlight reaches the ground. Such conditions prevail in young jack pine forests, and American Indians often set fire to older forests to create optimum habitat for blueberries. The herb layer, which is often well developed, is likely to be composed of Canada mayflower, blue-bead lily, big-leaved aster, bunchberry, wild sarsaparilla, feathermoss, and reindeer moss.

Other species likely to be found in jack pine forests include black spruce, northern pin oak, red oak, bur oak, red pine, white pine, large-toothed aspen, trembling aspen, paper birch, and balsam poplar. Over time the jack pine forest will be replaced by a forest dominated by such species as red oak, red maple, red pine, and white pine. As the soil accumulates more organic matter and nutrients, these trees are likely to be replaced by balsam fir, white spruce, and paper birch. This type of community is often considered the successional climax; however, it too is likely to be short-lived at a given location because of fire or windstorms. Thus, succession leading to climax is a dynamic process. All of the successional stages may be represented by individual stands within a single forest (Larsen 1980).

## Logging in the northern coniferous forest

Logging of forests was, and still is, an important part of the history and economy of Minnesota and has significantly affected the composition of the present-day forest landscapes (figure 5.13).

Logging of the pine forests resulted in natural reforestation by jack pine and aspen, both pioneer species. Now the mature aspen and jack pine are being logged, causing more changes in the forest.

At the beginning of the nineteenth century settlers began to move into the southeastern part of Minnesota. One of their first activities was to cut trees to provide firewood and logs for homes and barns. This cutting helped to clear the forest so that agricultural crops could be planted. Farther north, timber cruisers, working either for themselves or for the big timber companies, explored the wilderness to locate the best pine stands for logging. Most of these stands were purchased by timber companies, often at $1 or $2 per acre.

When a logging crew moved into a timber stand to begin harvest, they first had to build the camp in which they would live. Buildings and furniture were constructed from logs felled at the campsite. The sound of two-man crosscut saws and double-bitted axes could then be heard at ever increasing distances from the campsite.

Transportation of the huge pine logs was a problem. To solve this, much of the early cutting was done near lakes and rivers so that the logs could be floated to market in spring and summer. The first sawmill in Minnesota was constructed by the federal government at St. Anthony Falls in 1821 to provide timber for the construction of Fort Snelling. In 1837 the first commercial sawmill was built at Marine on St. Croix in Washington County to turn white pine logs into lumber. In the 1840s many new sawmills sprang up, and lumbering was the leading industry in Minnesota. The first railroad was established in 1862, and by 1870 it extended as far north as Duluth. Logs and lumber from the north could now be brought to Minneapolis and St. Paul, which were growing rapidly and needed an ever-increasing supply of building materials. By early in the twentieth century most of the pines had been cut, and the timber companies had moved their logging crews westward.

The loggers and timber companies showed little concern for the landscape that remained after har-

**Figure 5.13** *Modern logging with large, expensive machines can harvest a forest quickly, but care must be taken to maintain the landscape for a new forest and for wildlife. (Photo © Jim Brandenburg.)*

vest. Laws required that the slash left behind by the loggers be burned to clear the ground. Sometimes fires appeared to have been set for no reason at all. Many homesteads and some towns were destroyed by some of these fires.

During the harvest boom, replacement of the pine forests was not a concern of the loggers and timber companies. Minnesota's government, on the other hand, was beginning to consider replacement, and in 1871 a Tree Bounty Law was passed providing for the payment of a bounty for planting trees on the prairies. As a result, many farm windbreaks and woodlots were planted. This law, however, did little to aid in restoration of the logged pine forests. In 1876 the Minnesota State Forestry Association was organized, and in 1891 the Legislature established Itasca State Park in an effort to preserve part

of the state's great pine forests. Early in the 1900s tree nurseries were established, and fire prevention programs were initiated as both state and private organizations recognized the importance of forest preservation and restoration.

After most of northeastern Minnesota was logged, extensive pioneer forests of jack pine and aspen (sometimes called popple) developed naturally through succession on the bare mineral soil and now dominate the forests in which pine was once the dominant component. The red and white pine forests have not become reestablished, because of fire prevention and high numbers of deer.

Trembling, or quaking, aspen is by far the most common pioneer species (figure 5.14), but large-toothed aspen and balsam poplar are frequently present. Trembling aspen is a very adaptable

*110*

species, as evidenced by the fact that it is so widely distributed in North America. Aspen develop equally well from seeds or by suckering from roots, provided that they receive a large amount of sunlight. They grow quickly, are thinned naturally by competition, insects, and disease, and mature in 30 to 70 years, although some trees will survive more than 100 years.

A dense layer of shrubs is common in aspen forests. Beaked hazelnut, mountain maple, downy arrow-wood, bush honeysuckle, gooseberry, and raspberry are the most common shrubs. Characteristic herbs include big-leaved aster, wild sarsaparilla, bunchberry, Canada mayflower, sweet-scented bedstraw, twisted stalk, mountain rice grass, lady fern, and bracken fern.

The aspen forests that developed after logging are still present in many areas, although now most trees are mature. The mature trees create shade on the forest floor, which is covered with a layer of leaf and twig litter. These conditions inhibit the germination and growth of new aspen and pine. In some aspen stands, I have seen aspen saplings up to 3 or 4 feet in height with blackened terminal stems and leaves. This blackening, caused by the fungus

*Venturia tremulate,* is referred to as shepherd's crook. Usually these saplings die, but a few recover. This disease seems to be less common in open, sunny areas, where aspens thrive. It appears that the moist, shady conditions in the mature forest favor the growth of this fungus and impede the establishment of young aspen. Thus, the shade created by the aspen trees precludes the establishment and survival of their offspring, unless disturbance opens a gap in which the seedlings can survive.

In mature stands of aspen, small balsam fir and white spruce grow as understory trees. In contrast to aspen and paper birch, fir and spruce can reproduce in the dense shade of a mature forest and on top of leaf and twig litter. On less fertile sites, the aspen forest may be replaced by oaks, red pine, and white pine, and these, in turn, may be replaced by spruce and fir.

In the past decade, expansion in the wafer board and paper pulp industry in northern Minnesota has resulted in a marked increase in harvest of aspen and other species. Annual harvest climbed from 194 million cubic feet in 1976 to 251 million cubic feet in 1986 (Hahn and Smith 1987). This trend appears to be continuing, as indicated by

**Figure 5.14** *This dense stand of young trembling aspen is only a few years old. It developed on a site where mature aspen, such as those growing in the background, were harvested by loggers. (Photo by Jim LaVigne.)*

the harvest of aspen in 12 north-central counties. Aspen growth in these counties was 66.5 million cubic feet, and harvest was 76.4 million cubic feet in 1989 (Murray 1991). We cannot harvest more than we produce for very many years without destroying the resource. The impact of the timber industry on Minnesota ecosystems is currently receiving much attention from our legislature and interested businesses and organizations (Jaakko Pöyry Consulting 1992). It seems obvious that a management plan for wise use of forest resources is vital to the future ecology of our state.

### Ecosystem function

Nutrient cycling in northern coniferous forests is strongly influenced by low nutrient levels in the soil and prevailing cold temperatures. In older forests, high proportions of nitrogen and other nutrients are tied up in the litter because of slow decomposition (Van Cleve et al. 1983). Input of nutrients from precipitation and the atmosphere in stands with feathermosses is limited because the mosses absorb and hold nitrogen and other ions falling on the moss surface (Oechel and Lawrence 1985). Perhaps as an adaptation to such nutrient limitations, spruce, tamarack, and other species occurring in late-successional forests have shallow root systems, which enable them to obtain nutrients from the litter and upper soil horizons (Gale and Grigal 1987).

In general, values for both productivity and biomass of coniferous forests are lower than in deciduous forests, probably because of the cold temperatures characteristic of the northern coniferous forest biome, which influence both the rate of nutrient cycling and the rate of plant growth. Net primary production of about 7,200 pounds per acre per year (Table 4.1) is average for boreal coniferous forests (Whittaker and Likens 1975). Values in Minnesota's coniferous forests range from 3,000 to 8,000 pounds per acre per year. This range is probably related to the availability of nutrients, the successional stage, and the age of the stands studied. Early-successional stands are likely to be more

productive than late-successional stands (table 3.1). Stands of young trees are likely to have higher productivity than stands of old trees, in which most of the energy and nutrients are used for maintenance rather than production of new tissue.

In contrast, biomass is likely to be greater in late-successional and in old stands than in early-successional or young stands. Trees in the red pine stands studied by Ohmann and Grigal (1985), which had the highest biomass of 12 forest types examined, averaged 123 years old. Most of the biomass was in the tree trunks and branches of the large, long-lived red and white pine. White cedar stands, also composed of large, old trees, had the second highest biomass of the 12 forest types.

## Animals and community interactions

### Mammals

If asked to identify the most representative animals of the northern coniferous forest, most people would name beaver, deer, moose, and wolves. This is indeed an appropriate list. Beaver (see chapter 9) and deer (see chapter 4) are important species in most Minnesota forests. Browsing by deer and tree cutting and damming of streams by beaver can change the character of entire communities. Selective removal of certain tree and shrub species by these animals and the killing of entire stands by flooding sets back plant succession to an earlier stage, with all its associated changes in both plants and animals. Both deer and beaver are associated primarily with the early and middle stages of succession that follow logging and fire and are common in northeastern Minnesota. But as succession leads toward more mature forests of spruce and balsam fir, the numbers of these mammals tend to decline.

Moose (figure 5.15) are more restricted in their distribution than deer and beaver. The moose, the largest and most magnificent of all Minnesota mammals, may stand 6 feet tall at the shoulder and weigh up to 1,200 pounds. Its long legs enable it to move through deep snow, and its large hoofs and

112

**Figure 5.15** *Adult moose, like this cow, eat about 30 pounds of plant material daily. In 1994, 7,000 to 8,000 moose were esti-mated to live in northern Minnesota. (Photo by Bob Firth, Firth PhotoBank, Minneapolis.)*

dewclaws, horny projections on the back of the foot, aid it in crossing bogs and wetlands. Do not be deceived into thinking that such large animals cannot move quickly—they can travel at 20 to 30 miles per hour over rough terrain (Peterson 1955).

At the time of exploration and European settle-ment, woodland caribou (see chapter 7) and moose were common in parts of northern Minnesota and were used extensively for food by explorers, trap-pers, loggers, and settlers. An indication of their abundance can be gained from a report that in the spring of 1660, the exploration party of Radisson and Grosseilliers claims to have killed over 600 moose in Minnesota (Surber and Roberts 1932). Numbers apparently declined drastically in the 1800s. Herrick (1892, 272), in his book *Mammals of Minnesota*, stated, "A few specimens of this noble animal still may remain in the inaccessible regions of northern Minnesota."

Although few data are available before about 1946, estimates of moose in Minnesota suggest that numbers increased to 3,000 or 4,000 by about 1930 and to more than 6,000 in the early 1960s (Idstrom 1965). The population in 1994 is esti-mated to be 7,000 to 8,000 animals. In some years the population declines, probably because of severe infestations of winter ticks. As many as 25,000 to 40,000 ticks may be attached to a single moose. Infected moose rub their hides against trees to relieve the irritation caused by the ticks, and some moose have died of exposure due to loss of hair from rubbing (Mark Lennarz, Division of Fish and Wildlife, Minnesota Department of Natural Resources, personal communication).

Moose are generally solitary and secretive in their behavior and may be found in forest and wetland habitats. They are ruminants, like deer and cows, which means that the bacteria living in their digestive tract play a major role in digestion of food. Many of their preferred foods are characteristic of the early-successional communities that are interspersed with other successional stages throughout the north and northeast (Peek, Urich, and Mackie 1976). In summer, moose frequently feed on water lilies, wild rice, and other aquatic plants. Throughout the year they eat many kinds of browse, including willow, hazelnut, aspen, mountain maple, red osier dogwood, and even balsam fir. They do not eat white cedar except when preferred food species are unavailable (Peterson 1955), in contrast to deer, who prefer white cedar and rarely eat balsam fir. If a moose has eaten all of the browse within reach on a small tree, it will rear on its hind legs and bring its broad chest against the tree, pushing it to the ground, where the higher twigs and branches can be reached. Broken branches and trunks are characteristic of moose wintering sites.

With a daily consumption of over 30 pounds of food (Peterson 1955), a single moose will remove over 5 tons of vegetation per year. Their intensive foraging on early-successional trees and shrubs, such as aspen, alder, and willow, and their avoidance of most conifers result in less leaf litter and more needle litter in areas heavily browsed by moose (Pastor et al. 1993). Because needles are acidic and have a high lignin and resin content, they decompose more slowly than leaf litter, resulting in a slower return of the nutrients to the soil (Pastor et al. 1988). Thus, the feeding habits of moose can affect the environment of the coniferous forest. These connections among moose, bacteria, and coniferous trees remind us that the parts of an ecosystem are linked in many ways.

Wild animals are exposed to a wide variety of parasites and diseases, but fortunately most are well adapted to survive these potentially lethal fac-

tors. Although some old, weak, or undernourished animals may succumb to bacterial or viral infections or to parasites, most individuals survive. One disease, however, called moose sickness, has resulted in the death of many moose in Minnesota. The disease is caused by the small parasitic worm *Pneumostrongylus tenuis*, which damages the brain and central nervous system (Karns 1976). The asexual stage of the worm lives in a snail. Immature worms escape from the snail and may be ingested by a moose or deer when it drinks. Once inside the animal, the worm travels via blood vessels to the brain, where it burrows and reproduces. Its eggs eventually pass out of the deer or moose in feces, and those eggs that find their way into a pond develop into the asexual stage, which begins the cycle anew in the body of a snail. Although as many as 60% of the deer in some areas of Minnesota are infected with *P. tenuis*, they seem to be little affected by the worms. Infected moose, however, suffer lack of coordination and drooping of the head and ears and eventually die.

The abundance of deer that occurred in the early 1900s in the northeast following logging and fires provided many hosts for the parasite. The high number of infected deer resulted in a high probability of moose becoming infected with *P. tenuis*. A high rate of infection may explain why moose numbers were low in the early 1900s. A lower rate of infection in moose resulting from declining deer populations in the northeast may also explain why moose populations have increased in recent years.

Before European settlement, gray wolves, also known as timber wolves (figure 5.16), occurred throughout Minnesota and in most of North America. Wolves were at the top of the food chain: they fed on deer, moose, and caribou, and no animals except other wolves preyed on them. Hunting and the loss of wilderness due to logging and settlement probably combined to exterminate wolves over most of their range throughout North America. They are now found mainly in Minnesota, Alaska, and Canada. Wolves appear to be expanding their range to the southwest in Minnesota, and the 1994 state population was estimated at 2,000 individuals (W. Berg, Minnesota Department of Natural Resources, Section of Wildlife, personal communication; L. D. Mech, U.S. Department of the Interior, National Biological Service, personal communication).

Wolves live in social groups, called packs, of about five to eight individuals. Members of a pack are usually related, all being siblings or offspring of the dominant pair. Normally only the adult pair breeds, thus limiting the population in a given area. Mating occurs in late winter, and litters of four to seven young are born about two months later. The pups mature quickly and begin hunting with the adults in the following fall (Mech 1970).

Each pack travels over a territory of 50 to 120 square miles and marks its territorial boundaries with urine and feces. Howling is also believed to be a means of identifying an occupied territory (Harrington and Mech 1979). The combination of limited breeding and the maintenance of large territories appears to result in a population more or less in balance with its food supply.

Wolves prey mainly on white-tailed deer but also take moose, snowshoe hares, beavers, grouse, and other animals and at times eat fruits and berries. The amount of food required by an adult wolf during one year is estimated to be equivalent to about 20 deer. A single wolf would probably not be able to kill a healthy deer or moose, but pack members hunt as a team. They do not usually stalk or hide to ambush their prey, as mountain lions might, but rather search openly until prey is located. Then the chase begins, frequently ending in frustration as the deer or moose escapes, and sometimes ending in a kill. Often the strongest prey escape, and the weakest, the very young, the very old, and those debilitated by injury or disease are killed.

An interesting discovery has been made by scientists monitoring both radio-marked wolves and radio-marked deer in northeastern Minnesota. R. L. Hoskinson and L. D. Mech (1976) found

that deer may have a higher survival rate in the narrow zones where the wolf pack territories meet, presumably because the packs tend to spend less time in these buffer zones than within their territories. It is not surprising that deer quickly learn to avoid areas where they are likely to be attacked by wolves.

Although wolves eat other kinds of food, deer are the single most important prey species (Mech and Karns 1977). Thus, the future of the wolf population in northeastern Minnesota is largely dependent on the future of the deer population. The deer population, in turn, is dependent on the presence of communities in the early stages of succession, and these communities are dependent on land use and management that involves logging and fire.

Wolves have always played an important role in our culture, ranging from fairy tales to predation on livestock to real or perceived threats on human lives. Wolves do kill and eat farm animals, but S. H. Fritts (1980) reported that the total number of domestic animals taken each year in northern Minnesota is very small. Tales and folklore about wild wolves injuring people are plentiful, but there are no confirmed records of wolves injuring people in Minnesota. Regardless, the thought of wolves brings fear to the hearts of some people. To others, however, the beautiful and thrilling sound of wolves howling to a cold northern sky evokes a spine-tingling but also reassuring response that there is still wildness in a world so dominated by people.

Coyotes, or brush wolves, are also common in the northern coniferous forest and are often mistaken for wolves. Coyotes and wolves can be distinguished by their size and the way they hold their tails: coyotes usually weigh less than 30 pounds and often run with their tail between their legs, whereas wolves usually weigh over 60 pounds and run with their tail held horizontal. Before settle-

**Figure 5.16** (overleaf) Gray wolves, also known as timber wolves, are a valuable part of Minnesota's wilderness. (Photo © Jim Brandenburg.)

ment, coyotes in Minnesota were most common in the prairie biome, but today they are more likely to be seen in the northern coniferous forest (Hazard 1982). Coyotes and wolves apparently do not mix well. Data from radio-marked coyotes indicated that they generally avoided territories of wolf packs and that some were killed by wolves (Berg and Chesness 1978).

Mice, porcupines, deer, and snowshoe hares are common prey, but coyotes are also scavengers and will readily feed on carcasses of deer and livestock. Hunters, farmers, and ranchers who have often observed hairs from deer, sheep, and cattle in coyote feces blame them for killing these animals. Although coyotes certainly do kill some deer and livestock, they are frequently blamed for more damage than is justified.

Black bears are Minnesota's largest carnivores. The name is slightly misleading, as some have brown fur. In fact, both brown and black cubs may be found in the same litter. Bears are most common in the forests and wetlands of the northeast, but they sometimes wander as far south as Anoka County. Although they are classified as carnivores and feed on insects, mammals, and birds, black bears make heavy use of blueberries, cherries, raspberries, and hazelnuts. L. L. Rogers (1976) found that nut and berry crop failures in northeastern counties from 1974 to 1976 had a marked detrimental effect on the growth and reproduction of black bears, and that the population decreased by approximately 35% during these years. Bears also use garbage dumps and campgrounds as food sources, and although they are normally solitary in their habits, small groups may frequently be found at these feeding sites.

During winter, black bears den in hollow logs, under blowdowns or brush, or just in a pile of leaves. They are not true hibernators, but during winter dormancy their body temperature and respiration drop slightly. Females mate in midsummer, but the embryos do not begin to grow until early winter because of delayed implantation of the fertilized eggs (Hazard 1982). Two or three cubs

weighing about 10 ounces each are born in the den in January or February. They nurse and sleep while the mother continues her dormancy. By the time she emerges from her den in early spring, the cubs weigh 4 to 5 pounds and can travel and climb quite well. They usually stay with their mother until they are about one year old, denning with her during the winter. Females have litters once every two or three years.

Snowshoe hares (figure 3.7), at times perhaps the most conspicuous mammal in the northern coniferous forest, are well adapted to Minnesota's seasonal changes. Their large feet enable them to run on the snow, and their white coats in winter and brown coats in summer provide excellent protective coloration. They are browsers, eating mainly bark, twigs, and leaves.

Females may have three or four litters per year, with three to four young per litter. Adult hares are quite easy to see, but young hares are very difficult to find in the forest because they remain motionless most of the the time and have protective coloration. Orrin Rongstad and I suspected that mother hares spend very little time with their young. To learn more about parental care in snowshoe hares, we marked pregnant females with radio transmitters (Rongstad and Tester 1971).

Initially we had hoped that by plotting on a map the daily movements of female hares, we would see a pattern develop that would tell us when and where they had their young. We then intended to locate and radio-tag the young so that their movements could be recorded. Although we could predict the approximate date of birth of a litter, we were never able to locate the young. To solve this problem, we captured a pregnant female and put her in a small temporary pen in her normal home range. She had a litter of four young on the night of July 12. The next morning, all four young were huddled together under a small shelter in the pen. They remained close to each other most of the time during the first few days of their lives. On July 18, we placed transmitters on the female and her four young. The next afternoon we quietly raised the sides of the pen and allowed the hares to leave. On the day after release from the pen (day 8), all the young were located with a portable radio receiver. Each was in a different hiding place, separated from the others by as much as 60 feet. This spacing out may reduce the risk of a predator capturing all of the young in the litter.

Fixes obtained by telemetry were used to determine when the mother hare was with her young and how far away she went when she left them. The young did not remain together but did stay in the general vicinity of their birthplace. The most striking discovery was that the female was with her young only once each day for only five to eight minutes. This meeting happened every night at about 10:00 P.M. The female would often spend the rest of the day as far as 900 feet from her young.

After noting how little time the mother spent with her young, we examined the movement records of other females that had been monitored on a minute-by-minute basis. We found that there was one place to which an individual female returned at the same time each night. This place was undoubtedly where she met her young for nursing.

In another radio-tracking study in Hubbard County, we found that hares tended to avoid open habitats and concentrated their activities in lowlands with black spruce, tamarack, and alder and in edge habitats between alder and jack pine (Pietz and Tester 1983). Hares chose habitats with dense shrub cover and browsed jack pine, hazelnut, black spruce, and red osier dogwood heavily.

Numbers of hares fluctuate greatly, following about a 10-year cycle. In Minnesota densities as high as 500 hares per square mile were reported in the 1930s (Green and Evans 1940), and lows of less than 10 per square mile occur in some years. This cycle has been studied in detail in Alberta, Canada, where the fluctuations are believed to be linked to the availability and quality of browse and the numbers of lynx and other predators (Keith and Windberg 1978). Food supply and predation may be the cause of the cycles observed in Minnesota, but no studies have been done to confirm this.

Porcupines, also browsers, are often seen feeding near the tops of conifers, where the bark is thin and the twigs are small. They generally live alone, except for a brief mating period and sometimes in winter, when they may den in groups in hollow trees or in old buildings. Porcupines move slowly when on the ground, relying on their quills for protection. In summer they travel over 30 to 40 acres feeding on selected trees; in winter they may stay near a single tree for many days. Adults often have long lives because they have few predators. Fishers, members of the weasel family, prey on porcupines, but in most areas the number of fishers is too small to have much effect on the porcupine population.

The northern coniferous forest, like the deciduous forest, provides a home for many other species of mammals (see appendix E). Raccoons and red fox may be seen hunting along roads at night. Martens and fishers hunt in the trees for squirrels and other prey, and lynx, bobcat, and even mountain lions prowl the forest in search of small mammals, hares, and deer. Beaver can be observed gnawing on a tree or swimming in a lake or pond near their dam or lodge. Bats are common but not often observed unless one is standing by a pond or other opening in the forest near dusk. Then they can be seen catching insects in flight. Eastern and least chipmunks, gray, red, and northern flying squirrels, bog lemmings, red-backed voles, white-footed mice, meadow jumping mice, and a variety of shrews are also present. Some of these mammals, such as red squirrels and chipmunks, are readily seen, but others, such as flying squirrels, shrews, and bog lemmings, are rarely observed.

## Birds

The northern coniferous forest contains communities in all stages of succession. Each stage, from pioneer to climax, has characteristics that determine what species of animals are present. Many species of birds overlap several successional stages, but not all of the species listed as being present in the northern coniferous forest can be found in all communities. For example, mourning warblers, chestnut-sided warblers, brown thrashers, and American kestrels are the common birds in recent clear-cuts and grassy openings. Flickers, brown-headed cowbirds, and morning and chestnut-sided warblers are common in young aspen stands. Hermit thrushes, kinglets, and blackburnian warblers are characteristic of mature spruce and fir stands.

Many other species of birds can be observed in the northern coniferous forest. Gray jays eat a wide variety of food and often rob bait from trappers' sets and steal from campsites. Ravens, found primarily in wilderness areas, are carrion feeders. Purple finches, evening and pine grosbeaks, pine siskins, and red and white-winged crossbills obtain seeds from conifer cones. Sparrows and thrushes feed on seeds and insects on the forest floor. Warblers, kinglets, nuthatches, chickadees, and vireos find their insects on the trunks, branches, twigs, needles, and leaves of trees. Flycatchers pick flying insects out of the air.

Downy, hairy, pileated, and three-toed woodpeckers live year-round in northern forests. They benefit the forest through their foraging for insects under the bark of dead or dying trees. Holes made by woodpeckers may sometimes be used by swallows, flying squirrels, and other forest animals. Pileated woodpeckers are scarce, and one is most likely to be made aware of their presence by a pile of large chips of wood from the foraging holes they frequently make near the bases of dead trees.

Yellow-bellied sapsuckers, which migrate in spring and fall, make a series of evenly spaced holes in tree trunks to obtain sap. Pine, spruce, fir, and aspen seem to be the preferred species. Sap from these holes often attracts hummingbirds and flying squirrels. The holes are arranged in either horizontal rings or vertical rows, and the same tree is often used year after year. In some cases, direct injury to the trees may be severe, and fungus spores may enter the tree through the holes (Johnson and Lyon 1976).

Ruffed grouse (figure 5.17) are common and conspicuous birds of the northern coniferous

forest, especially in early-successional and mid-successional stages where trembling aspen is the dominant tree. The life history and ecology of the ruffed grouse are well known because of its importance as a game bird. G. Gullion (1990) studied ruffed grouse at the University of Minnesota Forest Research Center in Carlton County, and much of our knowledge about this species in Minnesota is due to his efforts.

Ruffed grouse use different habitats throughout the year to meet their requirements for food and cover. These varied habitats are found in aspen forests containing stands of various ages (figure 5.18). Male grouse seek out a log or raised site, usually protected by dense shrub or sapling cover characteristic of a 15- to 25-year-old aspen stand. Here they perform their courtship behavior known as drumming. This drumming log or mound serves as the focal point in the male's territory and may be used for many years. The drumming sound is produced by very rapid wing beats made while the male holds on to the log with its feet and braces with its tail. Air rushing in to fill the temporary vacuum actually creates the sound. Drumming communicates to nearby males that the territory is occupied and probably attracts hens that are ready to breed. Many males occupy their territories in fall, and some drumming can be heard at this time.

Radio transmitters that Herbert Archibald and Richard Huempfner placed on male grouse at Cedar Creek Natural History Area enabled them to monitor the exact timing of drumming, because the action of the wings made the whip antenna on the transmitter vibrate. At the peak of the breeding season in May, males sometimes drummed for more than 22 hours per day and spent the remaining time feeding. Males occupying adjacent territories appeared to respond to each other's drums, suggesting that the method of advertising an occupied territory was very effective. Each male maintained a territory of about 5 to 12 acres and

**Figure 5.17** (right) Ruffed grouse are popular game birds in northern forests. (Photo © Jim Brandenburg.)

121

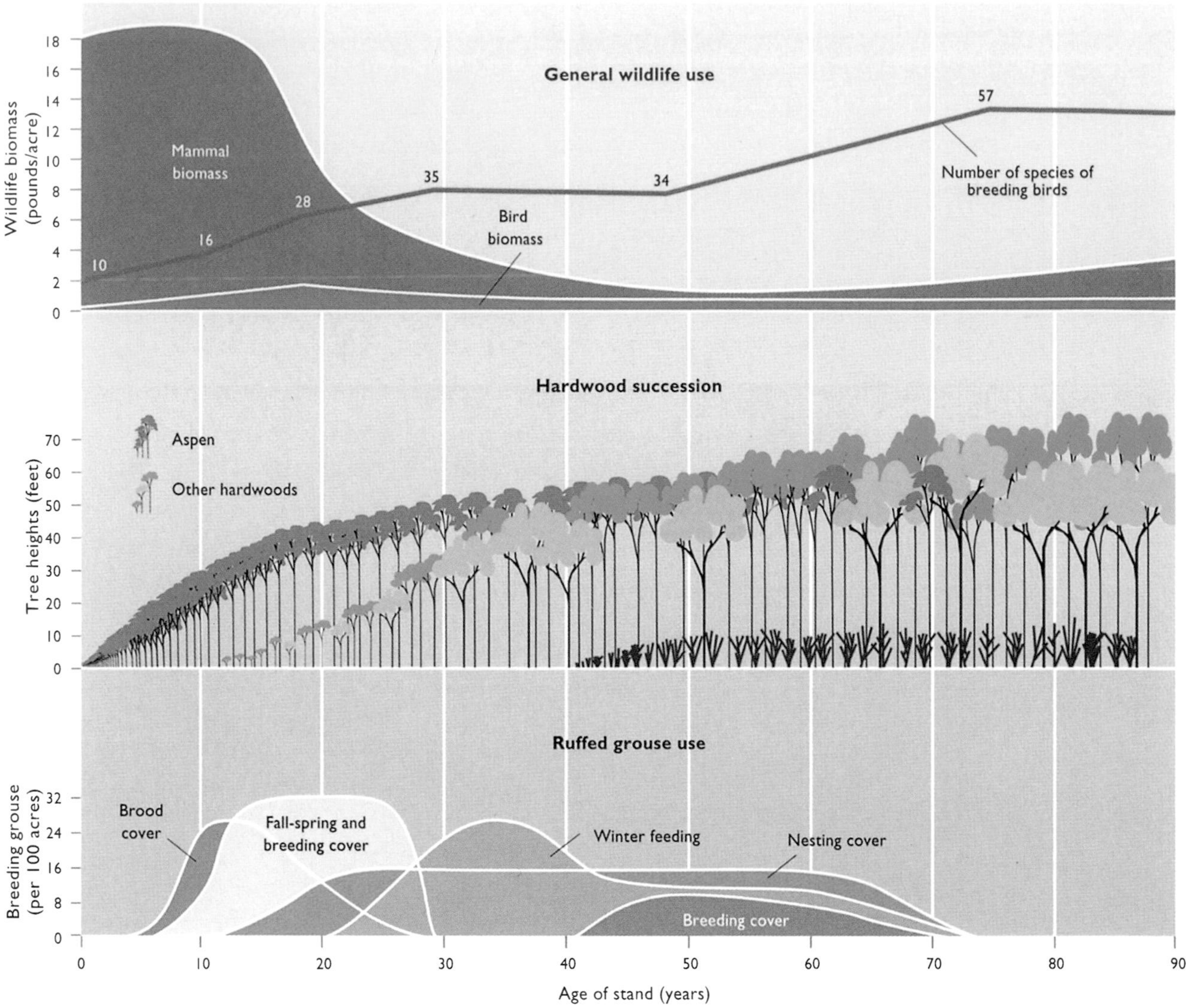

## General wildlife use

Wildlife biomass (pounds/acre)

18
16
14
12
10
8
6
4
2
0

Mammal biomass

57

Number of species of breeding birds

35
34
28
16
10

Bird biomass

## Hardwood succession

Tree heights (feet)

70
60
50
40
30
20
10
0

Aspen

Other hardwoods

## Ruffed grouse use

Breeding grouse (per 100 acres)

32
24
16
8
0

Brood cover

Fall-spring and breeding cover

Winter feeding

Nesting cover

Breeding cover

0      10      20      30      40      50      60      70      80      90

Age of stand (years)

**Figure 5.18** *(above) Ruffed grouse use different habitats to meet food and cover requirements throughout the year. The stages of succession in a northern forest, in which aspen is the most important pioneer species, provide the varied habitats needed by grouse. This figure also shows how the species diversity and biomass of birds and mammals changes as succession proceeds. (Modified from Gullion 1990, by permission.)*

**Figure 5.19** *(right) The ruffed grouse population increases and declines at about 10-year intervals, as shown by data from spring drumming counts in Minnesota. Population highs appear to occur near the end of each decade. The cause or causes of this cycle are unknown. (Data from Dexter 1993.)*

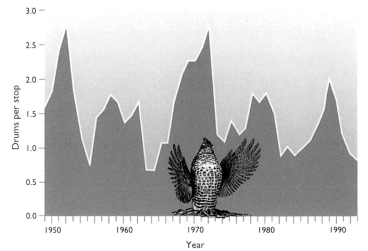

3.0
2.5
2.0
1.5
1.0
0.5
0.0

Drums per stop

1950      1960      1970      1980      1990

Year

seldom ventured far inside adjacent territories (Archibald 1976).

Hens do not form pair bonds but come to a male only for a short time for mating. Males do not participate in incubating the eggs or rearing young. Nests are located in aspen stands ranging from 15 to 65 years in age (figure 5.18). Incubation of the clutch, usually 10 to 12 eggs, lasts about 24 days. The chicks can run immediately after hatching and can fly in less than two weeks. During this period, they occupy young aspen stands and feed almost entirely on insects, which they capture themselves. By the time the young reach adult size, their diet has shifted mainly to plant food. Fruits and berries of dogwood and other shrubs, leaves of clover and strawberry, and buds and flowers of aspen, alder, and birch are important foods. Buds of male aspen are a very important winter food in Minnesota, and the size and nutrient content of the buds may influence grouse productivity and survival (Huff 1973).

Winter is the critical season for ruffed grouse, and the number that survive depends on the supply of buds in 30- to 40-year-old aspen stands and on the condition and depth of the snow. Goshawks, barred owls, and great horned owls are very effective predators on ruffed grouse in winter. If the snow is deep and soft, grouse will use burrow roosts. To form these, they fly directly from a feeding tree head first into the snow, disappearing from sight. Grouse in burrow roosts are quite safe from hawks and owls. When the snow is hard or crusty or not deep enough for burrows, however, the birds form a slight bowl in the snow for roosting (Huempfner and Tester 1988). Here they are highly vulnerable to hawks and owls.

Ruffed grouse populations appear to be cyclic and to follow the same time pattern as snowshoe hares. Data on population numbers in Minnesota from 1949 to 1993 (figure 5.19) show peaks about every 10 years, but obviously the cycle is not precise in either timing or numbers. The cause or causes of this cycle are unknown, although L. B. Keith and L. A. Windberg (1978) believe that

grouse are part of the snowshoe hare-vegetation-predator cycle discussed earlier. Other biologists, however, suspect that the quality of food, especially aspen buds, may be a major cause, and that food quality is somehow related to changes in weather or climate. In spite of many years of research and perhaps hundreds of published papers, the cause of the grouse and hare cycles is still not fully understood.

Spruce grouse also are found in northern coniferous forests and coexist with ruffed grouse in some localities, especially where jack pine uplands are interspersed with lowlands of spruce and tamarack (Pietz and Tester 1982). Spruce grouse eat a variety of herbs, shrubs, mushrooms, and insects in summer. In winter they feed on conifer needles. These appear to be poor quality food, but the bacteria and protozoa in the digestive tract convert the cellulose in the needles to sugars that the grouse can assimilate.

Numerous spruce grouse were radio-tagged in the Hubbard County study mentioned earlier. During winter the spruce grouse preferred jack pine stands, but in spring and summer most moved into the black spruce and tamarack bogs, where they nested and reared their young (Pietz and Tester 1982). In areas near the Canadian border with few upland sites, C. H. Haas (1974) found that spruce grouse carried out all phases of activity in lowland conifer forests.

Hawks and owls are common in all forests, but some of the most interesting and exciting species are associated with the northern coniferous forest. The largest and most spectacular is the bald eagle (figure 5.20), our national symbol. Although rare in many states, several hundred pairs of eagles nest each year in northern Minnesota. No one can mistake the majestic adults with their white head and tail as they soar over large lakes and rivers searching for food.

Their nest of sticks is often located high in a white pine and will usually be used by the same pair as long as they live. Each year more sticks are added. Nests may be 6 to 8 feet in diameter, 10 feet high, and may weigh over 2 tons. Eagles are

**Figure 5.20** *Several hundred pairs of bald eagles nest and raise their young each year in northern Minnesota. (Photo by Thomas D. Mangelsen/ Images of Nature.)*

quite secretive and usually nest in wilderness areas or forests with little disturbance from humans. The Chippewa National Forest is an especially good place to observe nesting bald eagles in Minnesota. During early winter eagles migrate south, and many can be seen during the winter along the Mississippi River and Lake Pepin south of the Twin Cities.

Survival of the young is high, and populations in Minnesota are increasing. Aerial surveys are conducted in April to determine nest occupancy and in late June or early July to determine production of young. In 1990, 437 breeding areas were occupied, and 467 young were produced (Miller and Pfannmuller 1991). These numbers are especially encouraging because the population nearly disappeared after the widespread use of DDT as an insecticide in the late 1940s and 1950s. DDT entered the food chain through insects and was concentrated in fish, birds, and mammals that fed on the insects. Bald eagles feed mainly on fish, most of which are dead or injured. They hunt by diving and snatching the fish out of the water with their talons. DDT in the body of the female eagles

interfered with calcium metabolism, resulting in thin-shelled eggs that frequently cracked during incubation. Since the ban of DDT in 1972, the amount of DDT in the food chain has decreased, and thinning of egg shells is now uncommon.

Although snowy owls nest in the tundra, they are abundant in Minnesota in some winters. In years when the lemming population crashes in the tundra, the owls migrate south in winter to find food but return to the tundra in spring to nest. Because they live in the far north, they are not familiar with humans and show little fear when approached. Snowy owls are large and can capture snowshoe hares, ducks, and grouse, but their main foods are small mammals such as mice and lemmings.

The boreal owl, common in Canada, has recently been found to breed in northeastern Minnesota. The first nest discovered in the United States was located in Minnesota in 1978. Nests are commonly in holes in mature aspen, spruce, or tamarack. Field studies in Lake and Cook Counties revealed that 22 of 23 cavities used by boreal owls were in aspen and one was in white pine (Lane

1990). This owl appears to be dependent on the presence of mature trees with cavities. Such trees are common in old-growth forests but are rare in young to medium-aged stands. Boreal owls feed on small mammals, especially meadow voles, red-backed voles, and bog lemmings, and on small birds.

### Amphibians and reptiles

Fewer kinds of amphibians and reptiles live in the northern coniferous forest than in the deciduous forest and the prairie. Two turtles, snapping and painted, live in the northeast as well as throughout the state. Both prefer slow-moving or quiet water with a muddy bottom and dense aquatic vegetation. Wood turtles, present only in the eastern counties, live in forests with clear, fast-flowing streams. Two snakes, the redbelly and the garter snake, are common and are also distributed statewide. These snakes feed on slugs, earthworms, toads, and frogs and are harmless to humans. Snakes can be found in deep woods and in meadows and clearings.

Several species of snakes may overwinter together in mammal burrows or in the tunnels of ant mounds. Jeff Lang (1971) discovered such a mound in October 1966 in a mixed conifer and deciduous woods in Clearwater County. He excavated the mound and found 155 redbelly snakes, 119 smooth green snakes, and 22 garter snakes, most buried from 4 to 7 feet deep. Some snakes were found nearly 9 feet deep.

Blue-spotted, redback, and tiger salamanders live in the northern coniferous forest but are rarely seen. American toads, on the other hand, are readily observed as they wait patiently for unsuspecting insects or other invertebrates to come close enough to be snapped up with a flick of their fleshy tongue, which is attached near the front of the lower jaw. Prey captured include cutworms, ants, beetles, butterflies, spiders, and earthworms. Toads are cold-blooded and escape the rigors of winter by burrowing in the ground to depths below the frost line with the aid of the hard, sharp tubercles on their hind feet. Toads may move as much as a quarter of a mile from their breeding pond and

summer range to a wintering site, and, amazingly, they return to within inches of the same wintering site each fall. Although biologists are not certain, they suspect that the toads use the position of the sun and the stars, and perhaps chemical cues, to navigate precisely to the same locations each year.

### Insects

Many kinds of insects live in the northern coniferous forest. Some, such as mosquitoes and deerflies, we call pests. Others, such as luna and cercropia moths, add beauty to the forest. Still others are predators on certain insect species, and all may serve as food for warblers, toads, and other animals.

From the standpoint of forest management, probably the most important insects in the northern coniferous forest are those that feed on conifer needles. Larvae of moths and sawflies damage many trees and sometimes destroy most of the trees in a community.

The spruce budworm, a moth larva, has been responsible for destruction of billions of board feet of balsam fir and spruce timber. Despite its name, the insect prefers feeding on balsam fir. Periodic outbreaks of the spruce budworm seem to be part of the natural cycle in ecosystems where these trees are common. Adult moths are mottled gray or brown and have a wing span of about 1 inch. Females deposit eggs on the tree needles. In about two weeks the eggs hatch, and the young caterpillars begin to feed and grow. In heavily infested trees, some caterpillars spin a thread and descend to the ground to search for another meal. Wind sometimes carries the thread and caterpillar to a new meal, but often the caterpillar is carried into a lake or onto an oak or maple that provides neither food nor shelter. In late summer the larvae, which have molted through several stages, spin cocoonlike shelters, where they overwinter. The following spring the larvae emerge and feed on needles and buds until early summer. They are about 1 inch long and are brownish with conspicuous white or yellow raised spots. The larvae then stop eating

125

and transform into pupas. About 10 days later the moths emerge, and the cycle begins again. During epidemics needles are severed at the base, and buds are damaged to the extent that the dead tops and webs full of dried needles seem to turn the forest gray. If defoliation continues for several years, many trees die, and the forest becomes susceptible to fire and blowdowns (Johnson and Lyon 1976).

Epidemics of spruce budworms are believed to be held in check in part by bird predation. Three warblers, bay-breasted, Tennessee, and blackburnian, have shown direct numerical responses to increasing numbers of budworms. In one study, the density of bay-breasted warblers increased from 10 to 120 pairs per 100 acres as the budworm population increased (Morris et al. 1958). At certain times, however, the budworm population grows so rapidly that the predators become satiated before they have an impact on the budworm population. Then the budworms may explode in an epidemic because the predators cannot reproduce fast enough to control the prey. The epidemic ends as the trees are defoliated and predation increases.

Pine sawflies are not flies at all but are nonstinging wasps with a sawlike appendage that is used to make slits in pine needles (Johnson and Lyon 1976). Females deposit as many as 100 eggs individually in rows of these slits. After hatching, larvae (figure 5.21) feed on the needles for about a month and then drop to the ground and spin cocoons. Winter is spent in the litter on the ground in the cocoon as a prepupa, a stage between a larva and pupa. Some prepupae complete development the following spring and emerge as adults. Some, however, may wait for a year or two before becoming adults. This mechanism, called diapause, has probably evolved to allow some sawflies to survive if all those that transform into adults in a given year should die. Early damage results in strawlike needle remains; later both old and new needles, and even bark, may be consumed.

Although insects that feed on plants may occasionally have a dramatic effect on a community, they are usually not so abundant that we notice their presence. On the other hand, there are several species in the forest that people would gladly avoid. Mosquitoes, deerflies, and blackflies are in this

**Figure 5.21** *Pine sawfly larvae consume needles and even bark of coniferous trees. Adult sawflies lay their eggs in pine needles, which they slit open with their sawlike appendage. (Photo by Don Rubbelke.)*

category, and when abundant, they will drive many visitors out of the woods and into their cabins or cars.

## Present status of the northern coniferous forest

In a few places in Minnesota, especially along the Canadian border, the northern coniferous forest still exists much as it did hundreds of years ago. The character of these communities is truly wilderness by almost any definition. An early surveyor, working to establish boundaries for the General Land Office, vividly conveyed his impression of the northern coniferous forest (as quoted in Ahlgren and Ahlgren 1984, 71–72): "In summing up the entire contract, I may truly say that it lies in the most forsaken country it has ever been my misfortune to encounter. There is apparently nothing (here) that would induce a sane person to enter within the unsacred domain of moose, wolves, bear, snow, rain, mosquitoes, flies, rocks, swamp, brush, and rapids."

Today these characteristics are considered to be valuable attributes of wilderness. In spite of increasing urbanization and technological advances, many humans are drawn to the wilderness of the northern coniferous forest. They are drawn not just to the trees, rocks, and water but to intangibles related to feelings and spirituality. Perhaps knowing that wilderness exists and that it can escape the impacts of civilization provides some assurance that we can live in harmony with our environment.

In contrast to these wilderness areas, much of the northern coniferous forest is in transition from its presettlement status (figure 5.22). The pine forests and the mixed hardwood-conifer forests have largely been harvested and replaced by aspen, paper birch, and second-growth maple forests, and by pine plantations (Stearns 1988). Most of these forest stands became established following logging and are 60 to 90 years old. Events such as wind, drought, and extreme temperatures cause natural disturbances in these forests, but their future will be determined by human management in the form of harvest and plantation establishment.

Many have found and many still find reasons for coming to the northern coniferous forest. American Indians, fur trappers, prospectors, loggers, hunters, fishermen, canoeists, hikers, and photographers have explored the landscape. Some left almost no trace of their presence, but some had widespread impact.

What does the future hold for the northern coniferous forest in Minnesota? To answer, we must address three quite different perspectives about the forest. One concerns economic values related to forest products and minerals, the second concerns aesthetic values related primarily to recreation, and the third concerns ecological values such as soil and water conservation and the preservation of rare species and old-growth forests.

Perhaps the best prediction is that communities will always be changing. Mature spruce and fir trees will burn, be logged, or will blow down, creating gaps in the forest canopy that allow sunlight to reach the ground. This in turn will lead to establishment of aspen or jack pine in a new successional sequence, which will eventually lead to a new spruce and fir stand. Wind, fire, and insects are to some extent beyond our control and so will continue to affect the northern coniferous forest. Logging for saw timber, pulpwood, and wafer board will continue and will very likely increase in the next few decades (Jaakko Pöyry Consulting 1992), causing further changes in the forest. Fragmentation may increase in some areas because of intensive harvest and may decrease in others because of reforestation.

Mining for copper and nickel and perhaps even gold may develop in Minnesota's northern coniferous forest, just as mining for iron ore developed many years ago. Surveys have indicated deposits of copper and nickel that could be mined, and several corporations have indicated interest in such ventures.

Recreational use will increase as the human population increases. But with management guided by knowledge of the ecological functioning of the for-

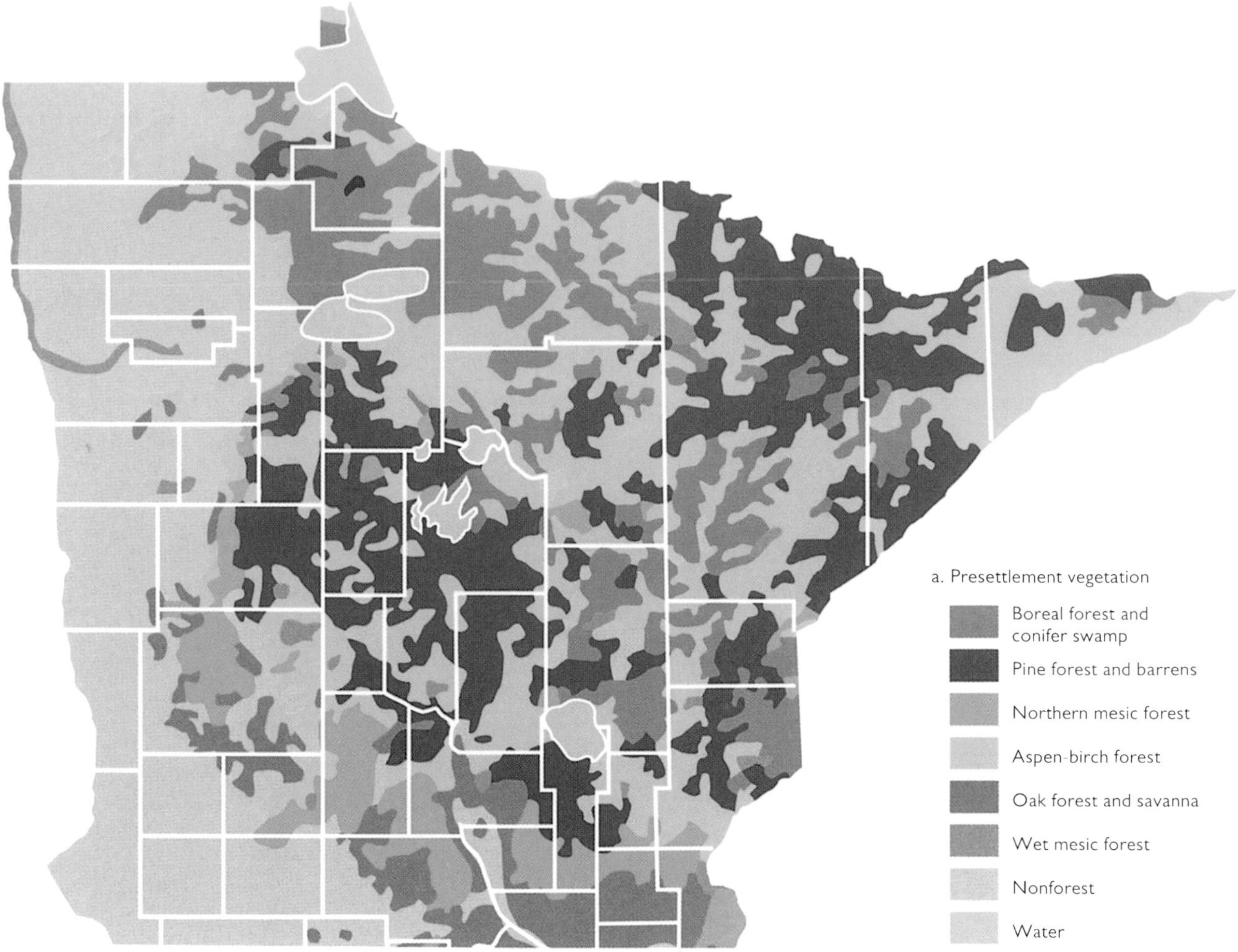

**Figure 5.22** (a) *Changes in the types of communities comprising the northern coniferous forest are strikingly revealed in these maps. Presettlement forest cover was determined from records made by government surveyors conducting the General Land Office survey in the mid-1800s. (Modified from Stearns 1988, by permission.)*

est, this use need not severely affect the forest. Do not be surprised if permits for recreational visits to popular wilderness areas such as the BWCAW and to parks and campgrounds are more and more difficult to obtain. Such rationing is necessary to protect the areas from overuse and to insure a quality recreational experience.

Old-growth forests are an important part of this experience for many people. These forests contain stands of large, old trees of such species as red and white pine, spruce, and white cedar that are relatively free of human disturbance. Maintenance

of old-growth forests is controversial: loggers are anxious to harvest the mature trees, and preservationists seek to save the remaining stands for their educational, scientific, and aesthetic values. Probably less than 1% of Minnesota's upland forests qualify as old growth, but these stands are currently receiving a great deal of attention (Rusterholz 1990).

The mixture of successional communities, the existence of wilderness, and the preservation of old-growth stands in the northern coniferous forest region have enabled most species to maintain sustainable populations. These forests contain

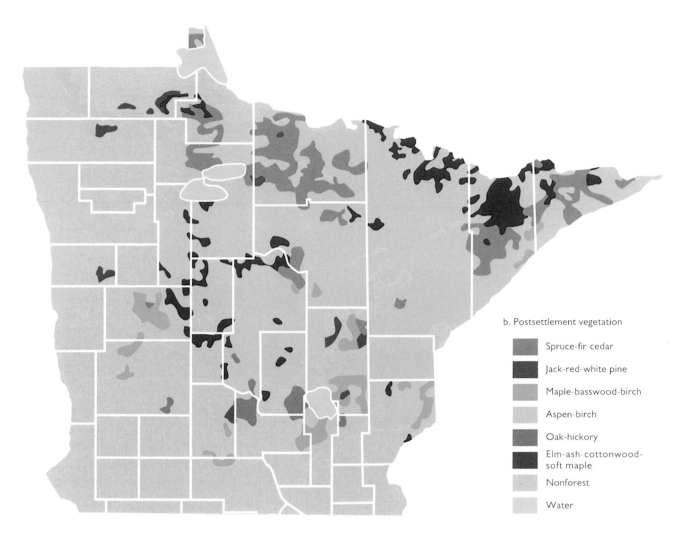

b. Postsettlement vegetation

- Spruce-fir-cedar
- Jack-red-white pine
- Maple-basswood-birch
- Aspen-birch
- Oak-hickory
- Elm-ash-cottonwood-soft maple
- Nonforest
- Water

**Figure 5.22** (b) *Composition of the present-day forest was generalized from cooperative forest surveys from 1977 to 1983. Reduction in the acreage of pine and the marked increase in aspen and birch reflect the impacts of logging and the development of second-growth forests on cutover areas. (Modified from Stearns 1988, by permission.)*

fewer endangered species than deciduous forests or tallgrass prairie. Of the 287 species of animals and plants listed as endangered, threatened, or of special concern in Minnesota, only 25 are found in northern coniferous forests. Eight of the 17 listed mammals, 2 of the 27 birds, and 15 of the 191 plants occur in these communities (Pfannmuller and Coffin 1989).

Interests in the northern coniferous forest range widely, from exploitation of its resources to complete protection. Compromises will surely be required. Private land will probably be used for the production of forest products, and public land will provide opportunities for protection. The Boundary Waters Canoe Area Wilderness, the Superior and Chippewa National Forests, Voyageurs National Park, and numerous state parks and forests lie in the northern coniferous forest biome. This large acreage in public ownership will, I hope, lead to preservation of important natural areas and sound management. It is essential, however, that these areas be preserved and managed with knowledge of the natural ecological processes that brought them into being.

# TALLGRASS PRAIRIE  6

A sea of waving grass stretching to the horizon, dotted here and there with wetlands and islands of trees, alive with herds of buffalo, antelope, and elk and with myriad waterfowl, godwits, bobolinks, and meadowlarks—this is the scene that greeted the early explorers of Minnesota's prairies. Reactions to the vastness of the prairie landscape were seldom indifferent. In *A Tour on the Prairies*, Washington Irving (1835, 271) wrote:

> To one unaccustomed to it, there is something inexpressibly lonely in the solitude of the prairie: the loneliness of a forest seems nothing to it. There the view is shut in by trees, and the imagination is left free to picture some livelier scene beyond; but here we have an immense extent of landscape without a sign of human existence. We have the consciousness of being far, far beyond the bounds of human habitation; we feel as if moving into a desert world.

Hamlin Garland (1928, 133) expressed a different reaction:

> Nothing could be more generous, more joyous, than these natural meadows in summer. The flash and ripple and glimmer of the tall sunflowers, the myriad voices of gleeful bobolinks, the

**Figure 6.1** *Remnants of the vast prairie still exist in Minnesota, where one can see gently rolling hills covered with grasses and punctuated with flowers. On this dry prairie in Lac qui Parle County, prairie coneflower, blue vervain, and daisy fleabane stand out. Flowering stems of sideoats grama grass can be seen in the foreground. (Photo by Blacklock Nature Photography.)*

meadow-larks piping from grassy bogs . . . made it all an ecstatic world to me. . . . The sun flamed across the splendid serial waves of the grasses and the perfumes of a hundred spicy plants rose in the shimmering air.

Some of the early settlers who originated from forested areas in Europe turned back into the forest when they came to the prairies. They believed that land that could not grow trees could not grow agricultural crops. They homesteaded in the forest and cleared trees to make fields. But by the middle to late 1800s, pioneers had learned that the soil of the tallgrass prairie was deep and rich, the climate was moderate, and the prairie sod could be broken. The prairies were quickly settled.

As the pioneers soon discovered, the prairie provided more than just fertile soil for crops. It gave its sod to build houses, but not easily—up to 10 yoke of oxen might be used to pull the sod-breaking plow. Wisps of dry prairie grasses were twisted and used as fuel, mattresses were stuffed with grass, and brooms were made from the tall grass stems.

Nearly 20 million acres of tallgrass prairie stretched over Minnesota, extending from the borders of Iowa and Wisconsin in the southeast to North Dakota and Manitoba in the northwest. Less than 1% of Minnesota's native prairie remains today, having been replaced by seemingly never-ending corn, soybean, and wheat fields. Thus, to understand the ecology of the prairie biome in Minnesota today, we must consider both the remaining native prairies (figure 6.1) and the fields, woodlots, shelterbelts, and farmsteads.

## Prairie-forest border

Differences in rainfall and temperature play a major role in determining the distribution of prairie across central North America and the location of the border between forest and prairie. Growth of trees on the prairie is, in general, inhibited by the low rainfall and high summer temperatures characteristic of the prairie biome. Change or transition from northern coniferous forest to deciduous forest to prairie is especially abrupt in northwestern

**Figure 6.2** *A few species of grass make up most of the biomass in the tallgrass prairie. Big bluestem, shown here, and Indian grass are dominant on moist sites. Root systems of these grasses may be three to four times as extensive as the visible part of the plants. (Photo © Jim Brandenburg.)*

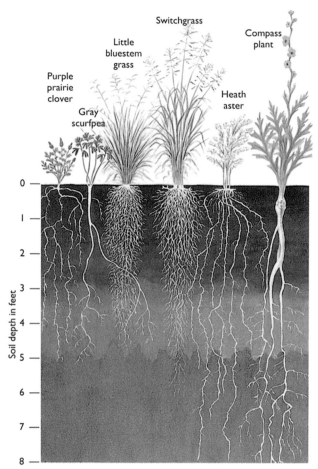

Purple prairie clover

Gray scurfpea

Little bluestem grass

Switchgrass

Heath aster

Compass plant

Soil depth in feet

0
1
2
3
4
5
6
7
8

**Figure 6.3** *Roots of prairie plants are highly variable. The dense, fine roots of little bluestem and switchgrass, for example, are able to absorb water quickly after a rain. Decomposition of these fine roots rapidly adds organic matter to the prairie soil. The longer, thicker roots of compass plant and scurfpea reach deeper for water and are not replaced as quickly as the finer roots of the grasses. (Illustration by Don Luce.)*

Minnesota in Becker, Mahnomen, and Clearwater Counties. In an effort to determine the reasons for this rapid change, I examined weather records from a number of locations within the transition zone. Average annual precipitation was 25.1 inches in the forests near Itasca State Park but 19.6 inches in the Red River valley about 50 miles to the west. January mean temperatures were similar, about 6.4°F, but August means were 65.1°F in the forest and 69.0°F in the Red River valley.

Prairie plants could live throughout central and northeastern Minnesota because adequate precipi-

tation occurs and the growing-season temperatures are favorable. However, this climate is also suitable for trees, which grow vigorously in response to the abundant moisture. The shade created by the trees inhibits the growth of prairie plants, which require high levels of light.

In fact, many kinds of prairie plants actually exist in some forested areas but do not grow to their normal size because of the deep shade. A striking example of this was observed in the Crex Meadows Wildlife Management Area near Grantsburg, Wisconsin, located close to the eastern border of Minnesota (Vogl 1964). Prescribed burning was being used to remove jack pine from the area in the hope that wildlife such as sharp-tailed grouse, ducks, and white-tailed deer would increase in numbers as the successional process brought about communities favorable to these species. During the growing seasons immediately following the fire, an unexpected and dramatic growth of certain prairie plants, such as big bluestem (figure 6.2) and Indian grass, occurred. These plants were in the jack pine forests all along but were so small that their presence had gone unnoticed. Once the trees were killed by fire, the prairie plants responded to the marked increase in light. I have seen seed stalks of these two grasses at Crex Meadows that were more than 7 feet tall.

## Tallgrass prairie ecology

Tallgrass prairies occurred in a broad sweep across Minnesota from southeast to northwest (figure 1.25). In a recent survey, over 900 species of plants were found on remaining prairies and in other communities in Lac qui Parle, Big Stone, Traverse, Wilkin, Clay, and Norman Counties (Wheeler, Dana, and Converse 1991), and some individual prairies support 200 or more species of plants. Some of these plants grow under a broad range of moisture conditions, but others are found only where certain moisture conditions exist. Root systems of prairie plants are extremely variable (figure 6.3). Some penetrate deeply as a tap root or as fine

133

rootlets, enabling the plant to use deep, subsoil moisture during droughts. Other roots are shallow and dense, enabling the plant to capture moisture from light rains that soak into the soil only a few inches. Thus, variations in the soil moisture, determined by the degree and direction of slope, the depth of the water table, and the soil texture from one location to another, result in variations in the plant species present. Ecologists have used these variations in soil moisture to classify tallgrass prairies into three types: mesic, dry, and wet (Wendt 1984).

### Mesic, dry, and wet prairies

Mesic prairie, present under conditions of moderate moisture, was the most common grassland type before settlement. The plant cover is made up primarily of five species of grasses: big and little bluestem, Indian grass, prairie dropseed, and porcupine grass (figure 6.4). Kentucky bluegrass, an introduced species, is often present, indicating that some type of disturbance has occurred. Sideoats grama, several species of panic grasses, muhly grass, and switchgrass are also common in mesic sites.

Prairie herbs not included in the grass family are referred to as forbs. Many species of forbs are present on mesic prairies in Minnesota, and their flowers often appear to be a dominant feature of the vegetation (figure 6.1). Although forbs are conspicuous, detailed studies have indicated that grasses usually dominate in terms of biomass and cover. On a typical remnant of mesic tallgrass prairie in Mahnomen County, grasses made up over 80% of the vegetative cover (Tester and Marshall 1961).

134

The species composition of mesic prairies differs from southeast to northwest. Wild false indigo, rattlesnake master, compass plant, wild quinine, and Indian plantain, for example, occur primarily in southeastern prairies whereas blanket flower, northern gentian, oat grass, and blunt sedge occur mainly in the northwest. Soft goldenrod, cut-leaf ironplant, locoweed, pliant milk-vetch, and red triple-awned grass are found only along the western edge of Minnesota. One woody species, leadplant

**Figure 6.4** *The seeds of porcupine grass have barbed tips and long awns, which curl when the seeds ripen and drop. The awns blow in the wind and twist and coil as temperature and humidity change. Movement of the awn pushes the barbed point deeper and deeper into the soil, until the seed is planted. (Photo of grass by Don Rubbelke; photo of seeds by Jim LaVigne.)*

(figure 6.5), is common on native prairie throughout Minnesota. It seems to be especially sensitive to disturbance; hence its presence indicates that the prairie has not been plowed or cultivated.

As the seasons progress, prairies exhibit marked changes in appearance due to the pattern of flowering of the various species of grasses, forbs, and shrubs. In early spring on a mesic prairie in west-central Minnesota, delicate pasque flowers are the first to blossom. Shortly thereafter, the tiny

**Figure 6.5** (right) Lead-plant, named for its grayish leaves, is a shrub on mesic prairies.

**Figure 6.6** (below) Purple coneflowers were called Thirst Plants by pioneers because the roots had a salty taste. Chewing the roots increases the flow of saliva, relieving thirst. (Moyle and Moyle 1977). (Photos by Richard Hamilton Smith.)

purplish white flowers of blue-eyed grass can be found hidden among the dead and decaying litter. Plants such as golden alexanders and lousewort are also flowering at this time. In early summer the pink flowers of prairie phlox, the brilliant orange trumpets of the Philadelphia lily, and the delicate slender florets of needle grass and porcupine grass dominate the scene. Grasses form an almost continuous carpet of green as a background for the wide variety of prairie flowers. As summer pro-

135

gresses, purple and white prairie clover, purple coneflower, and leadplant bloom. In August the yellows of the many species of goldenrod and sunflower, the bright purple spikes of blazing star, and the many white, blue, and purple asters take over the landscape. Now the purple and gold flowering heads of the bluestems and Indian grass begin to rise above the carpet of green. As fall approaches, the bluestems turn a rich reddish purple hue, and Indian grass turns from green to gold. Fringed and bottle gentians and the beautiful ivory-colored orchid called ladies tresses can be found blooming if one looks carefully in the litter formed by dead and dying grasses.

On dry locations, such as hilltops over bedrock, and on soils with sand or gravel, the composition of the prairie is somewhat different. Although some species from mesic sites, such as little bluestem and sideoats grama are present, big bluestem and Indian grass are replaced by plains muhly, blue grama, and hairy grama, grasses that grow to a maximum height of about 8 inches. Prairie dropseed, sand reed, June grass, and needle grass also occur on dry sites. Common forbs on these dry sites are pasque flower, prairie smoke, narrow-leaved puccoon, white-flowered beard-tongue, compass plant, purple coneflower (figure 6.6), and silky aster. Litter is about an inch deep, and the vegetation appears sparser than in mesic prairies.

In contrast, vegetation on wet prairies is often dense and tall. Sedges are abundant, and characteristic grasses include prairie cord-grass, switchgrass, mat muhly, blue-joint, and northern reed grass. Big bluestem may also be found on wet sites. Common forbs include numerous species of the mint family, wild licorice, white lady slipper (figure 6.7), New England aster, golden alexanders, and gayfeather. Wet prairies often contain wetlands with emergent plants such as sedges, cattail, bulrush, and arrowhead growing along the shore. The species of the wet prairie and the wetland tend to mix and blend,

and it is often difficult to tell where the prairie ends and the wetland begins.

In some wet prairies and along the shorelines of shallow lakes near the western edge of the state, the soil and water contain high levels of salt. These saline sites support a distinctive group of plants, including saltgrass, alkali grass, alkali plantain, spearscale, sea blite, and rarely, glasswort.

## Prairie fires

On October 2, 1835, at Lac qui Parle, George Featherstonhaugh ([1847] 1970, 354) described the spectacular character of wildfire on the prairie:

> Before going to my pallat I made another journey to the upland behind the fort, to see the prairies on fire. It is a spectacle one is never tired of looking at: half the horizon appeared like an advancing sea of fire, with dense clouds of smoke flying towards the moon, which was then shining brightly. Here I remained enjoying this glorious sight until a late hour.

Prairie fires (figure 6.8) set by lightning or by American Indians were common before European settlement (Higgins 1986), and important to the ecology of the prairie. Most of the prairie species are well adapted to fire and survive repeated fires with few ill effects. Many species are long-lived perennials with deep, extensive root systems that are highly adapted to surviving fire, as well as drought and grazing. Although fire may destroy all of the aboveground vegetation, the plants quickly produce vigorous new shoots from the growing point just at or below the soil surface. This method of growing from the base rather than from the tip enables grasses to withstand not only fire but also grazing, trampling, mowing, hail, and wind.

In addition, fires reduce the litter layer. The sun's rays strike the dark soil surface, resulting in higher temperatures and increased microbial activity, which are beneficial to the growth of prairie plants. Rainfall is absorbed directly into the soil rather than into the litter, and the nutrients

**Figure 6.7** *Small white lady slippers, only about 6 inches tall, grow in wet prairie sites. (Photo © Jim Brandenburg.)*

released by the fire are quickly recycled. Following most burns, production of tillers or roots, flowers, and seeds increases dramatically, and plants are often taller and more luxuriant than before the fire (Svejcar 1990).

Frequent fires are also believed to be responsible for preventing the spread of the deciduous forest westward into the prairie. As described in chapter 4, the position of the prairie-forest border before settlement was probably determined largely by wildfires set by lightning and by American Indians (Buell and Facey 1960). Because the prevailing winds throughout the summer and fall are from the southwest, fires that started on the prairies spread eastward and in many instances moved to the edge of the forest. Generally the environment in the forest was moister, primarily because of greater

precipitation and shade. Fuel was sparse, and the trees did not burn well. Thus, the fires quickly died out. These fires did, however, kill most of the aspen and bur oak trees that attempted to grow on the prairie, because the dense grass litter provided ample fuel for a hot fire.

Although fires did not burn every area every year, periodic fires at intervals as long as 10 or 20 years were sufficient to kill most invading trees and shrubs. The border between prairie and forest was likely irregular, with fingers of forests that extended westward into the prairie only to be killed back by fire (Buell and Cantlon 1951). Certain localities in the prairie, such as islands and peninsulas in lakes and wetlands, were protected from fires and often were covered by forests.

Early settlers in Mahnomen and Becker Coun-

138

**Figure 6.8** *Prairie fires occurred nearly every year, ignited by lightning or American Indians, until the time of European settlement. These fires had numerous beneficial effects on the prairie but were a serious threat to buildings and crops. (Photo courtesy of Dan Ruda, Minnesota Department of Natural Resources.)*

ties reported that fires frequently threatened their homes, buildings, and livestock (Buell and Facey 1960). With the continued settlement of the prairie came the development of railroads and highways with their adjoining ditches. These served as natural firebreaks and, along with efforts by the early settlers, served to control the spread of prairie fires. From then on, woody plants had an easier time invading the prairies. Today one can find many examples of groves of aspen and oak invading the few remnants of native prairie left in Minnesota (Buell and Facey 1960). Thus, the prairie-forest border is still changing. This change is a natural phenomenon influenced, to a certain extent, by human activities.

Fires in agricultural areas are usually controlled very quickly, and the few remaining tallgrass prairies in Minnesota have been protected from fire, doing them more harm than good. As the understanding of the ecology of prairie species has become more widespread, however, management agencies, such as The Nature Conservancy and the Minnesota Department of Natural Resources (DNR), have begun using prescribed burning to manage the prairies under their control. In addition to bene-fiting native plants by removing excess litter, fire exerts some control over non-native plants. Species that have invaded a prairie because of disturbance, such as overgrazing or burrowing by animals, are often highly susceptible to fire and will be reduced or eliminated by periodic burning. Quack grass and Kentucky bluegrass are good examples of such invaders. These grasses grow rapidly during the cool spring season. Because rapidly growing tissues are sensitive to heat, early spring fires are more destructive to these species than to prairie species that reach their maximum growth during summer (Glenn-Lewin et al. 1990). In addition, many prairie species will be stimulated by the burning and will increase in abundance more quickly than on areas that are not burned. My research showed that porcupine grass, switchgrass, azure aster, and rockrose increased markedly on frequently burned savannas in Anoka County.

## Successional communities

Once the prairie sod has been broken and the land has been used for agriculture for a period of years, succession leading to tallgrass prairie requires many years. Plants most likely to be pioneers on aban-doned agricultural land are those species that most consider to be weeds. They sprout readily, grow quickly, and soon cover the ground. Native plants, such as ragweed, and introduced plants, such as quack grass, foxtail, and goosefoot, are typical pio-neer species. They may persist for 20 years or more. Slowly, big bluestem, other prairie grasses, and forbs become established, provided that sources of seeds are available in nearby prairies. Perhaps after 100 years following the end of agriculture, the site may again resemble native prairie.

Intensive grazing of native prairie has a some-what similar effect in that succession is set back. The soil is not disturbed, as in the case of culti-vation, but is often highly compacted. Most of the typical prairie species will disappear or be greatly reduced in abundance. Increasers, as range managers refer to plants that increase under heavy grazing, include both native species, such as Canada goldenrod, and introduced species, such as Kentucky bluegrass, redtop, and Canada thistle. Heavily grazed pastures of Kentucky bluegrass or redtop with tall, conspicuous patches of thistle and goldenrod are common in parts of western Minnesota. If grazing is stopped, succession pro-ceeds very slowly because of the dense Kentucky bluegrass or redtop sod.

## Ecosystem function

Biomass and productivity values are much lower in prairies than in Minnesota forests (table 4.1). Measurements of biomass, however, commonly in-clude only the aboveground portions of the plants. In forests, tree trunks and branches constitute most of the biomass. In the prairie, most of the plant biomass is in the underground roots. Data from several investigations indicate that from 50% to 90% of the total plant biomass is underground

139

(Bray, Lawrence, and Pearson 1959; Ovington, Heitcamp, and Lawrence 1963).

Some of the roots die each year, and the entire root system is replaced by new growth about every four years (Dahlman and Kucera 1965). As a result, the prairie soil becomes richer and richer in organic matter from the decomposing root tissues. Approximately three to five years is required for decomposition of new litter and return of the nutrients to the soil. Thus, the litter will increase in depth for three to five years, and then the amount of new litter added each year will be approximately balanced by decomposition of the old litter. R. C. Dahlman and C. L. Kucera (1965) estimated that, under present rates of growth and decay, 110 years are required before the addition of organic matter is balanced by the leaching loss in the top 10 inches of the soil profile. In the zone from 10 to 18 inches, they estimated that 590 years are required to reach equilibrium. These estimates tell us that prairie vegetation must be present for many years before the organic matter in the soil reaches its maximum. These estimates also suggest that reestablishing a prairie and its characteristic soil after disturbance, such as plowing, may require 100 years or more.

As in other natural communities, most prairie vegetation is not eaten unless the site is a pasture. The vegetation becomes part of the detritus food web in which the plant tissues are broken down and nutrients are released. Nutrient cycling in prairie plants, and in many forest plants as well, actually begins before the annual foliage dies and becomes part of the litter. Leaching of nitrogen, phosphorus, and potassium from the green plants probably begins during early summer. By November in a bluestem prairie in Missouri, nitrogen in the standing stems and leaves had been reduced by about 75% (Kucera, Dahlman, and Koelling 1967). Some of the nitrogen was lost to the atmosphere, and some was carried into the soil.

Some prairie plants are especially well adapted to living in soil with low levels of nitrogen and actually appear to modify their environment. Big and little bluestem, for example, tie up nitrogen in

their slowly decomposing litter. They thus create a low-nitrogen environment in which they thrive and in which non-native grasses do poorly. This relationship between competition among plant species and nitrogen cycling, believed to be a critical process in succession, has been occurring for thousands of years. Now, however, stability of prairie remnants may be affected by high rates of deposition of nitrogen caused by air pollution (Wedin and Tilman 1992). About half of the nitrogen deposited comes from burning of fossil fuels and the other half from use of agricultural fertilizers.

The annual productivity of prairies, like crops and lawns, is strongly correlated with the availability of nutrients and with rainfall during the growing season. Burning also increases productivity, provided that adequate moisture is available, because some of the nutrients in the ash quickly leach into the soil and are then immediately available for uptake by plants (Kucera, Dahlman, and Koelling 1967). For example, net production on a mesic prairie in the Red River valley increased from about 3,000 pounds per acre in 1965 to 4,000 pounds per acre in 1966 following a prescribed burn in May 1966. Precipitation was similar during the two growing seasons (Hadley 1970). In Wisconsin, annual production was three times greater on burned than on unburned restored mesic prairie dominated by big bluestem (Peet, Anderson, and Adams 1975).

Estimates of the dry weight of aboveground biomass in Minnesota prairies (table 4.1) range from 400 pounds per acre in a dry sand prairie in Anoka County where nutrients are low (Ovington, Heitcamp, and Lawrence 1963) to 4,000 pounds per acre in a mesic prairie on rich soil in Polk County (Smeins and Olsen 1970). The most abundant species were the same in both prairies, big bluestem, porcupine grass, and Kentucky bluegrass.

Aboveground biomass does not provide an estimate of annual productivity, because it represents the weight of vegetation present at the instant in time when the sample is taken. Some species, such as pasque flowers, produce their growth early in the season and then wither, while others, such as big bluestem, reach their maximum growth in late summer. Therefore, estimates of annual net productivity must be based on repeated measures of standing biomass throughout the growing season. The annual productivity of the dry prairie in Anoka County was 830 pounds per acre, whereas the aboveground biomass at the time of sampling was only 410 pounds per acre (Ovington, Heitcamp, and Lawrence 1963).

The productivity of agricultural crops grown on Minnesota's rich prairie soils has been intensively studied by economists (Rock, Howse, and Wright 1993) and is also well known from experience to every farmer. Average productivity, measured as wet weight, of alfalfa hay (aboveground only) in 1992 was 6,100 pounds per acre (table 4.1). Alfalfa is cut three or four times each year, and these productivity values are the sum of the yields from each cutting. Yields of grain crops are commonly measured in bushels of grain per acre, which ignores the stems, leaves, and roots. Conversion of bushels to pounds, wet weight, provides a partial comparison to the productivity of prairie vegetation. In 1992, the average corn yield was 114 bushels or 6,384 pounds per acre. Soybeans yielded 32 bushels (1,920 pounds); wheat, 50 bushels (3,000 pounds); and oats, 70 bushels (2,240 pounds). Yields of root crops (underground only, wet weight) in 1992 were 24,800 pounds per acre of potatoes and 37,000 pounds per acre of sugar beets.

Grain and root crops are produced by annual plants, and their productivity is most comparable to pioneer prairie plants in a successional sequence. Pioneer species grow rapidly and are intolerant of stresses such as shade and competition for moisture (table 3.1). Rapid growth is beneficial for agricultural crops because it results in high yields; in a prairie, however, intolerance of stresses probably plays a role in the replacement of pioneer species by mid- and late-successional species. These plants have lower productivity but put more energy and nutrients into maintenance. Thus, the climax

prairie community is relatively stable and able to resist disturbances such as drought or disease. In contrast, many crops are sensitive to disturbances, and large quantities of chemicals may be needed to control insects and weeds.

## Animals and community interactions

### Mammals

Before European settlement, the plains buffalo (figure 6.9) was without a doubt the most important mammal on the prairie in terms of both numbers and impact. Records reveal that buffalo were common throughout much of western and southern Minnesota before the mid-1800s (Herrick 1892). Because Minnesota was on the eastern edge of the Great Plains, buffalo were not nearly as numerous as they were in the Dakotas, Nebraska, and Kansas. Nevertheless, early explorers reported seeing herds in Minnesota numbering in the thousands.

One need only imagine a herd of several thousand Hereford cattle roaming at will over present-day farmland to visualize the impact that a herd of several thousand buffalo might have on the prairie. The herd would consume nearly all of the vegetation in its path, would trample the remaining plants, litter, and soil, and would create muddy wallows out of many shallow wetlands. But, in contrast to domestic livestock, the buffalo herd might remain in one locality for a few days or weeks and then move on, not to return to the same spot for many years. Thus, the prairie had an opportunity to recover from the devastating impacts of the buffalo.

The last wild buffalo were observed in Minnesota in 1880 in Norman County (Surber and Roberts 1932). Although the buffalo are gone, evidence of their presence can still be seen in some of Minnesota's remaining prairies. Wallows still

141

**Figure 6.9** *(overleaf) The plains buffalo was the most important mammal on the prairie in terms of both numbers and impact. The last wild buffalo in Minnesota were observed in Norman County in 1880. (Photo © Jim Brandenburg.)*

exist in places that have escaped drainage and the breaking plow. Rubbing stones, polished smooth by the rubbing of itching hides and surrounded by trenchlike paths, can still be seen on some prairies.

Elk also were common on the prairies before settlement. Zebulon Pike, in his report of an expedition up the Mississippi River in 1805, mentions numerous observations of elk between Minneapolis and St. Cloud (Coues [1910] 1965). The town site of Elk River was very likely named for the presence of elk herds. Elk probably disappeared quickly as farms and ranches became established. Although the natural habitat of elk was grassland, they did not tolerate the disturbances caused by humans and quickly moved to mountainous areas as the prairies were settled. Wild elk were last observed in Minnesota in 1932 in the Northwest Angle near the Manitoba border (Swanson 1945).

In an effort to return elk to Minnesota, the DNR shipped 56 animals from Jackson Hole, Wyoming, to Itasca State Park in 1914 and 1915. Later 27 elk from the Itasca herd were released in Beltrami Country, and a few were released in the Superior National Forest (Fashingbauer 1965). Habitat in Beltrami Country is a mixture of open farmland, pasture, and willow and aspen brushland. This herd has persisted since the release, but numbers have probably never exceeded several hundred animals. Recently this herd has fluctuated from about 15 to 30 animals. These elk have become well adapted to the area and in winter frequently invade farmyards, where they feed with domestic livestock.

Today the burrowing mammals, such as gophers and badgers, have the greatest impact on the remaining tallgrass prairies in Minnesota. Pocket gophers feed almost exclusively on plant materials, taking some plants above ground and some from within their tunnel systems. Roots that protrude into the tunnels are often eaten directly, but sometimes pocket gophers grasp the roots and pull entire plants through the soil into the tunnel. Plant material that is not eaten immediately is stuffed into their large cheek pouches and carried to storage locations within the burrow system.

For some unknown reason, plants characteristic of early-successional stages, such as quack grass and goosefoot, are preferred over plants from later successional stages. High populations of pocket gophers appear, in effect, to "farm" their food resources. As gophers concentrate feeding in areas of dense vegetation, other areas accumulate plant biomass. A shift in feeding location to these formerly sparse patches then yields increased return from their foraging efforts (Behrend and Tester 1988; Zinnel 1992).

When gophers are excavating tunnels, they push soil onto the surface, forming the many small mounds that are commonly seen. Construction of the tunnel system and formation of the surface mounds influence soil structure and provide bare ground for pioneer plants to colonize. Digging loosens the soil, and the accumulation of feces, urine, and decomposing plant tissue in tunnels and dens increases nitrogen content (Zinnel and Tester 1992). Burrow systems, with their deep nest chambers, food cache sites, and numerous shallow feeding tunnels, require much energy to construct. As a result, they are more or less permanent features of the prairie and are often used year after year, but not necessarily by the same pocket gopher.

Several other species of gophers, or ground squirrels as they are more properly called, also live in the prairie. In Minnesota, the most common species is the thirteen-lined ground squirrel. Franklin's and Richardson's ground squirrels, which often live in the same areas, are larger and do not have stripes on the body. All three species, which are true hibernators, feed on plant and animal food taken above ground. Leaves, stems, fruits, and flowers make up a large proportion of their diet, but they are known to eat a variety of insects, eggs of ground-nesting birds, and even small frogs and toads. Burrow systems of these ground squirrels are not as elaborate as those of pocket gophers, because the ground squirrels obtain their food above ground.

Badgers, predators on gophers, ground squirrels, and other animals, make large holes and burrows.

Their strong front feet and long claws enable them to dig very rapidly, so rapidly in fact, that they can dig faster than the gophers and ground squirrels that are trying to escape. Gophers and ground squirrels have small home ranges, usually only a few tenths of an acre. Badgers, on the other hand, have very large home ranges, which may cover as many as 4 or 5 square miles.

During one spring and summer, a badger that was being radio-tracked at Cedar Creek Natural History Area moved about over an area approximately 2 miles wide and 2 miles long (Sargeant and Warner 1972). It foraged for food during the night and at dawn dug a den for itself wherever it happened to be. It slept in the den during the day and the next evening began moving about searching for food. At dawn it dug another den, and so the pattern continued with a new den each day for sleeping and a new area each night for foraging. In late August, the badger remained in a field of about 40 acres, feeding heavily on a large population of pocket gophers and using the same den each day for sleeping. During winter, badgers sleep much of the time, but they are not true hibernators; that is, their body temperature remains at its normal level during this period of semidormancy.

The major impact of badgers on the prairie is their extensive digging. Large holes often remain where a badger has tried to capture a gopher or other prey. Holes may be 8 to 12 inches in diameter and several feet deep, deep enough to cause a horse, cow, or buffalo to break a leg by stepping into it when running.

One of the most conspicuous animals on prairie and farmland today is the white-tailed jackrabbit. Populations of jackrabbits are usually low, only a few per square mile, but they can often be seen bounding across open fields and pastures in fall, winter, and spring, when they are not hidden by vegetation. Jackrabbits feed on a wide variety of plants, occupy a wide range of prairie habitats, and are present throughout southern and western Minnesota.

If we could observe the prairie at night, we would very likely see red fox, coyotes, striped skunks, and perhaps raccoon and weasel searching for food or patrolling the boundaries of their territories. During most of the daylight hours, these nocturnal animals are sleeping or resting, but their senses are so sharp that they can detect danger in the form of humans or other predators even while they are resting. It is extremely difficult, therefore, to find or observe them.

Fourteen species of mice, voles, and shrews, which ecologists collectively refer to as small mammals, occupied prairies before settlement. Today these species can often be found in roadside ditches, along the edges of wetlands, in areas where agricultural practices result in small patches of grassy habitat, and in the remaining tracts of prairie (Kalin 1976). The most common species are the meadow vole, deer mouse, short-tailed shrew, and the meadow jumping mouse.

Numbers of these species of small mammals fluctuate dramatically over the years, and much research has been devoted to analyzing the patterns of and reasons for these fluctuations or cycles. High populations of meadow voles seem to occur about every three to five years. Estimates of abundance range from a few per acre to literally thousands per acre depending upon the stage of the cycle. Explanations for these cycles are conflicting. Some researchers believe that they are related to changes in the quantity and quality of food plants caused by climate changes resulting from sunspot activity. Other researchers believe that predation is at least partially responsible for cycles in numbers, and still others think that the cycles are caused by changes in the genetic makeup of populations.

Meadow voles, which eat leaves and stems of grasses and forbs, are most abundant in thick vegetation with a well-defined layer of litter. Here they make their tunnel networks that serve as pathways to clumps of grass used for food. The nests and pathways are also used in winter, when they are insulated by the snow. The voles' network works well for them until a tunnel is found by an exploring short-tailed weasel, whose body diameter is about the same size as a vole's. The weasel searches

145

the tunnels, eats the voles, and occupies the nest as its own.

Deer mice, which feed primarily on seeds and fruits, prefer a more open type of vegetation and are likely to be most abundant in grazed areas or in places where fire has eliminated the litter layer (Tester and Marshall 1961). Their numbers fluctuate widely: high populations occur immediately following a fire, and very low populations occur in prairies protected from fires for four or five years.

Similarly, numbers of shrews are usually low; however, they tolerate many kinds of habitat, ranging from dense to sparse in terms of cover and wet to dry in terms of moisture. Shrews have a very high metabolic rate, which requires that they feed every few hours throughout the day and night. Poison glands in their mouths enable them to partially paralyze their victims. Thus, they often capture prey that are larger than they are.

Because fires were common on prairies, their impact on wildlife is of interest. The immediate effects of fire on animals are varied and depend on the duration and intensity of the fire and the type of shelter available to the animals during the fire. In a study of the effects of a prescribed burn on an oak savanna in Anoka County, I found that temperatures below the soil surface during the burn did not exceed 138°F and in most places were below 125°F (Tester 1965). Such temperatures for a few moments may not be lethal, but only 2 of 16 small mammals captured and marked before the burn were retrapped after the fire. This low return indicates low survival during burning, high predation following the fire, movement out of the burned area, or a combination of these results. When live traps were placed in the same locations 35 days after the fire, however, 62 small mammals were captured. This finding shows that the population increased rapidly, and suggests that burning a prairie probably has only a short-term effect on mammal populations.

It is not possible to discuss each of the species of mammals that occur on the prairie, or even to mention them all by name. I have discussed those that are important to the prairie ecosystem and those that are conspicuous or are most likely to be observed. White-tailed deer (see chapter 4) are also often abundant on prairie and agricultural land today. Their presence and abundance, however, are likely related in part to the availability of wooded habitats such as farm groves, shelterbelts, and forests growing along streams and rivers. Deer thrive in Minnesota's farmland, taking special advantage of the high nutrient content and ready availability of crops such as corn and apples.

Much of Minnesota's native prairie is now farmed. Except for buffalo and elk, all of the species of mammals discussed can be found on farmlands in Minnesota. Groves around farm buildings, shelterbelts, and forests along streams and rivers provide further diversity of habitat for mammals amidst row crops and pasture. Cottontails, which are not common on tallgrass prairie, are often abundant in these wooded habitats. They feed primarily on bark and twigs and sometimes cause severe damage to newly planted trees and shrubs. Cottontail numbers fluctuate from year to year, but the rabbits are likely to be most numerous in areas with dense, shrubby vegetation (Swihart and Yahner 1982).

These wooded sites also provide suitable habitat for small mammals not commonly associated with prairie. R. H. Yahner (1982) reports capturing red-backed voles, white-footed mice, red squirrels, and eastern chipmunks, all typical forest species, in shelterbelts. It appears that small mammals normally associated with forests are sufficiently mobile to occupy isolated wooded habitat, just as prairie species occupy grassland habitat in forests.

### Birds

Birds are likely to be much more conspicuous than mammals to an observer looking at prairie or farmland. Their significance in terms of the structure and function of the prairie and farmland ecosystem, however, is probably much less than that of mammals. Birds seem to be somewhat more selective than mammals with regard to choice of habitat. As a result, some species are found almost exclusively in the few remaining areas of native prairie, whereas other species have adapted well to living on agricultural land.

Although not the most abundant species, marbled godwits and upland sandpipers are probably the most conspicuous birds on native prairie. Marbled godwits, with their long legs and gently upcurved bill, are frequently seen walking and feeding on prairies in early spring before the vegetation has grown more than a few inches. Their loud, whistling calls and courtship flights provide dramatic action during the spring breeding season.

Upland sandpipers are often observed standing on a fence post or telephone pole. They are about a third the size of marbled godwits but are conspicuous because of their perching habit and their aerial courtship flights and calls. The clear, whistling call given in flight carries a long way over the open prairie. Both godwits and upland sandpipers nest on the ground and seem to prefer areas with only a small accumulation of litter. Fires and limited grazing create ideal habitat for these birds and probably played a key role in determining their abundance in the past.

Prairie-chickens, although not commonly observed, are certainly among the most spectacular birds of the prairie. In the early days of European settlement, they provided food for the pioneers. More recently they were highly sought after by upland bird hunters. The resonating calls of the cocks on their booming ground can be heard for over a mile in early spring, and their foot-stamping and flutter-jumping courtship displays have awed many an avid bird watcher (figure 6.10). Groups of about 6 to 10, but sometimes as many as 40, males gather in March, while snow still covers the prairie, and compete through displays and fights for the choice positions near the center of the booming ground. The grounds are commonly in pastures or

**Figure 6.10** *Prairie-chickens are among the most spectacular birds of the prairie. During spring courtship, males inflate their orange neck sacs to produce resonating calls that can be heard over a mile away. (Photo by Richard Hamilton Smith.)*

147

hay fields, because the birds prefer short grass or stubble for their performance. Courtship displays, which attract hens to the booming ground, reach their peak in April and May. Cocks erect their pinnae, stiff tufts of feathers on the neck, spread their wings, fan their tails, and inflate the orange skin sacs near the base of their necks. These sacs serve as resonating chambers, producing the booming calls, which sound as though they were coming from a kettledrum. Using both visual and audible displays, the cocks perform for several hours in the morning, beginning before sunrise, and again in the evening at dusk. Hens come to the booming ground for only a day or two, and mating is usually done by the cocks occupying the center territories. Nests are hidden in the surrounding prairie, which provides food for adults and young, cover for brood rearing, resting, and escape from predators, and protection from winter storms. Although large expanses of grasslands are essential, prairie-chickens feed extensively in farm fields.

The spread of prairie-chickens into southern Minnesota from Iowa and Wisconsin probably occurred in the early 1800s, following the beginnings of agriculture (Roberts 1932). Before this time, sharp-tailed grouse occurred throughout the prairie region of Minnesota. As the pioneers plowed the prairies for crops, the sharp-tails decreased because of the disturbance of their habitat, and the prairie-chickens expanded their range and increased in numbers. An equal mixture of grassland and cropland seemed to be ideal for prairie-chickens. As more and more of the prairie was broken, numbers of prairie-chickens declined. Presently a population of about 3,000 birds is concentrated along the eastern edge of the Red River valley in the northwest (Dexter 1993). A smaller population is present in Wadena and Hubbard Counties, and a successful introduction has recently established prairie-chickens in Lac qui Parle County. Prairie-chickens have been protected in Minnesota since 1935, and it does not seem likely that the population will increase sufficiently to permit hunting in the near future.

Marbled godwits, upland sandpipers, and prairie-chickens are dependent on native prairie for their existence. As I have already discussed, agricultural practices have eliminated most of the native prairie in Minnesota, and, as a result, numbers of these birds are low compared to presettlement times.

Sharp-tailed grouse, a close relative of prairie-chickens, are common in the brushy habitat characteristic of the prairie-forest border in northwestern Minnesota. Their main requirement is an interspersion of open areas such as cropland, pasture, sedge wetlands, or prairie with stands of willow, aspen, birch, or other shrubs. Their dancing ground, or courtship display area, is usually on a slightly raised meadow, pasture, or field several hundred feet away from shrubs or trees.

Since the 1940s, sharp-tail habitat has deteriorated as brushland has been cleared, natural succession has filled in once open areas, and wildfires have been suppressed. Management of land for sharp-tailed grouse involves setting back the process of succession to encourage the growth of brush. Burning is the natural method for perpetuating the brushland ecosystem (Ewing 1924), and both wild and prescribed fires are effective in creating and maintaining stands of brush. Prescribed burning is commonly used to improve habitat for sharp-tails.

Meadowlarks, bobolinks, and savannah, grasshopper, and clay-colored sparrows are common but less conspicuous on native prairies (Johnson and Temple 1986). In a few locations in northwestern Minnesota, where the vegetation is sparse because of lack of moisture, chestnut-collared longspurs can sometimes be found. Their distinctive fluttering courtship flight makes them conspicuous in the air, but they are extremely difficult to observe while on the ground. An especially good location to find chestnut-collared longspurs is in the grasslands located along the gravel ridges that parallel the Red River valley in Clay and Norman Counties.

Blackbirds of several kinds, red-winged, yellow-headed, brewer's, and rusty, are the most common birds in agricultural areas. In fall, waves of thousands of blackbirds can be seen rising and falling from corn and sunflower fields as they feed in preparation for migration. Red-wings, probably the most abundant birds in Minnesota (Henderson 1981a), and yellow-heads breed mainly near wetlands and other bodies of water but move to corn and other grains in late summer. Red-wings were found to be the most common nesting birds in fields removed from cultivation under the federal Conservation Reserve Program, followed by sedge wrens, bobolinks, common yellowthroats, and savannah sparrows (Johnson and Schwartz 1993). Brewer's and rusty blackbirds often nest in farm groves and windbreaks.

Blackbirds cause considerable damage to agricultural crops in the fall. Studies in the Dakotas and Minnesota revealed that blackbirds damaged about 1% to 2% of sunflower crops but that losses were two to four times higher in fields near wetlands (Hothem, DeHaven, and Fairaizl 1988; Otis and Kilburn 1988). Blackbirds sometimes congregate in farm feedlots in such numbers that their droppings in feed bunks make grain and silage unfit for cattle. The use of loud noises, distress calls, and other methods to frighten blackbirds away from crops and feedlots have generally not been very successful. Efforts to reduce the numbers of blackbirds through the use of poisons or chemicals causing lowered reproduction have also not been successful.

Two introduced game birds, the ring-necked pheasant, originally from China, and the gray partridge, originally from Europe, became well established on farmland in southern and western Minnesota in the 1930s and 1940s. Their numbers have fluctuated widely in response to such factors as changes in land use, drought, winter storms, temperature extremes, and predation. During the past two decades, pheasant populations have been low compared to the 1940s and 1950s. Wildlife biologists believe that the changes in pheasant numbers reflect in large part the more intensive use of agricultural land for row crops such as corn and soybeans. This type of farming does not provide

sufficient cover for nesting and escape from predators and severe weather. Gray partridge populations, on the other hand, appear to be increasing in many areas (figure 3.14).

In farmyards, house sparrows and rock doves, or pigeons as they are more commonly called, are likely to be found. They often nest in or on buildings. Mourning doves, common grackles, robins, gray catbirds, brown thrashers, and several kinds of sparrows nest in farm groves and shelterbelts (Yahner 1983). Meadowlarks and horned larks can often be seen along roadsides, in the short grass of pastures, and on plowed fields. Red-winged blackbirds, western meadowlarks, yellowthroats, and dickcissels often nest in alfalfa and grain fields.

Hawks and owls are important predators in the prairie food web. Red-tailed hawks and great horned owls nest in cottonwood and other trees along streams and in woodlots but capture ground squirrels and jackrabbits in open areas. Short-eared owls and northern harriers nest and hunt in the prairie, while burrowing owls actually nest in holes in the ground. They are able to dig but usually use abandoned skunk, fox, or ground squirrel burrows. Only a few pairs still nest in Minnesota, and the burrowing owl is listed as an endangered species (Coffin and Pfannmuller 1988).

## Amphibians and reptiles

Early in spring a loud, persistent peeping can be heard coming from prairie ponds and roadside ditches even before all of the ice has disappeared. These sounds are made by a tiny, inch-long amphibian called the western chorus frog. Chorus frogs probably spend the winter lying on or near the surface of the mud on the bottom of ponds and ditches, but some may spend the winter on the upland in a partially frozen condition. As soon as a bit of open water appears in the spring, they move to this spot and begin calling, which is part of their courtship display. Because they are tiny and very secretive, they are difficult to see unless one can patiently stand near the water and remain motionless for many minutes.

149

Leopard frogs, conspicuous with their dark spots against a greenish background, are easy to observe on the edges of most prairie wetlands and are often found in the uplands. They winter in the mud at the bottom of ponds, wetlands, lakes, and streams. This frog is the animal that most high school biology students become familiar with through dissection. It is also used extensively for medical research and at one time was used in pregnancy tests. Bass fisherman prize the frogs as bait.

One striking aspect of the behavior of leopard frogs makes them very conspicuous at certain times of the year. In summer when tadpoles have completed metamorphosis, as indicated by resorption of the tail, a major migration occurs as hundreds and sometimes thousands of young frogs move away from the breeding pond. Apparently they are seeking the shores of larger lakes or streams, where they spend the summer (Merrell 1977). Another migration occurs in September and October, when frogs move to their overwintering sites in lakes and streams. During these migrations, frogs may move several miles, often crossing heavily traveled roads. At these times thousands of frogs may be observed, and many are killed by traffic on the roads. In fact, highways may even become slippery from the frog carcasses.

Large commercial harvests of leopard frogs were common in Minnesota for many years. DNR records (Roy Johannes, Section of Fisheries, personal communication) show that in some years professional frog pickers harvested over 100,000 pounds of leopard frogs (figure 6.11). In the early 1970s, the adult population of leopard frogs in Minnesota began to decline, and in 1974 the DNR stopped the commercial harvest and marketing of leopard frogs. The season was reopened in 1987, but the number of leopard frogs harvested in recent years has been very low.

No explanation has been discovered to account for such a striking decline in numbers. A variety of causes for the decline have been offered, including overharvest, a disease known as redleg, caused by the bacterium *Aeromonas hydrophila*, increased use of herbicides and insecticides by farmers, and drainage of breeding ponds. Intensive study has failed to confirm the cause for the population decline.

My description of prairies in Minnesota would not be complete without a discussion of the Manitoba, or Canadian, toad, which can be extremely abundant on prairies in northwestern counties. Studies I conducted on a wet prairie in Mahnomen County (Tester and Breckenridge 1964a) suggest that the number of toads can be in excess of 64,000 per square mile in some years. In early spring, males gather to form breeding choruses in ponds and wetlands and even in roadside ditches (figure 6.12). During warm evenings their loud trills can be heard for long distances. These trills attract the females to the chorus, where eggs are laid and fertilized. The eggs develop into tadpoles, which metamorphose into small toads after several weeks. Small toads and the adults from the breeding chorus spend the remainder of the spring and summer along the shores of the wetlands and on the adjacent upland, where they feed on small moving organisms, mainly insects, worms, and other invertebrates.

In late August and early September the toads move away from the wetlands to a spot where they burrow and spend the winter. Most adult toads return to the same overwintering site every year. Young toads will return to the place that they selected to spend their first winter (Kelleher and Tester 1969). Many go to Mima mounds, a strange feature of the prairie landscape. These mounds, which number about one per acre on wet prairies, range from 30 to 100 feet in diameter and may be 2 to 4 feet above the level of the surrounding prairie (figure 6.13). Similarly shaped mounds in Washington were believed to be burial sites of the Mima Indians. The name Mima is still used to identify such mounds, although Indians had nothing to do with their origin.

Having observed that Manitoba toads were abundant on some prairie areas at certain times but difficult to locate at other times, W. J. Breckenridge and I decided to investigate their behavior and

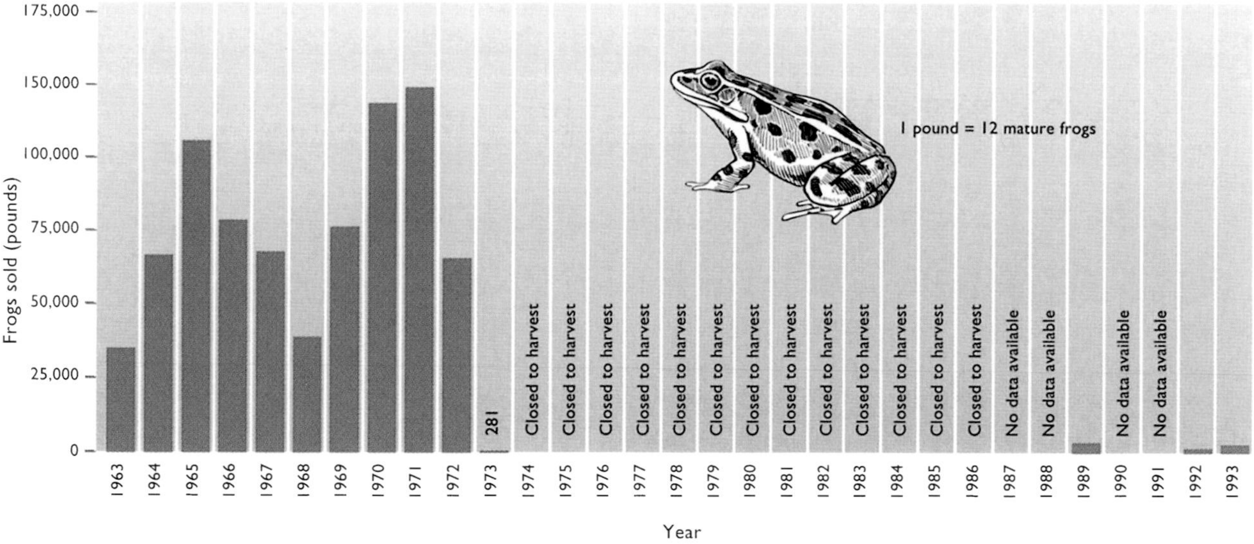

**Figure 6.11** *Commercial harvest of leopard frogs in Minnesota from 1963 to 1993. The large harvests in 1970 and 1971 reflect high market prices, not an abundance of frogs. Despite continuing high prices, only a few frogs were captured in 1973, because of a dramatic decline in the population. (Roy Johannes, Minnesota Department of Natural Resources, personal communication.)*

ecology and to try to determine where and how they survived the winter. We marked 20 adults with tags of radioactive tantalum by placing pieces of radioactive wire about ¼ inch long and ½₂ inch in diameter under the skin on the back of each toad (Tester 1963). We then used a scintillation counter, similar to a Geiger counter, to locate the toads throughout the summer and early fall. Every few days we would search the study area and determine the location of each marked toad. Overwintering sites were located when the toads disappeared into the soil on the mounds or on the surrounding prairie (Breckenridge and Tester 1961). Because most of our marked toads moved to the Mima mounds to overwinter, we felt that mounds must be playing an important part in the ecology of the toads.

The following spring before the frost was completely out of the soil, we placed a sheet-metal fence around one mound used by overwintering marked toads. To obtain an estimate of how many toads choose overwintering sites away from the Mima mounds, we placed a similar-sized fence around a patch of prairie next to the Mima mound. A trap was placed on the side of the fence nearest

the pond where the breeding chorus normally occurred. Our idea was that the toads would emerge from the ground in the spring and move toward the wetland, thereby ending up in our traps. This method proved to be very successful, and we caught 225 toads emerging from the mound but none from the fenced prairie. The following spring we repeated the same process, but that time we enclosed two mounds and one area of prairie. That year we captured 21 toads from the prairie enclosure, 1,594 from one mound, and 3,276 from the other (Tester and Breckenridge 1964a). We believe that the toads selected the mounds as wintering sites because the soil in the mounds was not compacted and provided a site where the toads could easily dig.

But how do these mounds form? Further study of 200 mounds on the prairie showed that their most striking feature was the disturbance of the soil by pocket gophers, Franklin's and thirteen-lined ground squirrels, badgers, toads, and other animals (Ross, Tester, and Breckenridge 1968). The small mounds of soil pushed onto the surface of the prairie by pocket gophers may have been the starting point for the development of the Mima

mounds. Continued disturbance by animals, including toads, probably caused the change in vegetation from grasses and forbs to shrubs, the lack of a soil profile, the looseness of the soil, and increased water permeability. Dr. Breckenridge and I calculated that the 3,276 toads that wintered in a single Mima mound moved 85 cubic feet of soil in a single year when they burrowed into and out of the mound. We believe that this volume of soil

**Figure 6.12** *(left) A male Manitoba toad in a breeding pond in Mahnomen County. The toad's inflated air sac is used to produce a loud, trilling call, which attracts female toads. (Photo courtesy of Bill Schmid.)*

**Figure 6.13** *(below) Mounds of soil, from about 30 to 100 feet in diameter and 2 to 4 feet high, are common on some prairies in northwestern Minnesota. These Mima mounds, as they are called, are formed by animal digging, wind, and frost action. They are used as overwintering sites by thousands of toads. The sheet-metal fences surrounding this mound were installed to enable us to capture and mark the toads as they emerged in spring. (Photo by author.)*

displacement for hundreds or even thousands of years is a significant factor in the formation of the mounds. Animal disturbance of the soil may also account for the lack of sod, the high organic content of the top soil horizon, and the presence of anomalies such as pockets of pebbles. Comparison of mounds and adjacent prairie soils suggests that the mounds are, in effect, "puffs" of soil formed primarily by animals disturbing and mixing the soil in a particular spot. Thus, our study of the ecology of the Manitoba toad led to the discovery of interrelationships between Mima mounds, toads, and other prairie animals.

We also studied the behavior of the toads during their winter stay in the mounds (Tester and Breckenridge 1964b). We followed the movements of marked individuals as they moved deeper in the soil throughout the winter to escape the frost line, which was also moving deeper (figure 6.14). Toads are not able to survive under water, however, and therefore cannot move deeper than the water table. In winters when the frost penetrated deeply, toads were trapped in the zone between the frozen soil above and the water table below. If these zones met, many toads died. We observed that the zones met when winter temperatures were extremely low and snow cover was light, resulting in deep penetration of the frost line.

This cause of mortality may be a significant control on populations of toads. Few predators eat them, perhaps because of vile-smelling secretions produced by the skin. These secretions are sometimes seen near the parotoid glands on the toad's back when the animal has been injured or frightened. Toads are eaten by birds such as herons and hawks, and badgers in captivity have been observed to eat toads after rubbing or rolling them in the soil. Presumably this removes the irritating secretions or masks the taste.

Of all the prairie amphibians, salamanders are probably the most secretive and least known. Adults spend the days hidden in litter or in burrows, and they do not call like toads and frogs during the breeding season. The tiger salamander may be

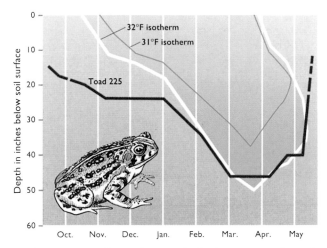

**Figure 6.14** *Relations between the depth of a Manitoba toad in winter and the position of the frost line in the soil. The frost line was between the 31°F and 32°F isotherms because of lowering of the freezing temperature by salts in the soil moisture. (Tester and Breckenridge 1964b, by permission.)*

familiar, however, because of its tendency to move close to the foundations of barns and houses in the fall. Perhaps the tiger salamander is attracted to the warmth of these places, or perhaps they simply encounter them in their movement to overwintering sites. In any case, they are frequently found in window wells and beside buildings, sometimes in large numbers.

This salamander reaches a length of about 8 inches, and its color is typically dark brown or black with many yellow or orange spots or blotches. Eggs are laid in ponds or wetlands in early spring, and the larvae develop into voracious predators that have an important role in the pond food web.

Adult salamanders live on the upland and have been found in many hiding places, including gopher burrows, rock ledges, basements, wells, and deserted foundations—anywhere they can find moist, undisturbed conditions (Breckenridge 1944). Winters are spent in similar locations protected from cold temperatures. The adults feed only on moving prey, such as insects and worms. In turn, they are eaten by snakes, hawks, owls, raccoons, skunks, and other predators.

A reptile shaped somewhat like a salamander, the prairie skink, can often be observed in dry

153

**Figure 6.15** *Several male plains garter snakes are courting a female by touching her body. The male that is successful in mating leaves a plug of semen in the female's cloaca to inhibit copulation by other males (Ross and Crews 1977). (Photo by Barney Oldfield.)*

prairies and savannas. Skinks are especially common on the gravel beach lines of old Glacial Lake Agassiz. The black stripes on the bluish gray body are conspicuous identification marks. Attempts to capture this skink often leave the predator with a squirming tail that has detached from the skink's body. A new tail will grow in a few weeks. In the meantime, the predator is left with a tiny bit of food that has served its purpose well by attracting the predator's attention while the skink escaped. W. J. Breckenridge (1944) reported that skinks eat a wide variety of insects, having a preference for grasshoppers, crickets, and spiders. Skinks, in turn, are preyed upon by a wide variety of animals including snakes, northern harriers, kestrels, shrikes, gophers, and raccoons. Young skinks are sometimes eaten by adults.

The first warm days of spring on the prairie signal the end of the winter season for the cold-blooded amphibians and reptiles who survived the rigors of winter. At this time the garter snakes and gopher snakes crawl out of the burrows where they spent the winter to find a warm spot in the sun.

The serious student of reptiles may find as many as six species of snakes on a Minnesota prairie: plains garter, western hognose, redbelly, smooth green, fox, and gopher. The most common is the plains garter snake. Garter snakes frequent the margins of ponds and wetlands, where they feed on toads, frogs, salamanders, and insects. Winters are spent underground, frequently in an abandoned gopher burrow. On rare occasions in early spring I have encountered groups of five or six garter snakes that appeared to be entangled or twisted together as they writhed along the ground (figure 6.15). This behavior is part of their courtship and breeding activity. About 20 to 60 live young are born to the females in July or August.

One of the most interesting snakes in Minnesota prairies is the western hognose snake, which is also called the puffing adder, blowsnake, or spread-head. These names all refer to the striking display that the snake makes when threatened by a predator. It draws its body tightly together in coils and inflates and spreads its head and neck and the forepart of its body. The head is raised like a cobra, and a loud hiss is given. As the enemy approaches, the hognose strikes repeatedly and vigorously. The snake is totally harmless, however, and the display is used only to frighten away potential predators. Hognose snakes will also play dead to escape a predator by turning belly-up. They give themselves away, however, by turning belly-up over and over if turned right-side up.

The gopher snake is the largest Minnesota snake, sometimes growing longer than 6 feet. It is totally harmless but because of its large size, frequently terrifies those who encounter it. It feeds on frogs, mice, gophers, and ground squirrels and frequently takes eggs of ducks and other ground-nesting birds. Gopher snakes have been observed crawling into gopher and ground squirrel burrows to capture adults as well as young.

### Insects

Thousands of toads and tens of thousands of buffalo sound like large numbers of animals. These numbers are minute, however, compared to the number of insects on a typical Minnesota prairie. Several thousand insect species may occur in a prairie community. I will describe a few examples of these species to illustrate the important role insects play in the ecology of the prairie.

In every prairie insects are important pollinators of plants. Seed production in many plants is totally dependent on transfer of pollen by insects. During feeding on pollen or nectar or hunting other insects on the grassland flowers, insects carry pollen from male flowers to female flowers, thus enabling plants to reproduce. This interrelationship, from which both the insects and the plants benefit, is a classic example of the process of coevolution. It is generally believed that in many cases a particular insect has evolved with a particular group of plant species so that both are dependent on one another and both benefit.

Moths and butterflies are perhaps the most conspicuous of the pollinating insects on the prairie. The larvae of some moths and butterflies are voracious feeders on the leaves and stems of the same plants that the adults cross-pollinate. Regal fritillaries, perhaps the most characteristic butterfly of the tallgrass prairie, satyrids, skippers, and monarchs (figure 6.16), which migrate southward to Mexico in late summer and fall, are the typical butterflies of the prairie.

Monarchs are sometimes called milkweed butterflies because they are so closely associated with this plant. Young monarchs that hatch early in spring in southern states migrate to Minnesota in May and June. The female monarch searches for a young, soft milkweed leaf and lays one egg, usually on the underside. The sticky egg adheres to the leaf, and the female flies off to a different leaf or to another milkweed plant to lay the next egg. She never lays more than one egg on a leaf but may lay as many as 400 eggs before she is finished (Graham and Graham 1976). In a few days a caterpillar hatches and immediately begins feeding on the tender leaf. During the next two weeks it molts its skin four times as it grows. Once the caterpillar has reached full size, it forms a chrysalis or cocoonlike structure around its body and hangs from the milkweed leaf until it develops into an adult butterfly and emerges.

During summer, monarchs may feed on nectar from a variety of flowers, but in August they seem to prefer blazing stars. After a brief period of mating and egg-laying, monarchs gather in preparation for migration, for they are one of the few insects that survive winter by flying south. Some migrate nearly 2,000 miles to central Mexico (Urquhart and Urquhart 1976).

Although not as conspicuous as the moths and butterflies, the hundreds of species of bees on the prairie are probably the most important pollinators.

*155*

Most of these bees do not store honey but rather live solitary lives in underground burrows. Unlike honeybees and bumblebees, which are social insects that live communally in nests with a queen, most of the prairie bees have individual nests, where eggs are laid on small pellets of pollen.

As I mentioned earlier in this chapter, different prairie plants flower at different times during the growing season. The species of insects that pollinate these plants while sucking nectar, seeking pollen, or hunting prey are active at the time the plants produce flowers. For example, in spring one group of bees visits flowers of wild plum, wolfberry, and meadow parsnip. In early summer, a different group of bees feeds from and pollinates spirea, violets, and clover. The bee species active in late summer visit a wide variety of goldenrods and asters (Costello 1969).

In some prairies, insects such as grasshoppers and crickets may be the most important grazers on the grasses and forbs. Actual consumption of vegetation on the prairie by insects, however, is relatively inconspicuous except when grasshoppers or crickets become so abundant that they literally eat every green stem. Historical records indicate that grasshopper plagues struck Minnesota numerous times in the 1800s and early 1900s. During these plagues, grasshoppers destroyed farmers' crops as well as all other vegetation in their path. So serious were some of these grasshopper outbreaks that a book bears the title *Harvest of Grief: Grasshopper Plagues and Public Assistance in Minnesota, 1873–78* (Atkins 1984).

An invasion of grasshoppers carried with it no less drama than a prairie fire threatening the fields of grain. As one pioneer reported in the early 1870s (Atkins 1984, 17):

> a large black cloud suddenly appeared high in the west from which came an ominous sound. The apparition moved directly toward us, its dark appearance became more and more terrifying and the sound changed to a deep hum. . . . We heard the buzzing; we saw the shining wings,

the long bodies, the legs. The grasshoppers—the scourge of the prairie—were upon us.

Hundreds of species of grasshoppers are known to exist in the North American prairies. About 135 species in the insect order Orthoptera, which includes grasshoppers, occur in Minnesota (Haarstad 1990). Any given prairie in Minnesota may have 50 or more species. Grasshoppers lay their eggs in small clusters 1 or 2 inches deep in the soil in late summer and fall. Eggs are able to survive the rigors of the Minnesota winter and hatch the following spring. The young nymphs go through five instars or molts before becoming adults in late summer. Each female will lay hundreds of eggs, and if a series of favorable hatching years occurs, grasshopper numbers approach plague conditions. In recent years, chemical insecticides have been used to control such outbreaks, with variable success.

Because of their small size and inconspicuous habits, one would not expect that insects play a major role in modification of the prairie soil, but in fact they do. Burrowing and digging by both larvae and adults loosens and stirs up the soil. Decomposing insect bodies and carcasses of dead birds and mammals hauled underground by burying beetles contribute to the soil nutrients and organic matter.

Among the more interesting and useful species on the prairie are the dung beetles. They shape balls of dung by tearing off pieces of manure and rolling them on the ground. When the ball of dung has been rolled to an appropriate spot, the beetle burrows around and under the ball until it disappears into the soil. Some beetles use the buried balls of manure as food. In one day a beetle may ingest manure juices equal to its own body weight. The beetle stays underground until the dung ball has been consumed and then returns to the surface to find more dung.

Some dung beetles use dung balls as a place to lay their eggs. When the larvae hatch, they feed on

**Figure 6.16** *(right) Monarch butterflies frequently gather on prairie plants before migration. (Photo © Jim Brandenburg.)*

the dung until they complete development and are ready to emerge as adult beetles.

Another important function of prairie insects is disposal of the dead. When an animal dies, its carcass is soon visited by blowflies, ants, carrion beetles, scarab beetles, yellow jackets, and many other species. This process of scavenging by the insects, as well as by snails, millipedes, and other invertebrates, greatly speeds the decomposition and disintegration of the carcass and the return of its nutrients to the ecosystem.

Lastly, insects play an important role in the food web of the prairie. Grazers, such as grasshoppers, crickets, and the larvae of moths and butterflies, convert plant material into animal tissue, which is then available to the third trophic level, the many species of birds, mammals, amphibians, and reptiles that feed, at least in part, on insects. In addition, dragonflies, walking sticks, and other predatory insects prey upon the grazing insects from the second trophic level. Thus, insects are extremely important components of the prairie ecosystem, contributing to the functions of energy flow, nutrient cycling, pollination, soil development, and decomposition.

When we consider the prairie species as a group, we see that the prairie, like all ecosystems, has many interconnections and pathways through which energy flows and nutrients cycle. Plants are being grazed, and animals are eating or being eaten, are competing for food and shelter, are searching for and interacting with mates, are producing wastes, are scavenging on dead plants, and are carrying out other apects of their daily lives. These interactions and adjustments have developed over a long period of time through the evolutionary process, resulting in the beauty and functionality of the prairie.

## Present status of the tallgrass prairie

Ole Rolvaag (1927, 110) foretold the fate of most of the prairie through the words of the Scandinavian pioneer in *Giants in the Earth*, who exclaimed: "Such soil. Only to sink the plow into it, to turn

over the sod—and there was a field ready for seeding."

Nearly all of Minnesota's 20 million acres of prairie were plowed under. Of the 150,000 acres of native prairie remaining today, about 48,000 acres (figure 6.17) are protected by the DNR, The Nature Conservancy, and other agencies in Minnesota (Natural Heritage Program, Minnesota Department of Natural Resources, unpublished data).

Many of the prairies acquired by the DNR were purchased as part of the Save Minnesota's Wetlands program, which began in the 1950s. Prairie wetlands are valuable waterfowl production areas, and the surrounding upland prairie is important nesting cover for ducks and other wildlife. Although the wetlands were the primary motivation for acquisition, extensive tracts of prairie were also preserved. Most of these prairies are open to the public and are readily accessible.

The Nature Conservancy, a private, nonprofit organization, has also played an important role in preserving Minnesota's prairie heritage. With the preservation of natural areas as its goal, The Nature Conservancy has used private funds to acquire

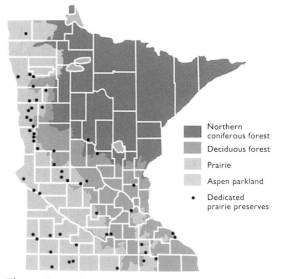

**Figure 6.17** *Some remnants of native prairie in Minnesota are protected as dedicated prairie preserves by government agencies and The Nature Conservancy. Most are less than 100 acres in size. (Adapted from an unpublished map, Minnesota Department of Natural Resources, Natural Heritage Program, by permission.)*

many tracts of native prairie throughout Minnesota. These prairies range in size from a few acres to nearly 2,000 acres and are distributed over most of southern and western Minnesota (Searle and Heitlinger 1980).

The size and location of prairie remnants are important to both animals and plants. Some animals show higher production of young on larger tracts and on those located farthest from forest edges (Johnson and Temple 1986). Isolated colonies of plants on small tracts of prairie may suffer from lack of genetic diversity and may eventually disappear. Current efforts by conservation organizations, particularly the DNR and The Nature Conservancy, are under way to establish large prairie preserves, greater than 2,000 acres, in the hope that such prairies will be adequate to maintain high species diversity and genetic diversity.

The effects of the loss and fragmentation of prairie due to agriculture and industrial development are evident in the status of some species of prairie plants. The prairie white fringed orchid, for example, was once common in southern Minnesota and in prairies in the Red River valley on the ancient beach ridges formed by Glacial Lake Agassiz. Since the 1970s, most of these prairies have been converted to farmland or to gravel mining, and the orchids are now restricted to one colony in Pipestone County and to several widely separated colonies in the Red River valley (Coffin and Pfannmuller 1988). The prairie white fringed orchid is now listed as state endangered and federally threatened.

Destruction and degradation of the natural environment is undoubtedly the most important factor influencing the decline and loss of native species. In view of the massive conversion of prairie to agriculture, it is not surprising that the number of animals and plants listed as endangered, threatened, or of special concern is higher for the prairie than for the forest biomes. Of the 287 species listed, 102 occur in Minnesota's prairie communities (Pfannmuller and Coffin 1989). Five of the 17 listed mammals, 19 of the 27 birds, 5 of the 17 amphibians and reptiles, and 64 of the 191 plants are associated with prairies.

What is the value of the few remaining acres of native prairies? Prairies, like hardwood and coniferous forests, are places where Minnesotans can enjoy and appreciate their heritage. In addition, valuable plant and animal genetic material may have disappeared as the prairies were broken. The remaining prairies may hold undiscovered genetic secrets of major importance to medicine and agriculture. For example, several years ago the Upjohn Company of Kalamazoo, Michigan, harvested seeds of tall meadow-rue from prairies in northwestern Minnesota for use in research on cancer (Robert Dana, Natural Heritage Program, Minnesota Department of Natural Resources, personal communication). In addition to values related to health and economics, scientific study of native prairies can provide basic information on soil formation and productivity, on relationships between prairie plants and animals, and on responses of the prairie ecosystem to stresses such as drought and flood. All Minnesotans owe a debt of gratitude to the state and federal governments and The Nature Conservancy for efforts in acquiring and maintaining these valuable nature preserves.

# WETLANDS

**7**

Swamps, sloughs, marshes, bogs, potholes—we know wetlands by many names. These words are often used interchangeably, however, because of the complexity of naming the many kinds of wetlands. To avoid this confusion, I will use the collective terms wetlands and peatlands to refer to Minnesota's water-saturated environments.

Wetlands occur in all three of the state's vegetation regions and can be grouped into three major types: prairie wetlands, peatlands, and forest wetlands. Most prairie wetlands (figure 7.1), once abundant in southern and western Minnesota, have been drained and converted to cropland. The remaining prairie wetlands have been studied intensively. Much of this research has been related to the value of these wetlands as habitat for breeding ducks and other wildlife.

In the northern coniferous forest region are vast acreages of vegetation-covered wetlands called peatlands. These wetlands have received much attention from ecologists in recent years because of their distinctive characteristics, their potential value as an energy source, and their high sensitivity to disturbance.

Forest wetlands are found on mineral soil in both coniferous and deciduous forests. Little is known about these wetlands, perhaps because they represent only a small fraction of the habitat in

**Figure 7.1** *A typical prairie wetland. The mixture of patches of emergent vegetation, such as bulrush and cattail, interspersed with open water in about equal proportions is considered optimum for the production of wetland birds and mammals. (Photo by Bob Firth, Firth PhotoBank, Minneapolis.)*

**Figure 7.2** *Open ditches and underground tile lines have drained most of Minnesota's prairie wetlands. (Photo courtesy of U.S. Fish and Wildlife Service.)*

Minnesota's forests. Still other wetlands occur along the shorelines of lakes and rivers and share some of the characteristics of both prairie and forest wetlands.

## Prairie wetlands

Just as the process of glaciation determined the general topography of our state, it also determined the location of prairie wetlands. As the glacial ice melted, basins or depressions in the moraines filled with water. In some places, blocks of ice broke off the glacier and were buried in till or outwash. These blocks of ice may have lasted for many years before finally melting, leaving depressions in the landscape, which often filled with water. We now refer to the deeper and larger of the areas that filled with water as lakes. The smaller, isolated, shallow bodies of water developed into what we call wetlands.

Before European settlement, wetlands could be found wherever prairies existed and were especially abundant in western and southern Minnesota (figure 1.25). The west-central portion of Minnesota lying in Otter Tail, Becker, Mahnomen, and adjacent counties was characterized by a mixture of large, deep wetlands interspersed with many

smaller, often temporary wetlands. Tens and even hundreds of wetlands could be found in each square mile.

Today only remnants of the prairie wetland habitat remain in Minnesota. Most prairie wetlands have been drained, and the land grows corn, wheat, and other crops (figure 7.2). Before settlement, Minnesota's prairie area contained an estimated seven million acres of wetlands. By 1980, only about 20% of this acreage remained. In the extreme southwestern corner of Minnesota nearly all of the wetlands have been drained. In Nobles County, 137,000 acres of wetlands existed before settlement. In 1984, less than 1,000 acres remained (Anderson and Craig 1984), and today only a few acres can be found.

### Wetland zones

Let us imagine that we are looking down on a typical Minnesota prairie wetland from a nearby hill. The first thing we are likely to notice is that the vegetation appears to be arranged in concentric rings surrounding a pool of open water (figure 7.3). These concentric rings, or zones, provide us with a way of describing the wetland.

Now let us examine these zones by walking from the hilltop into the wetland. As we approach

Shallow-marsh zone

Wet-meadow zone

Low-prairie zone

Seasonal pond or lake

Deep-marsh zone

Shallow-marsh zone

Permanent open-water zone

Wet-meadow zone

Low-prairie zone

Permanent pond or lake

**Figure 7.3** *Concentric vegetation zones are found in seasonal and permanent prairie wetlands. The number of zones is dependent primarily on water depth in the wetland.*

the wetland, we encounter the low-prairie zone. There is no standing water in this zone, and at times the soil does not even appear moist. During high water in early spring, however, this zone may be flooded for a short period of time. Cutgrass, blue-joint grass, and a variety of sedges, mints, and asters are characteristic plants of the low prairie.

As we move a few steps closer to the wetland, we enter the wet-meadow zone. Here we find very wet soil and perhaps some standing water. Spike rush, bur-reed, and several species of rushes belonging to the genus *Scirpus* are the most common plants. Water hemlock, the most poisonous plant in temperate North America (Voss 1985), grows 3 to 6 feet tall in the wet meadow. Eating only a small portion of the root or tuber, which

contains a poisonous resin, may cause convulsions and death in humans and in livestock. American Indians, however, smoked water hemlock leaves, and I have seen larvae of swallow-tail butterflies feeding on the plant.

Moving toward the center of the wetland, we enter the shallow-marsh zone, where the water varies from a few inches to a few feet in depth. Cattail and bulrush (figure 7.4) are the most common plants. Reed (figure 7.5), one of the most widely distributed plants in the world, is abundant in

**Figure 7.4** *(below) Bulrush, which grows in deeper water than cattail, may provide nesting cover for several kinds of ducks and other wetland birds. (Photo by Richard Hamilton Smith.)*

**Figure 7.5** *(overleaf) Reed* (Phragmites sp.)*, which may grow up to 10 feet tall, is one of the most widely distributed plants in the world. (Photo © Jim Brandenburg.)*

some wetlands. Its 4- to 10-foot stems have been used since ancient times for houses, fences, brooms, arrows, food, boats, fuel, sleeping mats, sandals, and writing tools (Lawrence 1972). In the shallow marsh we also find floating-leaved and submerged aquatic plants, such as pondweeds and bladderwort.

As we take more steps toward the center, water begins to creep over the tops of our hip waders. We have now come to the deep-marsh zone. Here there are few, if any, emergent aquatic plants but many submerged and floating-leaved species. Pondweeds, coontail, bladderwort, and yellow water-crowfoot are common here where the water ranges from a few feet up to about 8 feet in depth.

Sago pondweed, a top-rated duck food, often grows in abundance in the deep-marsh zone. Mallards eat its seeds, and diving ducks pull up and eat stems and tubers. Dense stands of sago form as root runners spread and tubers dislodged by ducks settle to the bottom and grow into new plants.

The open-water zone in the center of the pond is too deep for most rooted aquatic plants. However, the water contains algae and perhaps duckweeds, free-floating plants with their roots hanging in the water.

Not all wetlands contain all of the five zones described. The number of concentric bands of vegetation depends on the contour of the wetland bottom and the maximum depth of the water (Kantrud, Millar, and van der Valk 1989). Furthermore, as I will now discuss, we can expect to see striking changes in a wetland throughout the year and from year to year as the water level changes both seasonally and over longer periods of time.

## Water levels

Annual seasonal changes in the water level in the wetland ecosystem are inherent in its functioning. In late spring after snowmelt and heavy spring rains, the water level is at its highest. During this time many kinds of aquatic plants grow at their most rapid rate in response to the large flush of nutrients brought into the wetland by the water. Amphibians such as frogs, toads, and salamanders,

who spend most of their life cycle in the uplands, come to the wetland to mate and lay their eggs. Migrating geese and ducks rest and feed on even temporary wetlands flooded from snowmelt.

As summer progresses and dry periods become more frequent, the water level drops, reaching its lowest level in late fall; shallow wetlands may even dry up in the fall. Yet many of the plants and aquatic vertebrates and invertebrates that live in wetlands are able to survive this dramatic change in their environment. The massive root systems of perennials, such as cattails, enable the plants to survive the dry periods. The eggs of the amphibians hatch, and the young metamorphose and leave the wetlands before the water disappears. Many invertebrates are able to endure the dry periods and the winter cold by transforming into a cystlike dormant state. When the waters return in the spring, they again become active and reproduce.

Wetland water levels also fluctuate in response to long-term changes in the amount of rainfall, which occur over periods ranging from about 5 to 20 years. During drought periods, such as Minnesota experienced in the 1930s and to a lesser degree in the 1970s and early 1980s, many deep wetlands and nearly all shallow wetlands dried up. During periods of abundant rainfall, the wetlands reflooded, long-dormant seeds sprouted, and the wetlands were restored.

In many cases, the drying out and reflooding seemed to stimulate the wetlands to higher primary productivity, probably in response to nutrients released from the bottom sediments. Wildlife managers often increase the productivity of wetlands and thus their value as wildlife habitat by drawing down the water level to expose the bottom, sometimes cultivating the exposed bottom, and then reflooding (Harris and Marshall 1963).

## Classification

The dynamic nature of wetlands makes their naming and classification difficult. A number of classification systems have been developed, however, and are widely used. Key elements of three classification

**Table 7.1** *A partial comparison of three wetland classification systems*

| Major vegetation characteristics | Shaw and Fredine (1956) types | Stewart and Kantrud (1971) classes | Aaseng et al. (1993) classes or types |
|---|---|---|---|
| Wet prairie grasses or weeds | 1. seasonally flooded basins or flats | I. ephemeral ponds | wet prairie |
| Meadow sedges, rushes, grasses, and forbs | 2. inland fresh meadows | II. temporary ponds | wet meadow |
| Water-loving grasses and sedges, smartweeds, bur-reed | 3. inland shallow fresh marshes | III. seasonal ponds and lakes | emergent marsh |
| Cattail, bulrush, pondweeds | 4. inland deep fresh marshes | IV. semipermanent ponds and lakes | emergent marsh |
| Cattail and bulrush fringe | 5. inland open freshwater | V. permanent ponds and lakes | aquatic lake community |
| Alkali bulrush, wigeon grass | 10. inland saline marshes | VI. alkali ponds and lakes | saline wet prairie |

*Sources:* Modified from Weller (1987); Aaseng et al. (1993).

systems are presented in table 7.1. The systems are complex and depend primarily on the presence of zones with characteristic kinds of vegetation, which are influenced by water depth and water chemistry. Presence or absence of certain of the concentric zones is a primary criterion for classifying a given wetland in each of these systems.

The wetlands are also classified according to the salinity, or alkalinity, of the water, which is indicated by the presence of certain plant species. In Minnesota, the chemistry of the water in wetlands differs in different regions in the same way and for the same reasons that the water chemistry of lakes differs (see chapter 8). J. B. Moyle (1945, 1956) and E. Gorham, W. E. Dean, and J. E. Sanger (1983) found that the lakes and wetlands in the northeastern portion of Minnesota are generally acidic or neutral and that alkalinity, or salinity, increases toward the west (figure 7.6). This change is due in part to the higher concentration of calcium carbonate in the glacial till in the western regions of the state.

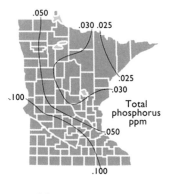
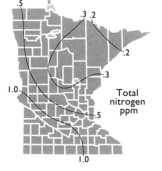
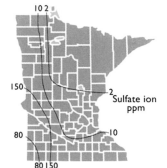

**Figure 7.6** *The chemistry of lakes and wetlands changes in a predictable pattern from northeastern to southwestern Minnesota. The map of aquatic floras shows a zonation of aquatic plants, reflecting the changes in water chemistry. (Modified from Moyle 1956, by permission.)*

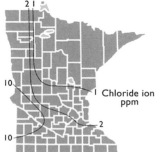
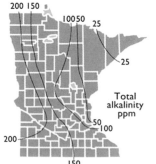

## Succession

The dynamic nature of wetlands also complicates our understanding of succession in prairie wetlands. We must look not only at the successional process over the 10,000 or so years since they were formed but also at the shorter-term changes that occur over about 5 to 20 years (Weller 1987).

In the newly formed pond the pioneer plants are algae and submerged and floating-leaved aquatics. Perhaps a few emergent plants grow near shore. M. W. Weller (1987) identifies this stage as the submergent stage. As the basin fills up with sediment from runoff and wind deposition and with decaying plants, it gets shallower over a broad area.

**Figure 7.7** *Succession in wetlands generally leads from the open-water stage through stages with submerged and emergent plants. Eventually the wetland basin fills with sediments, and a wet prairie or forest develops on the site. Succession has been ongoing in most Minnesota wetlands for about 10,000 years, since the end of glaciation. (Modified from Weller 1987, by permission.)*

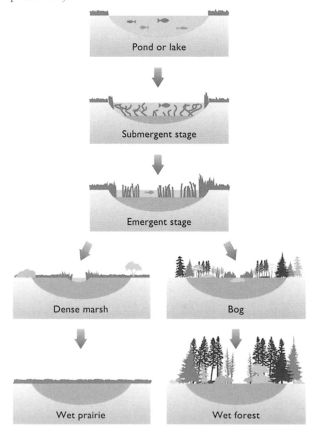

Emergent plants become more common, creating the emergent stage.

The sequence of succession from here on depends on the biome in which the wetland occurs. In succession in the prairie biome (shown on the left in figure 7.7), the emergent stage is followed by the dense marsh stage in which the basin becomes very shallow because of sedimentation. Emergents, sedges, and grasses cover the entire wetland. This stage is followed by the wet prairie stage in which the standing water is gone and sedimentation has essentially filled in the basin. The wet prairie is often invaded by native trees and shrubs such as willow and cottonwood, and by non-native Russian olive. This invasion may be due to the suppression of prairie fires.

Many examples of where this successional sequence has led to filling in of a wetland can be found in Minnesota today. There are, however, also many wetlands in each of the stages of succession, including the pioneer stage. Thus, the progress of succession is dependent on factors other than time. These factors are water depth, bottom type, climate, hydrologic changes, and the impact of human activity. For example, succession in some wetlands has been accelerated by human activities that have markedly changed the bottoms in these wetlands. When the wetlands were first formed, the basins consisted of glacial till, which varied from coarse gravel in morainic wetlands to fine sands and even silt in wetlands formed in outwash plains. Over the years, some wetlands in agricultural areas have received large quantities of fine particles of silt and clay carried in from the surrounding farmland by runoff during heavy rains and spring snowmelt. These sediments, plus the accumulation of dead and decaying plant material, created a soft, oozy bottom. In some areas, wetlands have filled in because of sedimentation, and no longer contain any standing water. In other areas with minimal erosion and with low productivity of aquatic plants, the bottoms remain sandy or gravelly, much as they were when the wetland was formed, and succession has proceeded at a slower pace.

Next let us consider the changes that occur in the vegetation of a wetland during a typical 5- to 20-year wet-dry cycle. These changes do not correspond to the succession sequence for wetlands just described; rather they occur repeatedly throughout the time that the wetland is proceeding through the stages of succession. The cycle has been divided into four stages: dry marsh, regenerating marsh, degenerating marsh, and lake. During periods of drought, when wetlands lose their standing water,

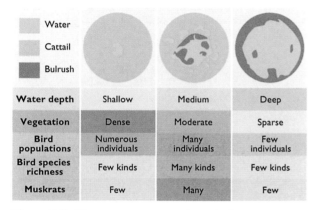

**Figure 7.8** *The 5- to 20-year wet-dry cycle that occurs in many wetlands causes marked changes in vegetation and animal populations. The changes are brought about primarily by the action of high water and the harvesting of vegetation by muskrats. (Modified from Weller 1987, by permission.)*

the bottom substrate is exposed, allowing germination of a wide range of plants. The seed bank of prairie wetlands is extremely rich and contains seeds of annual species, such as smartweed, beggarticks, and dock, that grow on the exposed mudflat, as well as seeds of emergent, submerged, and freefloating aquatic plants (van der Valk 1981; van der Valk and Davis 1978). Buried seeds of cattail, bulrush, reed, and arrowhead also germinate.

When the period of drought is broken by a period of normal rainfall, water returns to the wetland, and the mudflat species can no longer survive. Now the emergents thrive. In this regenerating marsh stage, the submerged and floating-leaved aquatic plants, such as pondweed, coontail, water milfoil, and chara, germinate under water. Emergent aquatic plants dominate this stage, and the

entire wetland basin may be covered so that water is not readily visible.

After a number of years, the emergent plants begin to die out, for reasons that are not well understood, and the wetland becomes a mixture of emergents and open water (figure 7.1). When this mixture is about half scattered tracts of open water interspersed with scattered tracts of emergents, the wetland is referred to as a hemi-marsh (Weller and Spatcher 1965). At this stage the wetland provides the greatest diversity of habitats and edges between habitats, and thus its animal population is at its greatest in both diversity and density (figure 7.8).

The degenerating marsh stage occurs next as the emergent plant populations continue to decline. Increasing numbers of muskrats consume more and more of the plants. The wetland now contains more open water than emergent vegetation.

Eventually, the open-water wetland stage is reached in which emergent vegetation is absent or limited to a narrow zone near the shore. Similarly, submerged and floating-leaved aquatic plants are sparse and are usually limited to the shoreline zone. In this stage, the wetland is very much like a pond or a shallow lake. This stage may continue for many years until the water level recedes because of drought and the bottom again dries out, allowing germination of the emergent and mudflat plants and thus beginning the cycle again.

Wetlands may repeat this cycle from dry marsh to lake over and over, depending to a great extent on periods of drought to initiate each cycle. The sequence is subject to change within a cycle in response to such events as flooding or to nearly total consumption of emergents by muskrats (Weller 1987).

## Ecosystem function

Plant growth is extremely luxuriant in wetlands because of the abundance of nutrients and water. The season of greatest primary production, when the maximum aboveground biomass is produced, occurs in late spring and early summer soon after stands of emergent vegetation become established.

169

Emergent plants, such as cattail and bulrush, have extensive roots and rhizomes that store the energy and nutrients used to initiate this rapid growth. The belowground production is thus an important part of the productivity of wetlands, in much the same manner as the roots are important in the productivity of the prairie. Furthermore, the belowground biomass provides a source of food for consumers active under the ice in winter, and the rhizomes provide the means for vegetative propagation or reproduction.

The annual productivity of a wetland depends to great extent on its stage in the 5- to 20-year wet-dry cycle described earlier. The production of plant material changes with stages in the cycle. As figure 7.9 shows, the mudflat species in a wetland in Iowa peaked in 1964 during the dry marsh stage. Both submerged and emergent species peaked in 1966 at the regenerating stage and then declined gradually until the lake stage was reached in 1970. The maximum total biomass was reached in 1966. (Note that these values represent only the aboveground portion of the plants.) Annual aboveground production changed 18-fold as the wetland proceeded through the different stages of the cycle (van der Valk and Davis 1978).

A comparison of the net primary productivity of a wide variety of habitats was summarized by C. J. Richardson (1979). The total of aboveground and belowground productivity was 24,200 pounds per acre per year for cattail marsh, 18,800 pounds per acre for reed-type marsh, and 9,200 pounds per acre for sedge marsh (figure 7.10). These values can be compared to the productivity of 4,600 pounds per acre per year of grasslands in the United States and to the aboveground productivity of only 13,200 pounds per acre per year of temperate coniferous forests and 9,000 pounds per acre per year of deciduous forests. Because of the availability of nutrients and moisture, wetlands are among the most productive ecosystems on Earth.

Net primary production in a cattail marsh was compared with the production of unfertilized and fertilized corn growing on sandy soil in east-central

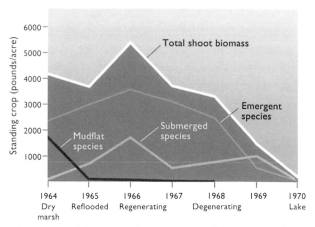

**Figure 7.9** *(above) Standing crops of mudflat species, submerged species, and emergent species change as prairie wetlands cycle through the stages leading from a dry marsh to a prairie lake. Total plant biomass declines through the cycle, except for a brief peak during the regenerating stage. (Data from van der Valk and Davis 1978.)*

**Figure 7.10** *(below) Net primary productivity of a variety of wetlands is higher than that of grassland. Unlimited water and abundant nutrients, which are carried into the wetland by runoff, create excellent conditions for plant growth. (Data from Richardson 1979.)*

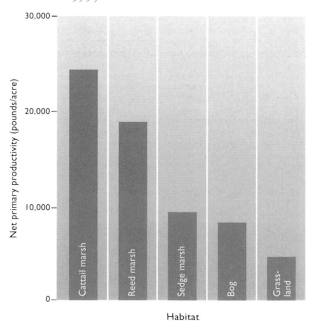

Minnesota by J. R. Bray, D. B. Lawrence, and L. C. Pearson (1959). They found that production of cattails was greater than production of fertilized corn and believed that this difference was due to the abundance of water and nutrients in the wetland and the straplike structure of the cattail leaves. Leaves with this shape standing vertically provide the maximum amount of photosynthetic surface exposed to the sun.

Although the emergent and floating-leaved aquatic plants are most important in primary production, the contribution by algae is also important. Algae do not have large standing crops, but the turnover or reproductive rate is very high. In some instances, this results in high annual production. Algae suspended in the water column usually make up only a small proportion of the annual production, whereas the diatoms, algae that grow on the surface of the sediments on the bottom of the wetland, may be quite important (Shamess, Robinson, and Goldsborough 1985). Mats of algae are conspicuous in many Minnesota wetlands in summer. Trails of birds, muskrats, and turtles can often be observed crossing the mats.

## Consumers and the prairie wetland food web

High productivity in wetlands results in complex food webs. D. E. Willard (1979) estimated that in the United States about 900 species of plants and animals use wetland habitats at some stage in their life history. Just as in the other ecosystems we have discussed, the nutrients and energy in the net primary production are made available to the consumers either through consumption of plant material by grazing or through decomposition in the detritus food web.

The most important grazer in prairie wetlands is likely to be the muskrat (figure 7.11), which uses cattail, bulrush, arrowhead, and many other species for food and to construct nests and lodges. Sometimes, they feed on clams, snails, and even fish. With this omnivorous feeding habit, muskrats fit into several places in the food web.

**Figure 7.11** *Muskrats are the most important grazers in many prairie wetlands. They sometimes remove nearly all of the cattails and bulrush, thereby changing the wetland from the emergent stage to the lake stage. (Photo by Blacklock Nature Photography.)*

171

High populations of muskrats, which develop during the regenerating and hemi-marsh stages, may consume a very large percentage of the emergent vegetation in a wetland, hastening the change toward the lake stage (Perry 1982). Muskrats also harvest roots and rhizomes of the emergent plants and have been known to dig as deep as 20 inches to uncover rhizomes (Lowery 1974). The large quantities of plant material harvested by muskrats are used for the construction of feeding huts, platforms, and winter lodges. Eventually, this material becomes part of the detritus food web.

Insects and other invertebrates also consume plant material, but their overall effect in terms of the percentage of primary production consumed is believed to be very small (Penko 1985). Some species of ducks and other birds feed on vegetation. Other species are omnivorous and also feed on insects and other animals at certain times in their life cycle or when prey is especially abundant. To further complicate the wetland food web, some consumers spend their entire life cycle within a single wetland, whereas others may be present for only a short period of time. Some obtain part of their food from the wetland and part from surrounding upland, and some species may move from wetland to wetland in their search for food.

At higher levels in the food web, insects and other invertebrates prey on insects (figure 3.5); frogs, toads, and salamanders prey on invertebrates; and mink prey on amphibians, muskrats, meadow voles, waterfowl, and other birds. Red fox from the surrounding uplands eat fruits and berries and prey on birds and mammals. Hawks and owls feed on a wide variety of animals from the wetland.

Although muskrats are considered to be the most important grazer, other grazers may significantly affect a wetland. Mallards and blue-winged teal completely eliminated the sago pondweed from a pond in southern Manitoba (Wrubleski 1984). Removal of the sago pondweed changed the structural character of the pond habitat, and this change influenced the chironomid, or midge, community in the pond. The smaller chironomid species that

normally grew attached to the pondweed were eliminated, and larger, bottom-dwelling species increased in abundance.

Little is known about the consumption of algae by grazers in wetland ecosystems. Invertebrate grazers such as copepods and cladocerans, or water fleas, are abundant in prairie wetlands and feed primarily on the algae growing in the open water. Cladocerans have been shown to control both the abundance and species composition of algae by selective grazing (Porter 1977). Some cladocerans feed on algae, whereas other species feed almost entirely on bacteria (Coveney et al. 1977). Chironomid larvae are also important consumers of algae (Murkin 1989). I have observed hundreds of toad tadpoles feeding on mats of algae in ponds in northwestern Minnesota, and H. R. Murkin observed coots, gadwalls, and mallards feeding on mats of algae in Manitoba.

In spite of the grazers in the food web, a prairie wetland is so productive that it would quickly be filled in with plant remains without the rapid decomposition that occurs in the detritus food web. The base of the detritus food web is made up of standing litter, fallen litter, and organic compounds that have leached from both standing and fallen litter (Murkin 1989). Individual plants die throughout the growing season, contributing gradually to the litter, but most death occurs from frost in the fall. Decomposition of litter involves leaching of soluble substances from the plant tissue, fragmentation due to animal activities and weathering, and the decay associated with bacteria, fungi, and other consumers (de la Cruz 1979). Muskrats may play a major role in the process of fragmentation.

Although some decomposition occurs near the water surface or on the bottom in aerated locations with adequate oxygen, much of the decomposition occurs under anaerobic conditions in the bottom sediments. The rate of decomposition is enhanced by warm temperatures, high oxygen concentration, high nutrient levels in the plant tissue, and low lignin and total fiber in the plant tissues (Brinson, Lugo, and Brown 1981).

Movements of nitrogen and phosphorus through the food chain have a strong influence on wetland productivity. Both enter the wetland through runoff, but some nitrogen input is from precipitation. These nutrients are incorporated into plant tissue during the growing season and are returned to the water and the bottom sediments through decomposition. During decomposition, plant tissues have been shown to accumulate nitrogen and phosphorus, actually doubling the amount present in the living tissue (Neely and Baker 1989). The increase may be attributed to the increase in populations of bacteria and fungi in the litter and to direct absorption of nitrogen compounds by the litter particles (Polunin 1984).

## Animals and community interactions in prairie wetlands

### Mammals

Many mammals use wetlands for cover and food. Most are primarily terrestrial, living in the grasslands or forests surrounding the wetland. The two most important species that live in wetlands are beaver and muskrat.

Beaver are most commonly associated with streams in forested areas, but they sometimes inhabit wetlands without flowing water. In such cases, beavers make heavy use of cattails, both for food and for lodge construction.

Muskrats use all types of wetlands at times, but their presence is usually dependent on water deep enough to sustain activity under the ice throughout winter. During summer and early fall, muskrats build lodges with vegetation, primarily cattails and other emergents, that they collect and carry to the building site. Burrow systems with nests and feeding chambers may also be dug into the banks of wetlands.

Muskrat houses may be over 9 feet in diameter and 6 feet high. They contain one or more chambers above the water level and have several entrances that provide access from under water. Smaller feeding platforms and houses are often located close to the nest house. Feeding sites, called push-ups, are formed in winter by muskrats carrying vegetation collected beneath the ice through cracks or holes in the ice surface.

During the breeding season from April to June, females remain more or less isolated in individual lodges and bear from two to three litters. The average litter size is 6 to 8, but occasionally 12 young will be born. Males remain associated with their females but do not play a major role in care of the young. Young muskrats begin establishing territories in the fall and are ready to breed the following spring. Fighting and aggression are common at this time of year, and muskrats that are unable to hold territories are likely to be killed by predators or in fights with other muskrats (Errington 1963).

Studies of radio-tagged muskrats at Cedar Creek Natural History Area by Paul Stolen found that in late summer they were active during both daylight and darkness building lodges and gathering food. After ice formed on the wetland in November, they reduced their total activity to about eight hours. All activity occurred at night, and was probably devoted entirely to feeding (Tester 1987; Tester and Figala 1990).

Muskrats are more social in winter, and several may live together in one chamber to conserve heat. Oxygen is generally adequate in the chamber, but carbon dioxide and other wetland gases often build to high levels (Huenecke, Erickson, and Marshall 1958).

Muskrats feed primarily on emergent vegetation but are known to use a wide range of other aquatic plants. During winter they burrow for the rhizomes of cattail and bulrush, and under extreme conditions will even eat the dead vegetation making up the walls of their houses.

The rapid growth of muskrat populations during the regenerating and hemi-marsh stages eventually leads to overpopulation and nearly total destruction of the emergent vegetation. At this time, muskrats are highly vulnerable to predation by mink, hawks, owls, red foxes, and raccoons.

Hemorrhagic disease and tularemia are also known to deplete dense populations.

Meadow voles, close relatives of the muskrat, are often found in the low-prairie and wet-meadow zones of prairie wetlands. Voles swim well and may live over water in the nests of diving ducks or in muskrat houses (Weller 1979). They are preyed upon by mink and raccoon, who hunt the edges of wetlands and often swim in the deeper zones searching for prey.

### Birds

Minnesota's wetlands are host to tens of thousands of waterfowl. On a bright spring morning, the stirring calls of migrating geese and swans announce the arrival of the new season, and the prairie wetlands seem to belong to the ducks.

Two main taxonomic groups of ducks use Minnesota's wetlands. Dabblers or puddle ducks, such as mallards, pintails, and blue-winged teal, feed on the surface or in shallow water and sometimes on land. Diving ducks, such as the canvasback, redhead, and lesser scaup, or bluebill, usually feed by diving underwater, are awkward on land, and seldom feed there.

Although different species of ducks use different types of wetland habitat to meet their varied requirements, all species seem to be attracted to areas with high primary productivity. West-central Minnesota provides ideal habitat for ducks because of its abundant wetlands of different types. Duck populations in the region vary in direct proportion to the number of ponds present during the month of May (figure 7.12; Batt et al. 1989).

Migrating ducks can be seen on shallow, temporary wetlands, which flood in early spring due to the water from snowmelt, and on nearly every permanent wetland and lake. Ducks may use several wetlands each day for feeding and resting (Gilmer et al. 1975). One might think that water is the critical factor in attracting ducks to wetlands, but it is actually the abundant food resources in the form of many kinds of invertebrates (Weller 1987). The wetlands provide food for energy for further

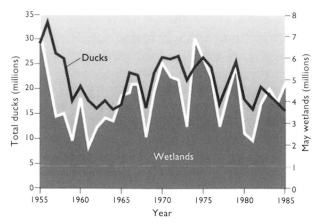

**Figure 7.12** *The number of ducks present in the prairie in any year is closely related to the total number of wetlands present in May of that year. (Modified from Batt et al. 1989, by permission.)*

migration and for maintaining fat reserves needed for reproduction. The birds eat invertebrates of all kinds and the seeds, leaves, and tubers of aquatic plants, both from the previous season and currently growing.

Ducks that remain in Minnesota to nest may already be paired or may be going through a period of courtship leading to pairing. Each pair requires a certain amount of space and privacy, called a breeding territory. Drakes defend their territories from other pairs of the same species. An abundance of small wetlands provides many opportunities for such territories and allows for high populations of breeding pairs. Similarly, wetlands in the hemimarsh stage provide visual isolation, allowing many territories to be established.

Nesting soon follows pairing. Diving duck hens often select a clump of emergent vegetation over water to support their nests. Dabbling duck hens usually place their nests on land. Most of these nests are near water, but some may be as far as 1.5 miles from the nearest water. Wood duck, goldeneye, hooded merganser, and bufflehead use holes in trees, and sometimes chimneys in vacant buildings, for their nests.

During the egg-laying period, hens consume a larger proportion of invertebrate food than drakes do to meet the high demand for protein and calcium required for egg production (Swanson and Duebbert 1989). Shallow wetlands are especially important to the hens because the wetlands produce large quantities of insects and other invertebrates.

Once the hen begins incubating her clutch of eggs, her mate leaves to join with other drakes on larger wetlands or lakes to undergo a complete change of feathers. Since the molt leaves the drakes flightless and vulnerable to predators, they usually molt on lakes that have large expanses of water and abundant emergent cover for escape.

Newly hatched ducklings are immediately led to a wetland by the hen, where they consume invertebrate food exclusively for the first days of their life (Swanson and Duebbert 1989). Mosquito, dragonfly and midge larvae, beetles, spiders, snails, leeches, and freshwater shrimp are eaten. Ducklings of most species require about six weeks to reach the stage of maturity at which they can fly. During this time they stay in wetlands and must move to deeper water as summer progresses and the shallow wetlands dry. The hen leads her ducklings from wetland to wetland, selecting those with abundant invertebrate food (Swanson 1977). After the young reach the flying stage, the hen abandons them and goes through her molt.

Young ducks learn to recognize landmarks as they are growing, and they migrate by instinct to their wintering areas in the south. Adult hens return in the spring and often use the same nest site as in previous years. Young birds are believed to return to their natal area to establish their first nest.

Tundra swans and snow geese migrate through Minnesota in spring and fall but do not nest here at present. Canada geese, however, are now common nesters in Minnesota. This has not always been the case. Before the 1920s, Canada geese nested throughout Minnesota, but during the 1930s and early 1940s none were known to nest in the state. In fact, the giant Canada goose was believed to have become extinct about 1920. Its existence went unknown for about 40 years until in the 1960s Harold Hanson, a biologist with the Illinois Natural History Survey, observed some of these geese

(figure 7.13) in the flock wintering in Silver Lake in Rochester in Olmsted County (Hanson 1965). Canada geese were probably present as migrants throughout this period but were simply not recognized. Due to the efforts of both government and private wildlife groups, the giant Canada goose is now a common breeder in the Twin Cities area, near Fergus Falls, on Heron Lake, and on wildlife management areas and refuges throughout the state.

The bond between a pair of Canada geese is permanent, and the male assists the female in taking care of the young. Family groups are often seen swimming or flying together. Some of these birds winter in Minnesota, and some travel farther south, especially during extremely cold winters when open water and food are scarce.

Many other kinds of birds use wetlands for feeding, nesting, or resting. It is estimated that about one-third of all the species of birds in North America use wetlands at some point in their life cycle (Kroodsma 1979). Bird usage appears to be dependent more on the pattern of vegetation and its structure than on the plant species present. Edges between vegetation and open water are especially important, and it is believed that wetlands with about 50% open water and 50% emergents are optimal for birds because of the diversity of habitat types (Weller 1987).

Coots, or mudhens as they are often called, are frequently associated with ducks on prairie wetlands. They have rather long legs, and their feet are not webbed. Although usually found in shallow water areas, they can also feed on dry ground. Coots are often seen feeding with redhead and canvasback ducks. I have observed coots on wetlands in Mahnomen and Becker Counties stealing parts of aquatic plants that the diving ducks have brought up from the bottom. Large numbers of coots migrate through Minnesota in spring and fall but are rarely seen because they move primarily at night.

Virginia rails and soras, which belong to the same family as coots, are also present in many wetlands but are seldom seen because they usually stay in heavy cover. They can swim but seem to prefer to pick their way through the emergent vegetation either by walking on the bottom in shallow water or by stepping from one piece of floating vegetation to another. Insects and seeds are their primary foods.

Common loons, white pelicans, and several species of grebes are fish-eating birds highly adapted for surface feeding or diving. Loons are more common on Minnesota's larger lakes but are often found on large prairie wetlands. Pied-billed grebes tolerate a wide range of conditions and are found in small numbers on most wetlands where food, nesting materials, and enough water for swimming are available. Western grebes (figure 7.14), the largest species in Minnesota, have been observed in concentrations as high as 500 on Lake Osakis in Todd County (Janssen 1987). White pelicans nested on Minnesota wetlands until 1878, when the last were killed (Henderson 1981b). In 1968, 90 years later, they resumed nesting in the state,

and in 1983, 1,450 nests were counted on Marsh Lake in Lac qui Parle County (Janssen 1987). Pelicans now nest on Marsh Lake and on several islands in Lake of the Woods.

A variety of large, long-legged wading birds are frequently observed searching the shoreline and shallow emergent zone for fish, frogs, and other foods. Great blue herons, great egrets, and American bitterns are often seen stalking fish, tadpoles, frogs, and other animals. The white plumage of the great egrets makes them highly visible to humans but may make them more difficult for fish and other prey to see against the bright background of the sky (Faaborg 1988). The mottled plumage of the American bittern, on the other hand, blends well with dead cattails and bulrushes in the emergent zone.

A large variety of smaller wading birds, commonly identified as shorebirds, use the shallow-

**Figure 7.13** *Thousands of giant Canada geese spend the winter on Silver Lake in Rochester in Olmsted County. Warm water from an electric power plant keeps the lake open. (Photo by Blacklock Nature Photography.)*

**Figure 7.14** *Western grebes are characteristic of prairie wetlands. This dramatic display occurs during courtship in the spring.* (*Photo by Blacklock Nature Photography.*)

water zones, the shorelines, mudflats, and wet meadows. They feed primarily on small invertebrates, seeds, and other parts of plants. This group includes the sandpipers, plovers, avocets, phalaropes, and snipe. Some of these birds turn over rocks looking for insects, some strain the water and sediments through their sievelike bills to obtain algae and microscopic invertebrates, and some probe in wet vegetation.

Other birds common in wetlands fly above the wetlands searching for food. Several species of terns nest in Minnesota wetlands and feed mainly on fish. Black terns will often attempt to chase a visitor away from their territory by flying at their head and calling loudly. Gulls, which feed on a wide variety of foods including dead animals and garbage, are common in many wetlands. Large nest-

ing colonies have been reported at several locations in Minnesota, including a colony of 100,000 on Heron Lake in the late 1930s (Janssen 1987).

Perhaps the most abundant species of bird using wetlands is the red-winged blackbird. Their nests are common in cattails, bulrush, and willows and other shrubs along the edges of wetlands throughout Minnesota. Red-wings arrive early in the spring and establish their territories throughout a wetland. The larger yellow-headed blackbirds arrive later and occupy zones of the wetland with emergent vegetation and deeper water by chasing the territorial red-wings away. In fall, the red-wings, yellow-heads, and other species of blackbirds form flocks of many thousands of individuals, which can often be seen streaming across a wetland from horizon to horizon.

*177*

*Amphibians and reptiles*

During the breeding season frogs and toads are conspicuous residents of nearly every wetland in Minnesota. The trills and calls coming from breeding choruses announce that winter is over for another year. Fourteen kinds of frogs and toads live in Minnesota, but all species are not active at the same time. Chorus frogs, spring peepers, and wood frogs establish their breeding choruses and begin calling first, followed by gray treefrogs and American toads. Green frogs and mink frogs begin their breeding later and are active even in midsummer. Some species are referred to as explosive breeders, completing their reproductive period in a matter of a few days, whereas other species continue breeding activity for as long as several months.

Wetlands are essential for breeding for all species of frogs and toads and for some salamanders living in Minnesota. It is likely that all but the most shallow and temporary wetlands harbor some species of amphibians even though much of their life cycle is spent in the uplands near water. Tiger salamanders may be especially abundant in prairie wetlands. Their life cycle allows for most young to metamorphose in summer into adults that can live either in the water or on land. In dry years some larval salamanders may keep their gills and stay beneath the water, thus surviving drought conditions on the uplands.

Frogs, toads, and salamanders feed on a wide variety of animals. They eat only moving prey and commonly feed on large quantities of insects, worms, and other small invertebrates. Often they can be observed waiting patiently in the grass or on a mudflat, watching the approach of a potential meal. When an insect approaches near enough, the frog or toad flicks out its sticky tongue and quickly snatches the food into its mouth. In turn, these amphibians make excellent meals for a long list of predators, including mink, hawks, herons, raccoons, skunks, snakes, turtles, fishes, and even other frogs and salamanders.

Wetlands in Minnesota provide homes for turtles and snakes. Turtles are more common and more widespread than snakes. Painted turtles, although more often found in lakes, can be observed sunning themselves on logs or rocks in a wetland. Snapping turtles spend most of their lives underwater but are sometimes seen laying their eggs on the upland.

Blanding's turtles require large expanses of wetland and sedge habitat with elevated sandy areas nearby for nesting. The Weaver Dunes in Wabasha County are reported to have the largest concentration of Blanding's turtles in the world (Personal communication from M. Pappas, reported in Henderson 1979). They feed on land as well as in the water, eating grasses, leaves, berries, fish, aquatic insects, and crayfish.

Blanding's turtles do not breed until they are at least 12 years old. The female digs a nest cavity in the soil with her hind legs and then lays her clutch of about 10 eggs. Given that they must be 12 years old before breeding and that they lay so few eggs per year, Blanding's turtles have the lowest reproductive potential of any turtle in Minnesota. Furthermore, dogs, raccoons, skunks, and other predators are reported to take over 95% of the eggs and hatchlings (Bury 1979). If the species is to survive, it must have large, undisturbed areas of suitable habitat.

*Fish*

The presence and abundance of fish in wetlands is limited to a large extent by the low oxygen content in the water, which often leads to the death of fish under the ice, referred to as winterkill. After a winterkill, fish may not be present in a wetland for a number of years, perhaps until embryos or fry are carried in by birds. High salinity levels in some wetlands are also detrimental to most species of fish. These physical factors, plus the fact that many wetlands exist in closed basins with no inlet or outlet, severely limit fish populations.

The most common fishes in Minnesota's wetlands are fathead minnows, brook sticklebacks, and bullheads. Fathead minnows and brook sticklebacks are tolerant of low oxygen and high salinity. They feed on copepods, cladocerans, and other crustaceans (Held and Peterka 1974). Minnows are a popular fishing bait and of value to the local economy. In 1978, income from fathead minnows sold for bait in Minnesota was over $12 million (Peterson and Hennagir 1980).

Small fish are important consumers of plant material and small invertebrates in the wetland ecosystem. They also provide excellent food for a variety of predators, including raccoon, mink, gulls, terns, and herons.

### Invertebrates

Most of the larger invertebrates in Minnesota's wetlands fall into four groups: snails and clams, worms, crustaceans, and insects. Many different kinds of snails and clams occur on the bottoms of wetlands or associated with vegetation. These mollusks feed on both living plant material and detritus and serve as a major source of food for some predators.

The most common worms in wetlands are annelids called oligochaetes. These aquatic earthworms live in the bottom sediments or on submerged aquatic plants. In some parts of the United States, oligochaetes are the most abundant large invertebrates found in bottom sediments (Clark 1979).

Crustaceans are found in bottom sediments, in open water, and in the aquatic vegetation. The nearly microscopic swimming forms, called plankton, include cladocerans, copepods, isopods, and amphipods.

Among the larger crustaceans, crayfish are probably the most familiar to Minnesotans. They possess a hard exoskeleton, which is shed as the body grows. The new exoskeleton is soft when the old one is shed but hardens in a few days. Their burrows in mudflats are often topped by chimney-like structures about 1 to 2 inches in diameter and 2 or 3 inches high. The crayfish I have eaten taste something like shrimp or lobster, but it takes many crayfish tails to make a meal.

Insects are the most diverse group of invertebrates in wetlands. Some species are aquatic for their entire life, but many species spend only a portion of their life cycle in water. Reproductive patterns and development rates of the various species are influenced by physical characteristics of the wetland such as water depth, temperature, water clarity, and salinity. Thus, annual and seasonal changes in these characteristics are important in determining the distribution and abundance of insects (Lammers 1976). These relationships have not been studied to a large extent, and the ecology of many species is virtually unknown.

One of the most spectacular concentrations of insects occurs when midges, belonging to the group of insects called Chironomidae, emerge from a wetland. Large clouds of adults can sometimes be seen swarming over the water, with flocks of swallows, terns, and gulls actively pursuing the abundant food supply.

Adult midges live only a few days, just long enough to reproduce. The eggs hatch, and most larvae mature in the bottom sediments. Some, referred to as bloodworms, are bright red in color. Wetlands may have as many as 50 or more species of midges and may contain more than 50,000 larvae per square yard (Coffman 1978). Midge larvae feed on algae and detritus, and, in turn, are eaten by fish, amphibians, grebes, and diving ducks. Adult midge swarms are often mistaken for mosquitoes, which are abundant in wetlands but rarely so abundant as midges.

Dragonflies and damselflies, close relatives, are among the most beautiful of the aquatic insects. They are highly efficient predators and capture mosquitoes and other insects in flight. The different species of adults fly at different levels above the wetland, feeding on different kinds of insect prey. In turn, they are often taken as food by birds and amphibians.

Larval dragonflies and damselflies, referred to as nymphs or naiads, live in the wetland attached to plant stems or in bottom sediments. They breathe through special gills in their rectum. Water is drawn in through the rectum and expelled through the gills, which take up oxygen. The expulsion can propel the nymphs through the water. Nymphs are effective predators on a wide variety of invertebrates

179

and small fish. Their lower lip is hinged and armed with hooks or spines. The nymph lies in ambush waiting for prey to move close. It then shoots out its lower lip, capturing the prey and bringing it back to its mouth.

Like other Minnesota ecosystems, wetlands have a very large number of species of insects. Only a few of the most important or conspicuous species have been discussed. Insects play a major role in the flow of energy and cycling of nutrients through their consumption of algae in the grazing food chain and of dead plant and animal material in the detritus food chain. In turn, many species of insects are consumed by predatory insects and by a wide variety of invertebrate and vertebrate predators.

### Non-native species

Most plants and animals have a remarkable ability to spread from one area to another. Seeds may be carried by wind, water, or animals, and animals move on land, water, or in the air. Species not normally present in natural communities that arrive from some distant place are referred to as non-native species or exotics. Often the arrival of non-native species is unexpected. In some instances, however, species are brought into an area by humans. Two such species, purple loosestrife and carp, have had severe impacts on Minnesota wetlands.

Purple loosestrife is an aquatic plant that was introduced from Europe in the early 1800s (figure 7.15). It spread quickly throughout the northeastern United States and adjacent Canada, and today many wetlands are totally covered by its purple flower spikes (Stuckey 1980). Individual plants can be 6 or 7 feet tall and produce tens of thousands of seeds each year. Seeds and pieces of broken stems, which root readily, are easily transported to new wetlands by flowing water or in mud on the feet of birds. Purple loosestrife was first discoverd in Minnesota in 1924. It was probably introduced for horticultural planting by a well-meaning garden club (Dunevitz and Bolin 1990).

The organisms that suppress or eliminate loosestrife either were not introduced along with the plant or were not able to survive in North America. As a result, loosestrife has few natural enemies, and individual plants grow rapidly, forming large clumps up to a yard in diameter. These clumps may eventually crowd out native plants such as cattail and bulrush (Reinartz, Popp, and Kuchenreuther 1987). When purple loosestrife displaces native plants, the wetland loses its attractiveness to many species of wildlife. Muskrats do not appear to use loosestrife for food and prefer cattail for building their houses. Ducks seem to avoid wetlands that have become dominated by loosestrife.

Once loosestrife has taken over a wetland, eradication is extremely difficult. Pulling or digging individual plants and spot applications of herbicide seem to be the only control methods that have proven effective (Reinartz, Popp, and Kuchenreuther 1987). Research is in progress, however, to test biological control methods by introducing several kinds of insects, including a European weevil that burrows into loosestrife roots and feeds on the soft tissue, eventually killing the plant. Before such an introduction is made, studies must be conducted to be sure that the weevils will not destroy native plants as well as purple loosestrife.

The other non-native species that has had an impact on wetlands is the carp. Carp are the largest members of the minnow family in North America, frequently weighing over 20 pounds. Prized as food by European immigrants, carp were stocked in many states. They were introduced from Europe in the 1870s, and the first stocking in Minnesota occurred in 1883, although carp were already moving into Minnesota via the southern part of the Mississippi River (Eddy and Underhill 1974). Carp are extremely hardy and have spread through most of the United States, occupying fertile, warmwater and muddy-bottomed lakes and wetlands.

**Figure 7.15** *Purple loosestrife, introduced from Europe for flower gardens, is invading many wetlands. It chokes out native wetland plants and creates a monoculture that has minimal value for wetland animals. (Photo by Richard Hamilton Smith.)*

Spawning occurs in late April or May, and a large female may produce over two million eggs. The eggs hatch quickly, and the young carp may reach a length of 8 inches or more during the first year (Eddy and Underhill 1974).

Young and adult carp feed on both plant and animal matter. They move along the bottom of the wetland, uprooting vegetation and stirring up a cloud of mud and debris. They then move through the area again, eating whatever is available. Carp are both grazers and secondary consumers, taking a large number of aquatic insects (Moyle and Kuehn 1964). Because they eat such a wide range of food, they are extremely efficient at producing biomass. S. Eddy and J. C. Underhill (1974) found that very fertile lakes and wetlands may support 1,000 pounds of carp per acre. This is far higher than the populations of game fish that our most productive lakes can support.

Today carp are considered detrimental to Minnesota's lakes and wetlands because they destroy aquatic vegetation and increase the turbidity of the water, which reduces photosynthesis and plant growth. These actions make the wetlands less desirable to ducks for breeding and feeding and interfere with the spawning and feeding of game fishes.

Intensive efforts have been made to remove carp by seining or trapping, but the effort is rarely totally successful. Migration barriers are sometimes constructed on inlets and outlets to prevent movement of carp into wetlands. Winterkill is common in shallow areas, and deeper wetlands of a few acres may be poisoned to eliminate carp. Although a number of anglers pursue carp for recreation and food, and large quantities have been harvested and sold by commercial fishermen, Minnesota's wetlands would probably be better off without any of these large minnows.

## Peatlands

"For peat sake, why is the ground so wet?" This title, which announced a meeting about the Red Lake peatlands in Beltrami County in northwestern

Minnesota, is an appropriate question to ask about peatlands. These wetlands are distinguished from other wetland types by the presence of water-saturated soil, called peat, that is composed of the partly decayed remains of plants. These remains accumulate because the decomposing action of bacteria and fungi is slowed by lack of oxygen in the waterlogged peatland environment. Such environments are found primarily in the boreal forest biome with its cool temperatures and short summers. When the cool, wet climate combines with poor drainage, precipitation exceeds evaporation, and the result is the wet condition favorable to the formation of peat.

In Minnesota peatlands occur throughout the coniferous forest region. They are estimated to cover about six million acres and contain the largest peat deposits in the United States outside of Alaska. The Red Lake peatland lying north and west of Upper and Lower Red Lakes covers an area about 50 miles long and 9 miles wide uninterrupted by streams or roads and represents the most pristine ecosystem in Minnesota (Glaser et al. 1981; Glaser 1987).

### Formation

After the glaciers retreated from Minnesota, the climate is believed to have been cool and moist for a few thousand years. Then it became warmer and drier, resulting in the spread of prairies and savannas to the north and east. During this time the bed of Glacial Lake Agassiz (figure 1.12) probably contained mostly prairie with scattered wetlands that dried out periodically, preventing the formation and accumulation of peat.

When the climate changed to a more cool and wet regime about 4,000 years ago, the prairie vegetation was replaced by forests, and wetlands grew larger and deeper. Poor drainage on the level basin of Glacial Lake Agassiz and the cool, moist climate slowed decomposition, resulting in the accumulation of peat.

Most peatlands in Minnesota formed by two processes: lake infilling and swamping. In the first process, peatlands formed in shallow ponds and ice-block lakes that were gradually filled in by the growth of floating mats of sedges and other plants and the accumulation of organic matter in the sediment. Although this successional process is very slow, as evidenced by the fact that the original ponds and lakes were formed following glaciation about 10,000 years ago, one can still find many examples of peatlands that illustrate the various stages in the successional process. A typical cross section of a northern Minnesota peatland (figure 7.16) shows how the floating sedge mat comes together with the accumulated organic matter on the bottom to eventually form a solid bed of peat. Sometimes a forest consisting primarily of black spruce and tamarack develops on the peat.

Peatland waters are typically cold, have low oxygen content, and are acidic. Under these conditions, decomposition of plant remains is very slow, and parts of individual plants may be identified in the accumulated peat. As shown in figure 7.16, sedge peat, composed mainly of sedge roots and stems, is produced under the sedge mat. Sphagnum

peat, composed mainly of moss, is found under the shrub zone, and woody peat, composed of roots and branches, occurs under the trees. Such an arrangement of layers of peat under a black spruce forest tells us that the site was once a pond or lake.

In the second process, swamping, or paludification as it is sometimes called, flat or gently sloping ground became swamped by mosses and sedges and accumulating peat. Like lake infilling, swamping depends on the rate of production of organic matter being higher than the rate of decomposition.

To understand paludification, imagine a shallow water-filled basin with a constant discharge through an outlet, a constant input of precipitation, and constant evapotranspiration. As long as the basin is full of water, the water table within the basin is flat. Once the basin is filled with sediments and peat, water can no longer flow freely to the outlet but must move through the peat and sediments. Locally the water ponds or increases in depth very slightly, until the pressure balances the resistance to its flow to the outlet. Decomposition slows there, more peat accumulates, and the ponding

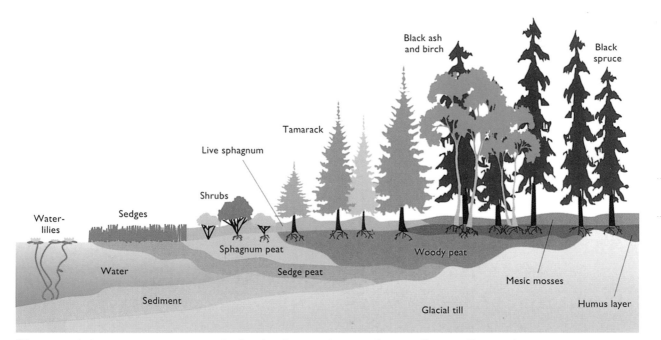

**Figure 7.16** *Stages of succession in a peatland with a floating sedge mat. The mat will eventually cover the open water, and the accumulation of organic matter on the bottom will form peat on which the forest will develop. Bog D, discussed in chapter 1, shows most of the stages diagrammed here.*

shifts to other parts of the basin. Over time, this process continues through the entire basin, resulting in a gradual raising of the peat surface (Edward J. Cushing, personal communication).

Several kinds of moss belonging to the genus *Sphagnum* (figure 7.17) grow on peat. *Sphagnum*

**Figure 7.17** *Mosses belonging to the genus* Sphagnum *are very common in Minnesota peatlands. These mosses, which can hold great quantities of water in their tissues, play a major role in the formation of peatland features. Insects are captured in the leaves of the carnivorous pitcher plant, shown in the foreground, where they are decomposed by digestive enzymes and bacteria. Thus nutrients from the insect tissue are made available to the pitcher plant. (Photo by Blacklock Nature Photography.)*

mosses have a slow decay rate, and their tissue consists of a mesh of large dead cells alternating with tiny live cells. The dead cells are empty and perforated and thus soak up and hold a great quantity of water. The absorbency of *Sphagnum* mosses is so great that they actually draw water upwards and retard the drainage flow of water.

In the mineral-soil upland around the edge of the basin, the water table also rises to keep pace with the peat growth, and peat-forming vegetation invades the new waterlogged soil. The effect is that the peat, as it spreads outward and maintains its slope, appears to grow uphill (figure 7.18). The sloping peatland adjacent to Cedar Creek in Anoka County and the very gentle slopes of the Red Lake peatland toward the rivers that drain it are examples of this kind of paludification.

These raised peat surfaces, or bogs, no longer receive runoff from mineral soil. Thus, the plants are no longer able to obtain nutrients from mineral soil and must rely solely on input from rain and snow. The water in bogs is ombrotrophic, meaning it is characterized by low levels of nutrients and relatively high acidity.

Peatlands that receive runoff from mineral soil are called fens. Their waters, referred to as minerotrophic, contain comparatively high levels of nutrients and are less acidic. These differences in water chemistry and flow between bogs and fens (table 7.2) are the keys to the patterns of vegetation types in some of the peatlands in northern Minnesota (Heinselman 1970). The complexity of these peatlands is dramatically illustrated in the aerial photograph of a portion of the Red Lake peatland (figure 7.19).

*Classification*

In a classification of peatland communities presented by P. H. Glaser (1987), ombrotrophic sites are classified as either nonforested bogs or forested bogs, and minerotrophic sites are classified as either rich fens or poor fens. The water table is very close to the surface in nonforested bogs, with pools forming in every depression. Water draining from

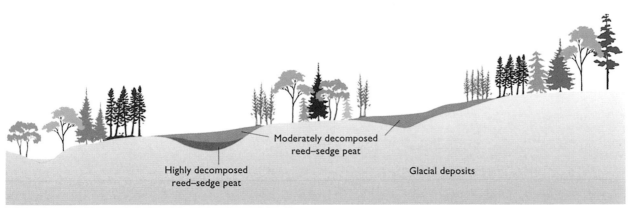

Initial paludification

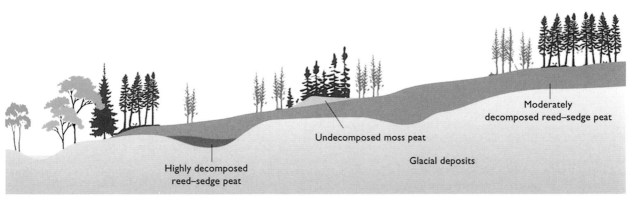

Advanced paludification: beginning accumulation of sphagnum moss peat

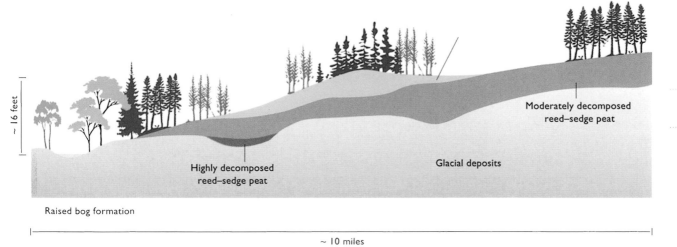

Raised bog formation

~ 10 miles

**Figure 7.18** *Raised bogs are formed by the process of paludification, in which peat and sediments retard drainage. Note that the vertical scale in this illustration is greatly exaggerated. (Modified from Minnesota Department of Natural Resources 1982, by permission.)*

**Table 7.2** *Characteristics of three peatland habitats*

|  | Fen | Poor fen | Bog |
|---|---|---|---|
| pH | 5.8–6.7 | 4.3–5.8 | 3.2–4.2 |
| Calcium (ppm) | 10–30 | 4–12 | 0.6–2.3 |
| Magnesium (ppm) | 1.1–2.8 | 1.2 | 0.2 |
| Peat depth (in) | 12–70 | 118–300 | 35–354 |
| Water movement | strong | sluggish | blocked |
| Water table | high | below surface | below surface |
| Depth to water (in) | -10 to +1 | +9 to +17 | +24 to +27 |
| Plant species diversity | moderate to high | moderate to high | very low |
| Dominant species | white cedar | tamarack | black spruce |
|  | balsam fir | black spruce | bog laurel |
|  | paper birch | bog birch | leather-leaf |
|  | alder | bog rosemary | Labrador tea |
|  | red osier dogwood | cattail | cotton grass |

Sources: Data from Heinselman (1970); Glaser (1987); and Glaser et al. (1981).

the raised bogs downslope to *Sphagnum* lawns results in the formation of bog islands in the midst of water tracks, producing some of the patterns shown in figure 7.19. The number of species of plants in nonforested bogs is lower than in other peatland sites; *Sphagnum* mosses and a number of shrubs such as leather-leaf, bog laurel, and Labrador tea are the most common. Small clumps of stunted black spruce also occur.

On forested bogs the water table is deeper, providing a drier site for tree growth. Black spruce and tamarack (figure 7.20), tolerant of the acid conditions and low levels of nutrients, dominate the forested bogs.

Rich fens occur in minerotrophic water tracks. Because the fens are less acidic and contain more nutrients, they are much richer in species diversity. Three landforms in rich fens are described by

**Figure 7.19** *Water tracks and forested islands show the complexity of habitats in peatlands. (Photo courtesy of Minnesota Department of Natural Resources.)*

Glaser and co-workers (1981). Elongated pools characterized by sedges occur oriented across the slope and across the path of water flow. The ridges between the pools, where bog birch, bog rosemary, and leather-leaf are common, are called strings.

A third landform, long, narrow islands shaped like teardrops, may develop from sedge meadows, or they may be isolated by the spreading out of a water track (figure 7.21). Teardrop islands have their blunt end pointing upstream with a tail of bog birch extending downstream. H. E. Wright Jr. (1984) has aptly described areas containing teardrop islands as resembling a great armada of ships cruising on a wavy sea. Tamarack and black spruce, surrounded by hummocks of sedges and mosses, grow on the islands. These hummocks often cover fallen trees, which are thus incorporated into the peat.

A third major category, the poor fen, is described by Glaser and co-workers (1981) as occurring where minerotrophic and ombrotrophic sites meet. Vegetation is more diverse in these transition areas than in fens or bogs. Characteristics of fens, poor fens, and bogs are given in table 7.2. Note the striking differences in pH and in depth to the water table in these three peatland habitats.

### Succession

The question of whether succession occurs in peatlands has received considerable attention but is not yet resolved (Conway 1949; Heinselman 1970; Glaser 1987; Glaser et al. 1981; Gorham 1990). Certainly, the buildup of organic matter eventually changes the physical conditions under which the original plant community developed. When the peat deposit on a rich fen is raised above the groundwater table, mineral nutrients from the fen do not reach the surface of the peat. The site is then ombrotrophic. A common sequence in Minnesota is from a sedge-dominated fen to a tamarack-dominated fen to a *Sphagnum*-dominated bog (Gorham 1990). A forest of black spruce may then develop on the bog (Glaser and Janssens 1986). The number of vascular plant species is reduced from 195

**Figure 7.20** *Tamarack, the only Minnesota conifer that loses its needles every fall, is one of the dominant trees in forested bogs. (Photo by Jim LaVigne.)*

in rich and poor fens to only 23 in ombrotrophic bogs (Wheeler et al. 1983).

This process, which can occur in a timescale of decades to a few centuries, seems to fit the concept of succession. M. L. Heinselman (1970), however, avoids the use of the terms succession and climax. He suggests that rather than a trend toward establishment of a terrestrial forest, there has been a general swamping of the landscape brought about by a rise in the water table. This swamping has resulted in a deterioration of tree growth and a diversified landscape. He cautions, however, that there are limits to paludification brought about by the permeability of the peat and the steepness of the water-table slopes.

### Vegetation

Black spruce and tamarack, the most important tree species in peatlands (Glaser 1987), are well adapted to growth on the rapidly rising peat surface. Their lower branches develop roots when they contact the wet moss, a process called layering,

187

**Figure 7.21** *Islands of black spruce and tamarack dot the vastness of Minnesota's Red Lake peatland. (Photo courtesy of Dan Ruda, Minnesota Department of Natural Resources.)*

and new roots grow from the trunk as the moss increases in depth. These roots then replace the deeper roots, which may be killed by the rising water table. Black spruce also may sprout new trees from its roots and even from a blown-down trunk. Trees growing in a row may have all originated as sprouts from a fallen trunk (Glaser 1987).

Many other peatland plants have adaptations to deal with the acidic, waterlogged substrate in which they grow and with the extreme temperatures and low nutrients in their environment. Most shrubs, members of the heath family, have thick, leathery leaves densely covered with hairs to help prevent water loss and to conserve nutrients (Glaser 1987). Although water conservation would seem unnecessary in a peatland environment, frequently the top portion of the plant is growing while its roots are locked in ice. Even when the ice has thawed, the acidic, oxygen-poor water may be difficult for plants to use.

The steep temperature gradients in the bog in summer, when the surface layers are warm and deep layers are frozen solid, are due in large part to the insulating properties of *Sphagnum* mosses. *Sphagnum* mosses are responsible also for increasing the acidity and reducing the nutrient content of bog sites through their ability to absorb calcium, potassium, and other positive ions from their surroundings (Curtis 1959). The negative sulfate, chloride, and nitrate ions then combine with hydrogen to form acids. *Sphagnum* mosses help to perpetuate an environment favorable to their growth: low temperatures and high rainfall are favorable to their development, their insulating properties enhance the temperature gradient, and their physiological processes reduce the nutrient and oxygen content of the water, making the site even more favorable for *Sphagnum* growth.

A lichen, called old man's beard, is often found hanging like Spanish moss from dead black spruce and tamarack branches. Old man's beard grows best in moist environments and obtains nutrients from rain, snow, and dust in the atmosphere. This lichen is very sensitive to sulfur and nitrous oxides, which are common components of air pollution. Thus, the lichen serves as a biological indicator of air quality.

Some plants growing in the nutrient-poor bog are able to capture and digest insects, enabling them to gain additional nutrients. The carnivorous pitcher plant (figure 7.17), with its long trumpet-shaped leaf, is a common and conspicuous member of the bog community. Fringes of downward-pointing hairs inside the leaf prevent the unwary insect that is attracted to the color and smell of the sweet solution in the bottom of the cup from crawling out. Translucent spots on the leaf appear as windows through which the insects try to escape, only to fall back into the pitcher.

When an insect falls into the fluid at the bottom of the leaf, digestive enzymes and bacteria rapidly decompose its body and make the nutrients in the insect tissue available to the plant (Mellichamp 1978). Although this process is commonly believed to be important to the plant, the actual need of the pitcher plant to obtain nutrients by carnivory is in question. One of the closely related species of pitcher plants in the southeastern United States showed no increase in productivity when the plants were fertilized (Eleutarius and Jones 1969). In another study, N. L. Christensen (1976) fed insects to pitcher plants and found no increase in lithium, magnesium, or potassium in plant tissues. Levels of nitrogen and phosphorus, however, were higher than in unfed controls. It is possible that carnivory may be important under conditions of nutrient stress or for providing trace elements that might be lacking in the bog environment.

The interior of the pitcher leaves creates a microhabitat. Compared to the surrounding environment, the cavity within a pitcher is typically higher in relative humidity, lower in light intensity, and somewhat less variable in temperature. In addition, pitchers contain a decomposing mass of trapped prey, which is a potential food source for other organisms. A number of species, such as mosquitoes, flies, moths, aphids, and mites, are able to live in this environment without being either trapped or digested.

### Ecosystem function

Bogs receive all of their nutrients from precipitation, whereas fens obtain nutrients from mineral soil as well as from precipitation. The average aboveground productivity of nine northern non-forested fen sites ranged from about 1,000 to 9,000 pounds per acre per year, and root production ranged from 1,300 to 4,600 pounds per acre per year (Reader 1978). Annual production of *Sphagnum* mosses in six Minnesota bog forests ranged from 3,000 to 4,600 pounds per acre (Grigal 1985). These values fall within the range of productivity of other kinds of forests in Minnesota but are much lower than the productivity of sugar beets (table 4.1). Although adequate moisture is available in most peatland forests, nutrients are often limited.

Studies of energy flow in peatlands indicate that only a small percentage of the net primary production is consumed in the grazing food web. The remainder enters the detritus web, where decomposition is slowed by the cold temperature, acidic conditions, and low levels of oxygen in the water-logged substrate (Moore and Bellamy 1974). In a study of the uppermost 4 inches of the sedge mats in bogs in Hubbard and Clearwater Counties, G. A. Leisman (1953) found that about 5,500 pounds of organic matter were deposited per acre every year. The slow decomposition of this organic matter leads to further peat accumulation and slows the recycling of nutrients in a system that is already nutrient limited. Carbon, hydrogen, oxygen, and nitrogen may be sequestered in peat for thousands of years, far longer than their normal cycles (Gorham 1990).

Decomposition in peatlands has implications for the greenhouse effect and global warming. Methane, a greenhouse gas, is currently produced from the activities of methane-forming bacteria living in the saturated peat. If global warming occurs and the peatlands begin to dry out, methane emissions will decline, but oxidation of the plant remains will release carbon dioxide, also a greenhouse gas. Peatlands will thus contribute to greenhouse gases in the atmosphere regardless of climate change.

## Animals and community interactions in peatlands

### Mammals

Many of the animals associated with the northern coniferous forest are likely to be observed at times in the various peatland habitats. Home ranges of many of these species, however, are considerably larger than the individual communities making up the peatland mosaic. The most diverse mammal populations are found in peatland habitats with the greatest diversity of vegetation. Mammal diversity ranged from 10 species in tamarack swamps to 6 in forested bogs (Nordquist 1992).

Woodland caribou were common in northern Minnesota before European settlement but are extremely rare today. The last resident caribou in Minnesota inhabited the Red Lake peatland in the late 1930s (Manweiler 1938; Nelson 1947). Shooting, logging, fires, and settlement all played a role in the decline of the population. Plans to reintroduce woodland caribou into northeastern Minnesota are currently under consideration by the Minnesota Department of Natural Resources (DNR) and other groups. However, the parasitic worm *Pneumostrongylus tenuis*, which is common in deer, is likely to limit the success of such an introduction.

Caribou have large feet with dew claws and with tufts of long hairs between the hooves, enabling them to move easily over snow and peatland. For most of the year they eat woody browse supplemented by lichens, sedges, and fungi.

Moose are common in some peatland habitats. Willow, alder, and birch growing in fens are used

for food, and both upland and lowland forests provide cover.

Wolves and mountain lions are known to make extensive use of peatlands. Although sightings of the secretive mountain lion are increasingly frequent, the number of mountain lions in Minnesota is probably very low.

Shrews and voles are common small mammals in peatlands. The water shrew has stiff hairs on its hind feet that aid in swimming. It prefers flowing water and feeds on aquatic invertebrates but also uses slugs, earthworms, and spiders taken on land (Whitaker and Schmeltz 1973).

Southern bog lemmings inhabit wet wooded and open areas and are likely to be found in tamarack or spruce forests with a moss understory. They are primarily herbivorous, feeding on leaves and stems as well as on fruits and roots. They nest in clumps of moss and may have several litters of one to five young per year. Bog lemming populations are known to fluctuate markedly, but no population estimates are available for Minnesota peatlands.

### Birds

Many species of birds use bog and fen habitats in Minnesota. In a variety of communities in the Red Lake peatland, 70 species were observed during the breeding season (Warner and Wells 1980). Only 4 species were observed in nonforested bogs, compared to 32 species in cedar and spruce bog forests. The numerous layers, or strata, in the forests provide more habitats for birds than the uniform shrub layer in the nonforested bogs.

Sandhill cranes, which once nested in wetlands throughout the state, now nest only in areas where they are isolated from human disturbance. Many peatlands provide this isolation, and sandhill cranes nest in them every spring. Several bird species, such as the yellow rail and the great gray owl, which are classified as threatened or rare (Coffin and Pfannmuller 1988), use peatlands during certain seasons.

Two of Minnesota's warblers, Connecticut and palm, breed primarily in bogs and fens. Their nests of fine grasses and sedges are located in the tops of

*Sphagnum* hummocks in tamarack and spruce forests. LeConte's, savannah, and Lincoln's sparrows are likely to be found in more open, shrub peatland habitats (Niemi and Hanowski 1992). Spruce grouse, ruby-crowned kinglets, gray jays, Swainson's thrushes, and Tennessee, Nashville, yellow-rumped, and Cape May warblers are common in peatland forests.

### Amphibians and reptiles

Because of the toxicity of acid water, only a few amphibians and reptiles live in peatlands, primarily in fens. Wood frogs are the most tolerant of high acidity and are able to metamorphose in poor fen sites (Karns 1984). Eggs of the other five species studied by D. R. Karns—blue-spotted salamander, American toad, leopard frog, chorus frog, and spring peeper—did not hatch in bog water but did hatch in fen water. Larvae of all of these species except spring peepers had 0% survival when experimentally transplanted into bog water. Only 7% of the spring peeper tadpoles survived, whereas 100% survived in fen water.

Common garter and redbelly snakes and painted and snapping turtles are the only reptiles reported from Minnesota peatlands (Karns 1992). None of these species are restricted to peatland habitats.

## Other wetlands occurring in Minnesota

Wetlands are found along shorelines of many lakes and rivers and in forests. Bays and shoreline zones of lakes, with zonation patterns extending from shallow to deep water, often appear similar to prairie wetlands. Stands of bulrush, cattail, sedges, and wild rice often characterize these wetlands, but a wide variety of aquatic plants and animals are usually present. Although these wetlands appear similar to those on the prairie, their water chemistry may be quite different. Wind and wave action frequently mix the lake waters and wetland waters. Thus, the wetland waters have the same concentrations of chemicals as the lake.

Linear-shaped wetlands sometimes occur along the shorelines of rivers, and isolated, looping segments of meandering rivers, called oxbows, often have wetland characteristics. Periodic flooding by the river results in rapid changes in water levels in these wetlands, and sediments deposited by the floodwaters provide important nutrients for wetland plants. Trees, such as elm, willow, silver maple, and black ash, may become established in river bottomlands, and sometimes border rivers for many miles.

Forest wetlands (figure 7.22) are considered to be wetlands that occur in deciduous or northern coniferous forests on mineral soil or on shallow peat. Little research has been carried out on forest wetlands in Minnesota, perhaps because they are considered to be of minor importance for wildlife or because they have low commercial value.

Many forest wetlands originated as depressions in glacial moraine or outwash topography, and they are therefore similar in age to prairie wetlands. In Minnesota, much of the deciduous forest was cleared for agriculture. Wetlands existing in these forests were drained to create cropland. In the northern coniferous forest, however, little drainage has occurred, and the distribution of wetlands is probably similar to what it was before settlement, except for the natural changes that have occurred because of succession.

The hydrology of these wetlands is similar in most respects to that of prairie wetlands. Surface water is present all year in some, but many dry up in late summer. Aeration then takes place in the upper soil layer. In addition to this annual pattern, long-term changes in water levels occur because of drought and years of abundant precipitation.

Zonation is not usually as well defined in forest wetlands as in prairie wetlands. As a result, the commonly used wetland classification schemes are not particularly helpful in discussing these habitats. The successional pattern in Minnesota's forest wetlands, as presented by N. E. Aaseng and co-workers (1993), will be used to describe the communities that can be observed today.

Wet meadow, the pioneer stage, is an open com-

**Figure 7.22** *Little is known about forest wetlands in Minnesota, although they are common in both deciduous and northern coniferous forests. (Photo by Bob Firth, Firth PhotoBank, Minneapolis.)*

munity dominated by sedges or sedges and grasses growing on wet mineral soil, muck, or shallow peat. Wide-leaved sedges, blue-joint grass, and bog reed grass may form dense, closed stands. Herbs likely to be present are joe-pye weed, mint, and swamp milkweed. Willows may grow around the edges of the wet meadow, but mosses are rare. Fires will maintain the wet meadow stage, whereas drought, drainage, or absence of fire will probably lead to an intermediate stage of succession, the shrub wetland.

Two types of shrub wetlands, alder swamp and willow swamp, are considered to be midsuccessional in Minnesota. Speckled alder is the dominant shrub in alder swamps, and several species of willow are the most abundant shrubs in willow swamps. However, both alder and willows, plus red osier dogwood, are likely to be present in shrub wetlands. The shrubs may form a continuous canopy over the wetland, or the canopy may be interrupted by light gaps. Alder has the ability to take nitrogen from the atmosphere. This occurs through the action of the nitrogen-fixing bacteria in the bright orange nodules on the shallow roots. Sedges, blue-joint grass, cattails, marsh fern, and jewelweed are the common herbs.

Shrub wetlands will persist for many years in the absence of fires, drought, or floods. Such disturbances are relatively common in Minnesota, however, and may set back succession to the wet meadow stage. If the shrub wetland is not disturbed, trees slowly begin to invade the community around the periphery, signaling the beginning of the next stage, the swamp forest.

In the deciduous forest region, the swamp forest trees include black ash, paper birch, yellow birch, American elm, slippery elm, trembling aspen, green ash, and red maple. Tamarack, white cedar, and white pine may also be present. Alder is often found in the shrub layer, and sedges and ferns may be abundant. Windthrow is common in these communities, probably because of the saturated soil and shallow root systems of some of the trees.

In the northern coniferous region, the swamp forest is dominated by black spruce, tamarack, or white cedar. Standing water may be present, but often the water table drops in late summer and the soils become aerated. On sites protected from fire and with high levels of nutrients, white cedar will form extensive stands. Sites with lower levels of nutrients and a more acidic substrate are characterized by dense stands of tamarack or black spruce.

Standing crop biomass and productivity of some forest wetlands are high because of the availability of water and the accumulation of nutrients in runoff from surrounding uplands. W. A. Reiners (1972) reported aboveground biomass of 90,000 pounds per acre and net primary productivity of 6,500 pounds per acre per year in a swamp forest in Anoka County (table 4.1). Black ash, white cedar, American elm, yellow birch, and speckled alder were the dominant species in this late-successional community.

Forest wetlands are usually small in size, but some in the northern coniferous forest region may be several hundred acres. As a result, animals present are likely to be those found in the surrounding forests. Deer, moose, coyote, wolf, red fox, bobcat, lynx, several species of weasels, skunk, raccoon, bear, and porcupine are large to medium-sized mammals characteristic of forest wetlands (Berg 1992). Small mammals include several species of shrews and voles, bog lemming, deer mouse, and

jumping mouse (Nordquist 1992). Many species of birds may use forest wetlands for brief periods, but the predominant species are sedge wren, common yellowthroat, and swamp, LeConte's, and clay-colored sparrows (Niemi and Hanowski 1992). All species of frogs and toads resident in forest communities might be found in forest wetlands during the breeding season.

## Present status of wetlands

Throughout history, humans have placed little value on wetlands. Americans have repeatedly passed legislation encouraging drainage and the economic development of wetlands, particularly for agricultural use. Only forest wetlands seem for the most part to have escaped the pressures of development, perhaps because they have little commercial value. Today the DNR (Tom Landwehr, Section of Wildlife, Minnesota Department of Natural Resources, personal communication) estimates that over 90% of the wetlands in the prairie have been lost, from about 40% to 60% in the deciduous forest, and less than 5% in the northern coniferous forest (figure 7.23).

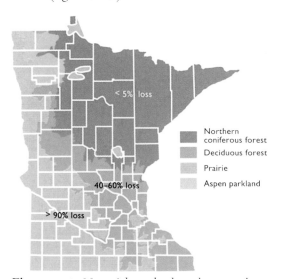

**Figure 7.23** *Most of the wetlands in the prairie biome have been drained. Many deciduous forest wetlands were also drained after the forests were cleared for agriculture. (Tom Landwehr, Section of Wildlife, Minnesota Department of Natural Resources, personal communication.)*

*Prairie wetlands*

Like the prairie uplands, many of Minnesota's original prairie wetlands no longer exist, and most of those that remain in agricultural areas are threatened by drainage. From the farmer's perspective, drainage has usually been beneficial. From the perspective of many others—the nature photographer, the bird-watcher, the trapper, the duck hunter—the decrease in wetlands has been detrimental.

Until recently most efforts to protect wetlands have been led by people interested in waterfowl. Government agencies, private landowners, and sportsmen's groups have implemented programs to protect wetlands from drainage and to manage wetlands to enhance production of ducks and geese (Thompson and Sather 1970). Fortunately, these people have been joined by others in recent years, and wetlands are now recognized as being valuable resources with many benefits. The federal Swamp-buster and Clean Water Acts have aided in wetland preservation efforts.

Although it is difficult to assign a dollar value to wetland benefits, they are vitally important in evaluating the contributions of wetlands to our society. R. M. Darnell (1979) divided the environmental values of wetlands to society into four categories: goods, services, esoteric values, and maintenance values. Wetlands have provided civilizations with oxygen, water, fish, wildlife, fibers, pharmaceuticals, and other goods. Recent studies in Minnesota, initiated in response to the depletion of fossil fuels, have examined the use of cattails as an energy source (Dubbe, Garver, and Pratt 1988).

Wetlands assimilate heavy metals, pesticides, and other wastes (Kadlec and Kadlec 1978), remove carbon dioxide, sulfur dioxide, and nitrous oxide from the atmosphere, and add oxygen. Wetlands mineralize sewage effluents, remove nitrogen

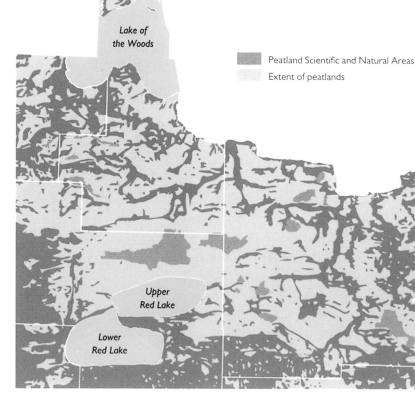

**Figure 7.24** *The major concentration of peatlands in Minnesota lies in the region north and east of Upper and Lower Red Lakes. Only a small amount of disturbance due to human activities has occurred in these peatlands, and 22 sites are permanently protected as Scientific and Natural Areas. (Modified, by permission, from an unpublished map, Minnesota Department of Natural Resources, Natural Heritage and Nongame Research Program; see page 324 for complete permissions information.)*

Lake of the Woods

Peatland Scientific and Natural Areas

Extent of peatlands

Upper Red Lake

Lower Red Lake

and phosphorus from runoff, trap sediments, augment stream flow during droughts, recharge aquifers, and store groundwater, thus reducing flooding. Flooding along the Minnesota River and the Red River of the North is undoubtedly more frequent now than in the 1800s because so many wetlands have been drained. Runoff now moves quickly to the rivers rather than being held in wetlands.

The esoteric values of wetlands include their high species diversity and their enhancement of the quality of life. They provide opportunities for recreation, education, research, and aesthetic pleasure (Errington 1957).

Darnell's fourth category refers to the fact that wetlands are self-maintaining. They do not require energy or human input to provide their values.

Public attitudes toward drainage of wetlands appear to be changing. The Swampbuster provisions of the Food Security Act of 1985 eliminated indirect federal subsidies for drainage by denying farm program benefits to farmers converting wetlands to croplands. The concept of "no net loss of wetlands" is being spoken of in national political circles and by Minnesota legislators.

In the final analysis, the advantages of draining wetlands for development must be weighed against their many other values. The remaining wetlands in Minnesota must be treated as valuable resources and must be managed on the basis of sound knowledge of the structure and function of the wetland ecosystem.

## Peatlands

Drainage ditches running east-west and north-south, without regard to the natural drainage pattern, crisscross many of northwestern Minnesota's peatlands every mile or two. These ditches were constructed in the early 1900s in an attempt to encourage farmers to homestead (Soper 1919). Advertisements placed in newspapers to entice settlers to come to the peatlands described the area as needing little or no clearing and as having an ideal climate and rich soil. A few homesteads were established, but most of the settlers soon abandoned their farms when the drainage ditches proved to be ineffective more than a few feet from the ditch bank. Although the attempt at drainage and settlement was a financial fiasco, little damage was done to the peatlands. Today the ditches are used by beaver, ducks, and other wildlife, and most peatlands remain undeveloped.

Interest in Minnesota's peatlands increased markedly in the 1970s in response to the world energy crisis. Conversion of peat to synthetic natural gas and use of peat as fuel were given serious consideration by industries and state agencies. In addition to its fuel value, peat contains organic and inorganic molecules that can be used to make metallurgical coke, activated charcoal, ethyl alcohol, waxes, oil, and other products (Botch and Masing 1983). Although peat is used to produce a variety of products in other parts of the world, in North America peat has been used primarily for horticultural purposes. Other peat products are as yet unable to compete economically with similar products derived from other materials.

Because the peatlands are the least disturbed of Minnesota's ecosystems, they provide viable habitat for their native plants and animals. Nevertheless, 43 peatland species are listed as endangered, threatened, or of special concern. Three of the 17 listed mammals, 7 of the 27 birds, and 25 of the 191 plants are associated with the peatlands (Pfannmuller and Coffin 1989).

In a strong move toward peatland protection, the Minnesota legislature passed the Wetland Conservation Act in 1991. This act designated 146,239 acres in 18 peatlands as Scientific and Natural Areas (figure 7.24). Only four peatlands in Minnesota had previously received protection, due to their location in wildlife management areas or refuges. Passage of this act represents one of the most significant efforts to protect peatlands worldwide. A valuable part of Minnesota's peatland ecosystem, with its rare plants and animals and complex patterns, is now preserved for the future.

195

# L A K E S

⑧

Cool, clear water, swimming, walleye fishing, the haunting cry of a loon—Minnesota's many lakes provide recreation and aesthetic pleasures for many people (figure 8.1). Nearly 25% of the outdoor recreation hours of Minnesotans are spent fishing, swimming, or boating (Minnesota Department of Natural Resources 1990). Our lakes also attract thousands of visitors each year and are a source of pride for both urban and rural residents. Lakes are plentiful in Minnesota, but less then 0.1% of the world's water is in lakes (Cole 1983). This fact explains in part why lakes are a highly valued resource, and underlines the importance of preventing their degradation.

## Formation and distribution

Lakes in Minnesota were formed primarily by erosion and deposition during glaciation (Wright 1989; Zumberge 1952). Ice gouged and scraped the surface of the earth, leaving many depressions that eventually filled with water. Buried ice blocks melted long after the glacier retreated, and the resulting depressions also became lakes. In some areas, glacial till, moraines, or the ice itself blocked natural drainage pathways much like a modern dam.

The action of rivers can result in the formation of lakes. As a river winds its way along, it sometimes creates oxbow lakes by pinching off a loop

**Figure 8.1** *Many Minnesota lakes are lined with cabins, homes, and resorts. Storm clouds, reflected in the calm water, predict imminent rain. (Photo by Bob Firth, Firth PhotoBank, Minneapolis.)*

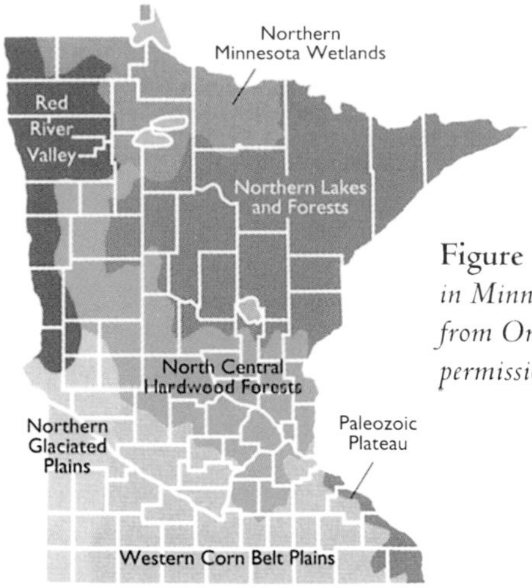

**Figure 8.2** *Ecoregions in Minnesota. (Modified from Omernik 1987, by permission.)*

in its channel. Oxbow lakes are quite common on meandering rivers flowing through relatively flat topography, such as along the Mississippi River in Aitkin County. Rivers may also form lakes when blockage occurs in their channel. Lake Pepin was formed when the delta of the Chippewa River partially blocked the main flow of the Mississippi River. The water backed up from this point northward and formed the lake.

Lakes are also formed by human activities. Because people think that lakes enhance the landscape, developers sometimes create them. Other lakes are created or enlarged when dams are constructed to provide flood controls on major rivers. For example, water levels in Lake Winnibigoshish increased when the flow of the Mississippi River was impeded by the construction of the Federal Dam.

Beavers also create lakes by constructing dams, but most of these lakes are temporary, lasting only until the dam breaks. Many are only a few acres in size, but some exceed 30 acres and may be 6 to 8 feet deep (Zumberge 1952).

Minnesota's many lakes are not evenly distributed throughout the state. Bordering the northeastern corner of the state is Lake Superior, the largest freshwater lake in the world. Numerous other lakes are located in the northern coniferous forest biome. Here many of the lakes are long and narrow with steep, rocky shorelines. In the Boundary Waters Canoe Area Wilderness lakes are so close to each other that we can travel by canoe

throughout this area. The deciduous forest biome contains thousands of lakes, many with gently sloping, marshy shorelines. Fewer lakes are present in the unforested part of Minnesota. Most of these lakes are quite shallow, but some are very large. In the southeast, few lakes exist because the last glacier did not cover the area and no moraines, with their lake-forming depressions, are present. In the Red River valley in the northwest, lakes are rare because few water-holding depressions exist in the bottom of former Glacial Lake Agassiz.

A lake's origin and its location in the state are only two of the many variables that influence the characteristics of Minnesota's lakes—we cannot describe a typical Minnesota lake. We can, however, categorize the lakes in a general way according to their location in the state. Recent assessments of many lakes in Minnesota have been based on an ecoregion approach (Heiskary and Wilson 1989; Heiskary and Lindbloom 1993). Ecoregions were mapped for the entire United States by the U.S. Environmental Protection Agency (Omernik 1987). The map divides the state into seven relatively homogeneous regions, based on a combination of factors including land use, land surface form, potential natural vegetation, and soils (figure 8.2). Of Minnesota's 12,034 lakes greater than 10 acres in size, 98% occur in four of the seven ecoregions: Northern Lakes and Forests, North Central Hardwood Forests, Northern Glaciated Plains, and Western Corn Belt Plains (Heiskary and Wilson 1989).

I will describe characteristics of a number of lakes in each of these four ecoregions. But first, an understanding of Minnesota's lakes requires a basic knowledge of the lake ecosystem.

## Physical and chemical characteristics

### The watershed and lake chemistry

The ecology of a lake is largely determined by its water chemistry and the physical characteristics of its basin. Water chemistry is determined by the characteristics of the drainage basin, or watershed, in which the lake lies, especially the parent rock material from which the soil developed. The watershed is the area from which water, often called runoff, drains into the lake and includes the drainage area of all the streams and rivers that flow into the lake. Groundwater entering the lake comes from its own watershed, which may or may not correspond to the runoff watershed. Groundwater is especially important to water chemistry because it is more concentrated in many chemicals than is runoff. Water chemistry is modified by other characteristics of the watershed, including topography, size relative to the lake's size, soils, vegetation, groundwater hydrology, land use, and pollution. Thus, much of the variation in Minnesota's lakes is attributable to variations in their watersheds and in the size, depth, and shape of their basins.

Ecologically important chemical parameters in lake water are hardness (calcium and magnesium), alkalinity, pH, phosphorus, sulfate, and dissolved organic matter. With the exception of dissolved organic matter, most chemicals in lake water are derived from the parent rock material in the watershed and are correlated with each other. Especially rich sources of these chemicals are calcium-bearing sedimentary rocks such as limestone and dolomite. Limestone and dolomite dissolve easily and are present in Minnesota both as bedrock and in the glacial tills spread unevenly across the state.

The concentrations of these chemicals in the lake are influenced by other characteristics of the watershed. The topography, soils, and vegetation of the watershed all influence how much of these chemicals dissolve in the runoff, how much of the runoff reaches the lake, and how quickly. Typical watersheds in northeastern Minnesota, for example, have steep slopes and contain igneous and metamorphic rocks, which are resistant to solution and erosion. Thus, water runs off rapidly carrying only small quantities of dissolved minerals to the lake.

The size of the watershed relative to the size of the lake also influences the concentrations of chemicals. Lakes with a large volume of water and a small watershed may have low concentrations of chemicals compared to lakes with a small volume of water but a large watershed. Climate may also affect concentrations. Chemicals may become more concentrated in lakes in areas such as southwestern Minnesota that have a low precipitation rate and a high evaporation rate.

Water residence time, the speed at which water moves through a lake's basin, may affect the length of time required to remove pollutants from the lake after the input of pollutants has ceased. Water residence time varies substantially in Minnesota lakes, ranging from an extremely low time of 0.05 year, or 19 days, in Lake Pepin, part of the Mississippi River in Goodhue and Wabasha Counties, to a more moderate time of 4 years for Lake Calhoun in Minneapolis to more than 25 years for lakes with small watersheds relative to lake size and volume. Residence time is about 25 years in Ten Mile Lake in Cass County and on the order of 30 to 40 years in Lake Mille Lacs in Mille Lacs and Aitkin Counties. These lakes have watershed to lake ratios of approximately 3:1 and 2:1, respectively. In contrast, Lake Pepin has a ratio of 1,225:1 (Steve Heiskary, personal communication).

Because the amounts of chemicals naturally present in a lake are correlated, scientists who study lakes, called limnologists, can get a rough idea of a lake's chemistry by measuring one of the chemical parameters, for example, hardness. Lakes with high amounts of calcium and magnesium are called hard-water lakes, and lakes with low amounts of these chemicals are called soft-water lakes. Hard-water lakes usually have high alkalinity, that is, high content of carbonates, associated with the hardness. They also typically contain more phosphorus than soft-water lakes. Thus, the terms hard-water and soft-water are a shorthand way of characterizing a lake's chemistry.

Similarly, alkalinity is sometimes used as a measure of nutrient availability. Alkalinity is also important because it determines the ability of water to buffer acids that may be introduced to the lake from organic soils in the watershed and from acid rain. Because the acidity of water affects aquatic life, limnologists also measure the pH of water, which is the concentration of hydrogen ions. Lakes with pH values below 7.0 are acidic, and those with values above 7.0 are alkaline.

### Phosphorus and nitrogen

Plants will grow until they run out of a limiting resource. For terrestrial plants, limiting resources are water, space, light, phosphorus, nitrogen, or potassium. For algae and other aquatic plants in lakes, the limiting nutrients are usually phosphorus and nitrogen. Although the concentration of phosphorus and nitrogen in plants is low, they each play a key role in growth.

Phosphorus, in the form of phosphate, enters a lake primarily by way of runoff and rainfall on the lake surface. Human activities often increase the amount of phosphorus that enters the lake from its watershed. Point sources, those discharged at a single point, usually a pipe, include discharges from industrial facilities and municipal wastewater treatment plants. Other sources, called nonpoint sources, include urban stormwater (runoff from roads, lawns, and parking lots) and runoff from agricultural areas (cultivated land, pastures, and feedlots). Phosphorus is also transferred up into the surface layer of water from the lake bottom during mixing of the lake by the wind.

Algae and other aquatic plants grow more rapidly as more and more phosphorus enters the lake. At a certain point the rooted plants decline, and the algae take over until they actually dominate the lake by shading out the other plants (Burgis and Morris 1987). High populations of algae may result in green scums, which are undesirable for swimming and boating. Efforts to decrease the growth of nuisance algae are usually directed to decreasing the amount of phosphorus entering the lake.

Nitrogen plays a somewhat similar role as phosphorus in the productivity of lakes. Efforts to control nitrogen amounts are not common, however, because nitrogen can be obtained from the air by some species of algae. Thus, decreasing the amount of nitrogen entering the lake may not reduce the growth of algae but may only change the dominant species in the lake. Nitrogen can exist in the form of ammonia, nitrate, nitrite, and organic nitrogen in plant and animal tissue.

### Light

Because plants form the base of the lake food web, light for photosynthesis is essential to the functioning of the lake ecosystem. Where light is blocked by algae or sediments, fewer rooted plants will grow. Three factors concerning light are of special importance. The first concerns the quantity of light that falls on the lake surface. Light is greatly reduced in winter because of the shorter days and because snow cover both reflects and absorbs light.

The second factor concerns the depth to which the light penetrates. The transparency, or clarity, of water is a measure of the actual depth of light penetration in a lake. This depth is influenced by the amount of algae present in the water (Schindler and Comita 1972), by the amount of nonalgal suspended particles, and by the amount of organic matter, which absorbs light. Transparency is measured with a Secchi disk, an 8-inch disk, usually all white or with alternating black and white quadrants (figure 8.3), which is lowered into the water until it is no longer visible. This depth is referred to as the transparency of the lake. The fewer algae in the water, the higher the transparency in most cases. Thus, variations in the transparency of a given lake are a good estimate of the changes in the abundance of algae, if other factors such as color and suspended sediment do not vary.

Highly colored lakes often have low transparency but low algal quantities because the color intercepts the light and reduces productivity. This brown color in lakes, often referred to as bog staining, is usually from waters that have drained through organic soils in the watershed, and not from iron as is commonly believed. Color is not indicative of poor water quality. Examples of highly colored lakes are Big Sandy in Aitkin County and Lake Vermilion and Island Lake in St. Louis County.

**Figure 8.3** *The transparency, or clarity, of water is measured by lowering an 8-inch black-and-white Secchi disk into the water until it is no longer visible. This measures the depth of light penetration in the lake. (Photo by Jim LaVigne.)*

The third factor concerns how light affects organisms. Plants need at least 1% of full sunlight for photosynthesis. At and near the lake surface, plenty of light is available. As depth increases, however, more and more light is lost by scattering and by absorption by the algae and other organisms. In clear lakes, high light at the surface may actually inhibit photosynthesis, however, and in such lakes the main algal populations are often in the deeper water.

## Oxygen

The amount of oxygen in lake water, critical to fish and all other organisms, is primarily determined by exchange with the atmosphere, the amount given off by photosynthesis, and the amount used in respiration, including decomposition. Because light is necessary for photosynthesis, oxygen is produced in the lake only as deep as the light penetrates. Below that level, decomposition and respiration use up oxygen and produce carbon dioxide.

In winter, ice may decrease the amount of light penetration. When heavy snow cover blocks the sunlight, little photosynthesis occurs. At such times the oxygen concentration may become so low that nearly all of the game fish and many of the rough fish in the lake die. This kind of winterkill is most common in shallow lakes when oxygen is consumed by a large amount of dead vegetation undergoing decomposition by bacteria.

## Stratification

Most deep lakes stratify, forming layers of water with different temperatures. The stability of this stratification is a function of the depth and size of the lake. Stratification occurs because of the changes in the density of water as it warms and cools. As the lake water cools in the fall, it becomes more and more dense, reaching its maximum density at about 39°F. Surface water that has reached this temperature sinks because it is denser or heavier than the other water in the lake. This process of cooling and sinking continues until the entire lake is at 39°F. As the surface water continues to cool below this temperature, it starts to get lighter again, and it floats at the surface until the temperature drops sufficiently to cause ice to form. The physical change from liquid to solid at 32°F results in the ice being 8.5% less dense than liquid water at the same temperature (Burgis and Morris 1987).

Ice formation and subsequent cooling during winter create a layer of water just under the ice that is colder and lighter than the remaining water in the lake. The water immediately beneath the ice is 32°F, and the temperature increases with increasing depth until it reaches 39°F. All water below this level is at about 39°F.

In spring, when the ice melts and the water temperature reaches 39°F, the wind mixes the entire body of water, resulting in a period of circulation called the spring turnover. As the warmth of spring and summer continue to raise the temperature of the surface water above 39°F, consequently making it lighter, the wind continues to mix the upper layer. In small, relatively deep lakes, this process leads to stratification, or layering, because the wind is not able to mix the entire depth. The amount of energy the wind can input for mixing is proportional to the surface area of the lake. The more wind can blow across the surface, the deeper the lake will mix. Lake Superior, for example, mixes to depths of 30 to 50 feet during summer (R. Megard, personal communication).

A layer of warm water, called the epilimnion, which is continually mixed by the wind, develops as the upper layer (figure 8.4). Just below the epilimnion is a layer where the temperature drops markedly. This layer, called the metalimnion, may be from a few to many feet thick. The temperature gradient in the metalimnion is called the thermocline. Here the temperature decreases rapidly with a small change in depth. In lakes that are deep enough, a third layer, called the hypolimnion, forms beneath the metalimnion. Here the temperature decreases slightly with depth.

As lower air temperatures in the fall cool the surface waters, they sink and are replaced by warmer waters until the entire lake is again at a uniform temperature. At this time, mixing is complete for the second time in a year. This mixing is called the fall turnover.

Stratification and the mixing of lake water by wind action combined with the fact that cold water holds more oxygen than warm water produce changes in the distribution of oxygen in the lake profile through the seasons. In a stratified lake in summer the oxygen concentration is generally high in the epilimnion and low or even near zero in the hypolimnion (figure 8.4). During the spring and fall turnovers, the oxygen concentration is the same from top to bottom.

This pattern is typical for many of the lakes in the North Central Hardwood Forests and in the Northern Lakes and Forests ecoregions. Exceptions occur in lakes such as Mille Lacs, which have a very large surface area but a shallow depth. Here the wind is able to mix the entire lake throughout the summer, and any stratification is temporary. Many of the lakes in southern Minnesota are shallow, and here again the wind is able to mix these lakes throughout the summer, preventing stratification.

Sometimes a small but relatively deep lake sheltered by surrounding hills or trees can remain continually stratified; that is, the wind seldom inputs enough energy to mix the deepest waters. Examples

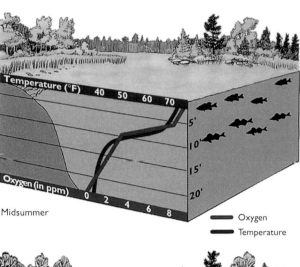

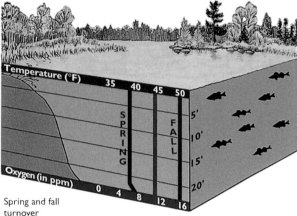

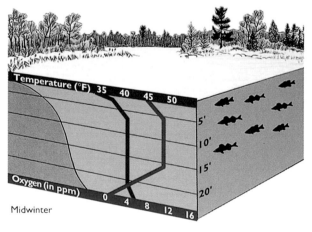

**Figure 8.4** *Gradients in temperature and oxygen, decreasing from surface to bottom in summer, cause the water in many lakes to stratify, or form layers. Note the thermocline, or sharp temperature change, in the middle layer in summer. In spring and fall, when the water temperature is the same from surface to bottom, the entire body of water mixes. (Illustration by Don Luce; modified from Smith 1990, by permission.)*

of small sheltered lakes that do not mix are tiny Deming Lake in Itasca State Park and Brownie Lake in Minneapolis.

*Water movement*

On windy days the water in large lakes is sometimes pushed by the wind and piled up on the downwind shore. This water movement may expose sand or mudflats on the upwind shores of lakes. When the wind stops blowing, this water flows back. Such tidelike effects are referred to as seiches. Seiches, although seldom seen, can be important because, along with internal currents, they mix water in the lake, transporting and circulating heat, dissolved gases, and nutrients.

A more visible type of water movement caused by the wind is Langmuir circulation (Cole 1983).

Wind creates long spirals of water oriented parallel to the direction of the wind. These spirals rotate in alternating directions, and where they come together, foam and floating debris accumulate in windrows. In these streaks of convergence, which are marked by parallel lines of foam, water moves downward carrying debris and mixing nutrients and organisms.

## Lake communities

Limnologists divide lakes into zones, or communities, based on physical characteristics and the type of vegetation present. The shallowest part of a lake, where rooted aquatic plants grow, is the littoral zone. The open water too deep for rooted plants to grow but receiving adequate light for growth of

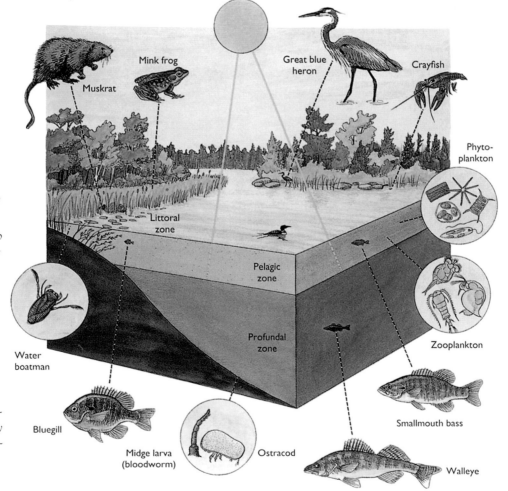

**Figure 8.5** *Lakes can be divided into zones based on physical characteristics and vegetation. The area in which sunlight penetrates to the lake bottom is called the littoral zone. The pelagic zone is the layer of open water receiving sufficient sunlight for algal growth but too deep for growth of rooted plants. The profundal zone is too dark for photosynthesis to occur. Bottom-dwelling bacteria and other organisms, collectively called benthos, occupy the sediments in the profundal zone. (Illustration by Don Luce.)*

203

algae is the pelagic zone. In most lakes that stratify, a third zone, the profundal, occurs beneath the pelagic zone. This zone includes the water to which light does not penetrate and the sediments on the lake bottom. The proportion of the lake in each of these zones is dependent on the surface area of the lake, the shape of the shoreline, and the contour of the basin. Some lakes have all zones, but many may have only one or two. I will discuss these communities according to their position in the lake, beginning near the shore and proceeding to the deepest area (figure 8.5).

## Littoral community

The littoral community occupies that portion of a lake from the shore outward to where rooted plants no longer grow. Vegetation usually exists in bands: emergent plants are closest to shore, floating-leaf plants, such as pondweeds and water-lilies, occur in somewhat deeper water, and submerged aquatic plants occur as deep as light penetration allows.

The kinds of aquatic plants growing in a lake are strongly tied to water chemistry. J. B. Moyle (1945) studied the distribution of aquatic plants in Minnesota lakes and divided them into three categories, soft-water, hard-water, and alkali, according to the alkalinity, sulfate concentration, and pH of the lakes. Soft-water flora occurred in lakes with alkalinity below 40 parts per million, sulfate below 5 parts per million, and pH below 7.4. Common soft-water species were quillwort, floating bur-reed, spiral pondweed, water bulrush, pipewort, and water lobelia. Soft-water lakes are found primarily in the northeast.

The hard-water flora described by Moyle is typical of lakes in the central portion of the state occurring on glacial drift rich in calcium carbonate. Common plants in hard-water lakes are a large alga called *Chara*, pondweeds, slender naiad, Canada waterweed, coontail, water milfoil, yellow water-lily, white water-lily (figure 8.6), reed (figure 7.5),

**Figure 8.6** *White water-lilies are common in many lakes. (Photo by Richard Hamilton Smith.)*

bulrush (figure 7.4), and cattail (Moyle 1945). In the northern half of the state wild rice forms dense stands along the shores of many lakes (figure 8.7).

An alkali-water flora occurs in some lakes in southwestern and western Minnesota that are characterized by high concentrations of dissolved carbonate and sulfate. Plants in this group include widgeon grass, large naiad, prairie bulrush, and a few other species.

Probably the best known of all the aquatic plants in the littoral zone of Minnesota's central and northern lakes is wild rice, which is harvested for eating. This annual grass, which grows from seeds in the same manner as wheat and oats, is usually found in water less than 3 feet deep. Seeds germinate in the bottom mud in May and June and produce ribbonlike leaves that grow and float on the surface by late June and July (figure 8.7). Fruiting stalks began to emerge from the water in July and continue to grow until the grains are ripe in late August and September. By this time, the fruiting stalks extend several feet above the water, and the rice beds resemble fields of wheat or oats. Most plants have one stalk, but some will produce five or six.

The distribution of wild rice is strongly influenced by water chemistry. Wild rice tolerates a wide range of alkalinity but appears to grow best in water with alkalinity greater than 40 parts per million and with a sulfate content less than 10 parts per million (Moyle 1944). The best stands seem to develop on mud bottoms underlain by gravel or sand and in lakes with moving water from inlets, outlets, or strong wind action. Nett Lake in Koochiching County, Big Sandy in Aitkin County, and Bowstring in Itasca County are some of the best producers of wild rice in Minnesota. Years of low rainfall with receding water levels often lead to the best crops.

Wild rice growing in Minnesota lakes must be harvested by hand. Harvesters use two tapered wooden sticks about 18 inches long, called flails. Ripe grains are harvested into a canoe by pulling a group of heads over the canoe with one flail and

205

tapping with the other. The unripened grains remain on the head. Thus, the stand can be harvested several times at about two-day intervals. A typical harvest is 100 pounds of green rice per acre, which is then processed into about 40 pounds of rice, ready to be prepared for eating (Moyle and Krueger 1964). Harvesting does little damage to the rice stands and, in fact, may leave most of the grain for reseeding and as food for ducks and other birds.

In recent years, farming of wild rice has been carried out in paddies, where water levels can be controlled. Commercial production has developed in northern Minnesota, and about four million pounds were produced in 1989 (Wright 1990). Some gourmets, however, claim that naturally grown wild rice from Minnesota's lakes and rivers is far superior to rice from commercial paddies.

American Indians have harvested Minnesota wild rice for hundreds of years, and it was one of their staple foods. Estimates of Indian harvests in the late 1800s suggest that about 350,000 pounds were taken annually (Jenks 1898). No estimates of the current harvest of wild rice are available, but during the 1970s, before the development of rice farming in paddies, the annual harvest ranged from about 400,000 to 1,750,000 pounds. Probably less than half of these annual harvests were made by Indians (Gerald McHugh, Enforcement Division, Minnesota Department of Natural Resources, personal communication).

Aquatic plants in the littoral zone are grazed by a variety of organisms, including snails, adult insects such as leaf miners, and several kinds of immature insects. Muskrats use these plants for

**Figure 8.7** *Wild rice is one of the best-known aquatic plants in the littoral zone of many lakes. It is easily recognized in late summer when its seed heads rise up from the water. For much of its growing season, however, the leaves lie flat on the water surface, as shown here. (Photo by Richard Hamilton Smith.)*

feeding and for lodge construction. Carp root up entire plants but generally feed only on small particles of roots and shoots that break off from the uprooted plants.

Stands of aquatic plants in the littoral zone influence the open water of the lake by intercepting silt and nutrients in runoff. Nitrogen, phosphorus, and other nutrients in this runoff may be used directly by the plants. Seeds and portions of leaves and stems from these plants are often transported to deeper waters after they partly decompose and there contribute to sediment accumulation.

The plants also serve as refuges for organisms that do not actually feed on the plants directly. A large group of plants and animals spend all or a portion of their lives attached to larger plants (figure 8.8) or to stones or other objects on the bottom. This mixture of plants and animals forms a slippery, gelatinous coating, which becomes especially noticeable when we grasp a submerged plant stem or step on a slippery rock.

Small fish and invertebrates use plant cover for protection, and the term nursery is frequently given to such habitat. Adult fish establish nests and lay eggs in littoral areas. Adult fish also forage here on other fish, insects, and a wide variety of other organisms.

Both immature and adult stages of many kinds of insects live in the littoral zone. Caddis fly larvae build cases of sticks, sand, or stones held together by a sticky thread secreted by glands. Some species carry their case with them when they move, and some species attach the cases to stones or other objects on the bottom. The water boatman collects a bubble of air in the hairs on its underside. It carries this bubble when diving, thus giving itself a source of oxygen. Predaceous diving beetles, which feed on other insects, worms, and even fish, are also common. Water boatmen and diving beetles are hunting carnivores that move through the aquatic plants searching for their prey. Other species such as dragonfly larvae and hydra, a small anemone or jellyfish relative with tentacles armed with stinging cells, hunt by attaching themselves to plants or

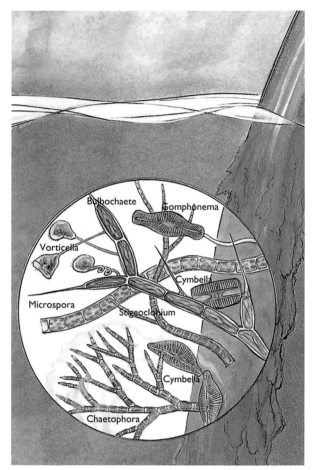

**Figure 8.8** *A large group of plants and animals spend all or a portion of their lives attached to plant stems, stones, or other objects on the bottom. They form a gelatinous coating that makes the plant stems and rocks slippery. (Illustration by Don Luce.)*

hiding among the leaves, waiting for prey to come near.

*Pelagic community*

If we travel by boat toward the center of the lake, we eventually leave the zone of rooted aquatic plants and enter the open water, the pelagic zone. Here the primary producers are phytoplankton, a group of free-floating plants called algae.

Algae, many of which have no common names, only Latin ones, represent a large and varied group of many thousands of kinds of plants. Most are aquatic, living in freshwater or saltwater, and most are microscopic. They come in a variety of shapes and sizes and generally are composed of a single cell only a few thousandths of an inch long (figure 8.9). Some, however, form threads or chains, and others are united in globular colonies large enough to be seen without a microscope. They occur in a wide range of colors because of their photosynthetic pigments. These algae are grouped into categories such as blue-green, green, and golden-brown. Diatoms, members of the golden-brown algae, have cell walls made of glasslike silica. They accumulate oils as a nutrient reserve and thus are a valuable food source for many aquatic animals. Often diatoms are attached to plants or even to other algae.

Algae are the primary producers in the pelagic zone, using the sun's energy, carbon dioxide, and nutrients in the water to produce plant biomass. Rapid reproduction of algae sometimes leads to blooms, which cover the water with a thick green, brownish, or blue-green scum, sometimes consisting of just one species. Certain species of blue-green algae, which some authorities classify as cyanobacteria, may be toxic to animals.

That so many species with somewhat similar life histories can live in the water environment seems to contradict the ecological principle that two species cannot occupy the same niche. In fact, algae species are occupying different niches by living at different depths, reproducing at different times, using different types and amounts of nutrients, and growing at different rates.

Because most algae have short lives but high rates of reproduction, they provide excellent examples of how the abundance of different kinds of organisms changes with the seasons. In the spring a pulse of diatom growth often occurs. This may be followed by pulses of other golden-brown and then green algae in the summer. In late summer, the blue-greens sometimes seem to explode and cover lakes with a thick, sometimes smelly scum. Diatoms again may show a pulse of productivity in the fall.

A number of ideas have been offered to explain these changes. One is that diatoms grow rapidly in the spring until the silicate in the water, which

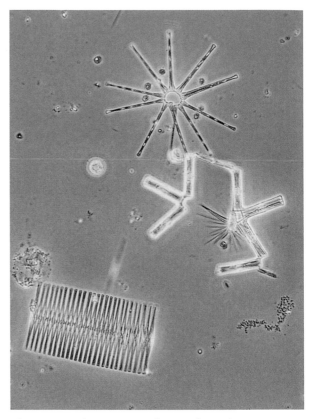

**Figure 8.9** *Algae represent a large and varied group of plants. Most are microscopic and live in water. Many kinds are composed of a single cell only a few thousandths of an inch long. Others, like the larger diatoms pictured here, form chains or clusters. Green algae, blue-green algae, and diatoms are commonly found in Minnesota's lakes and wetlands. (Photo courtesy of Robert Megard.)*

makes up their shells or cases, is used up. The silicate is not replaced in the water until the fall turnover occurs and silicate from the sediment is mixed into the water. At this time diatom growth again increases (Moss 1988). It is also believed that some algae are better at growing when nutrients such as phosphorus are low, producing a succession of species as nutrients vary in concentration. Water temperature also plays a role; cool temperatures favor diatoms, and warm temperatures favor green and blue-green algae.

Algae in Minnesota lakes are consumed by a wide variety of smaller aquatic animals called zooplankton. These zooplankton are an important link between algae and the larger consumers such as fish. The zooplankton population in a typical lake is diverse but not as diverse as the algal population. Three groups are common in most lakes: cladocerans, copepods, and rotifers.

Cladocerans (figure 8.10), or water fleas as they are often called, are tiny shrimplike invertebrates called crustaceans, which have a hard shell, or exoskeleton. Many kinds of cladocerans filter algae from water by using the fringe of fine hairs on their legs to collect their algal food supply. Reproduction in cladocerans is unusual in that eggs develop without fertilization in the females.

Cladocerans may become extremely abundant in lakes and, as a result, are important in the transfer of nutrients and energy through the food chain. They constitute from one-third to more than one-

half of the zooplankton and are the most important food item for many carnivorous fishes (Reid and Wood 1976). When cladocerans are very abundant, they consume large quantities of algae. As a result, the lake water can become extremely clear. This happens during late spring in many lakes, when water transparency suddenly increases.

Copepods (figure 8.11), another type of crustacean, make up the second group of zooplankton. Most of the 6,000 species are marine, but many freshwater species are present in Minnesota lakes. Although they are very small—most are less than 0.1 inch in length—they are vigorous swimmers. The species in the pelagic zone are filter feeders, consuming small algae and other organisms carried to the mouth by currents created by their leg movements. The density of copepods in lakes may be greater than 1,000 per quart (Reid and Wood 1976). In such high densities, copepods may also play a very important role in lake food webs.

Rotifers (figure 8.12), the third group of zooplankton, are tiny animals with a wheel of fine hairs, called cilia, around the mouth. These cilia beat rhythmically, directing water and food into the

**Figure 8.10** *(below, left) Cladocerans, or water fleas, are small crustaceans that feed on algae. Although most are very tiny, about 0.01 inch long, they are important food for many carnivorous fishes. The starlike clusters in the background are colonies of diatoms. (Photo courtesy of Robert Megard.)*
**Figure 8.11** *(below, right) Copepods, another crustacean, are about 0.1 inch long. They feed on algae and other organisms and are eaten by many fishes. (Photo by Don Rubbelke.)*

**Figure 8.12** *(right) This entire colony of rotifers is about 0.04 inch in diameter. A wheel of fine hairs around the top of each rotifer directs food into the mouth. (Photo by Don Rubbelke.)*

**Figure 8.13** *(below, right) Sonar instruments, such as those used by anglers, are used to examine distribution patterns of zooplankton and fishes in lakes. This image was recorded along a transect across Elk Lake in Clearwater County at noon on August 19, 1991, by using an echo sounder. The scale on the left is depth, measured in meters. The bottom scale is distance along the transect, also measured in meters. The lake bottom is shown by the curving red surface. During the day the density of cladocerans, copepods, and other zooplankton in the warm epilimnion is low (blue layers interspersed with green). In the cold hypolimnion the density of zooplankton (yellow layer) and phantom midges (orange layer) is high. Schools of small fish (yellow comet-shaped images) feed on the zooplankton. (Photo courtesy of Robert Megard.)*

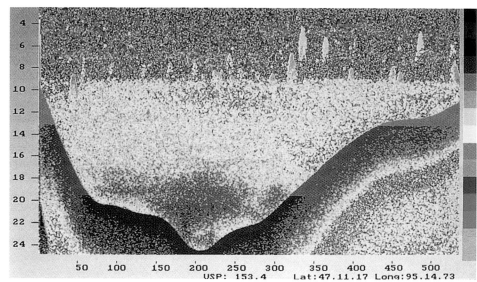

210

mouth and providing a means of locomotion. In some species, no males exist, and the females are able to produce young without fertilization. In species in which males are present, some young may be born without fertilization. Fertilized eggs are also produced, however, which overwinter and survive drought (Moss 1988). Rotifers are widely distributed in freshwater, and with over 1,500 species known, many kinds are likely to be present in a typical Minnesota lake (Reid and Wood 1976).

A number of larger zooplankton, which are somewhat stronger swimmers, are also present in the pelagic zone. These animals, which include the fairy shrimp, scuds, and amphipods, are common in most Minnesota lakes. They are important in the transfer of nutrients and energy through the food chain from the plankton to larger consumers such as fish and birds.

One of the most interesting animals living in lakes is the phantom midge. This insect, which has a transparent body, actually lives in three worlds, the bottom sediments, the water column, and the air. Phantom midge larvae spend the day in the sediments on the lake bottom. At about dusk, they rise to feed on the zooplankton nearer the surface. When the larvae reach maturity, they develop into pupae, which rise to the surface and metamorphose into a mosquito-like fly. The reason for the daily movements of phantom midge larvae from the sediments to the epilimnion, documented by using sonar interfaced with a computer (figure 8.13), is not well understood. These diurnal migrations may represent a strategy to escape predators.

Another group of organisms is associated with the film at the interface between the water surface and the air. This film is created by a phenomenon known as surface tension. Duckweeds, tiny floating plants with their roots hanging in the water (figure 8.14), and some algae live in this zone. Among the animals present are cladocerans, small spiders, and insects such as springtails, whirlygigs, and water striders.

### Profundal community

In the hypolimnion, the temperature is cold, currents are at a minimum, and light is greatly reduced. Plants do not grow in the hypolimnion of most lakes, because of the low light level. Because no living plants are present, the animals that live in this zone are detritivores, such as bacteria, protozoa,

**Figure 8.14** *Masses of duckweed, tiny free-floating plants with roots hanging in the water, may cover sheltered, shallow areas in the littoral zone and can be found floating in the pelagic zone. (Photo by Bob Firth, Firth PhotoBank, Minneapolis.)*

and some invertebrates. Fish that prey on these invertebrates may also be present. Food for the detritivores comes from the sediments, from the rain of fragments and feces from organisms living in the epilimnion, and from materials carried into the lake from its watershed basin. This so-called rain includes dead plants and animals, which simply sink to the bottom, materials leaked from plants or excreted by animals, particles of plants that simply break off and settle, and particles blown in by the wind or carried in by runoff.

Many of the organisms that live in the bottom sediments are wormlike. Larvae of chironomids, another group of midges, may be extremely abundant and are conspicuous because of their blood red color. Members of the earthworm group Annelida present in the sediments include leeches and aquatic earthworms called oligochaetes. Only a few of the leeches are bloodsuckers. Some are scavengers, feeding on many kinds of dead organic material, and some are predators. Most of the oligochaetes are tubelike worms with one end stuck in the sediments and the other waving gently in the water. They may be especially abundant in lakes containing high levels of nutrients.

## Ecosystem function

### Productivity

Four factors determine the productivity of a lake: basin morphology, watershed characteristics, nutrient supply or fertility, and climate. Depth, surface area, and transparency determine the volume of water in which photosynthesis, and thus productivity, can occur. For example, rooted aquatic plants may grow throughout clear, shallow lakes. In deep lakes, however, rooted plants will be limited to the shoreline zone where light can penetrate to the bottom.

Weathering of rock and soil within the watershed makes nutrients available to a lake via runoff. Human activities may introduce nutrients from wastewater treatment, soil erosion, stormwater

runoff, and other activities. Additional nutrients may enter a lake through precipitation and in the form of wind-carried dust and particles.

Climate, especially temperature, also affects productivity. Photosynthesis can occur over a wide range of temperatures, but maximum productivity is usually achieved under warmer temperatures. Temperature may affect productivity in Minnesota, but the effect is combined with the effects of basin characteristics and fertility.

Limnologists often classify lakes according to their productivity into four groups. An oligotrophic lake is relatively low in nutrients and contains few planktonic algae. The water is clear and deep and has a high level of oxygen in the hypolimnion. A eutrophic lake is rich in nutrients and algae, tends to be shallow, and has limited oxygen in the hypolimnion (table 8.1). Mesotrophic lakes are intermediate between the two. Lakes with the highest productivity are hypereutrophic. Minnesota has lakes in all four groups: approximately 15% are estimated to be oligotrophic, 33% to be mesotrophic, 35% to be eutrophic, and 17% to be hypereutrophic (Heiskary and Wilson 1990).

Three measurements are generally used to estimate the trophic status of a lake (Carlson 1977): phosphorus concentration, the transparency of the water, and the amount of algae as measured by the concentration of chlorophyll *a*, a green pigment in plant cells that captures sunlight in photosynthesis. Calculation of phosphorus concentration, or loading, taking into consideration mixing of the lake water and the form in which the phosphorus exists, allows prediction of the amount of algal growth that will occur. Transparency determines the depth to which photosynthesis will occur. The actual amount of algae present can be estimated by measuring the quantity of chlorophyll *a* in a sample of water.

Actual field studies to determine the productivity of Minnesota lakes are rare; however, much information on water quality and trophic status has been collected by the Minnesota Pollution Control Agency, and fish populations have been monitored

**Table 8.1** *Characteristics of oligotrophic and eutrophic lakes*

| Oligotrophic lakes | Eutrophic lakes |
|---|---|
| deep with steep banks | shallow with broad littoral zone |
| epilimnion small compared to hypolimnion | epilimnion to hypolimnion ratio greater |
| high transparency | low transparency |
| blue or green water | green to yellow or brownish green water |
| low nutrient content | high nutrient content |
| organic matter content low in sediments | organic matter content high in sediments |
| oxygen abundant at all levels at all times | oxygen often depleted in hypolimnion |
| production to respiration ratio low | production to respiration ratio high |
| rooted plants limited | rooted plants abundant |
| low phytoplankton population | high phytoplankton population |
| algal blooms rare | algal blooms common |
| diverse but sparse bottom fauna | few species but abundant bottom fauna |
| trout and whitefish in hypolimnion | no trout or whitefish in hypolimnion |

*Source:* Data from Thienemann (1925).

by the Minnesota Department of Natural Resources (DNR) Section of Fisheries. These findings and data from studies of production in lakes in other areas were used by Tom Waters (unpublished manuscript) to prepare general estimates of production for Minnesota lakes. Primary production of algae and aquatic plants ranged from about 3,000 pounds per acre per year in the northeast to 22,000 pounds per acre per year in the southwest. Production of zooplankton and bottom-dwelling organisms was estimated to range from about 350 to 7,000 pounds per acre per year from northeast to southwest.

Waters (unpublished manuscript) approximated the production of fish in two categories: first-level carnivores, such as perch, sunfish, and smallmouth bass, that feed primarily on zooplankton, insects, and other invertebrates; and top carnivores, such as lake trout, walleye, and largemouth bass, that feed

mainly on smaller fish. The productivity of first-level carnivores ranged from 9 to 200 pounds per acre per year and that of top carnivores from 4 to 20 pounds per acre per year from northeast to southwest. These values can be interpreted as representing one 4-pound walleye produced per acre per year in an oligotrophic lake. It is important to remember, however, that the biomass of walleye per acre may be much greater.

Many Minnesota lakes that are classified as hypereutrophic (Heiskary and Wilson 1990) contain bullheads, carp, redhorse, and other rough fish. Much higher production, approximately 700 pounds per acre per year, of these species is likely to occur in these lakes.

*Succession*

As a lake becomes older, both the physical and biological characteristics of the lake change as its basin is gradually filled in by sediment and vegetation. The lake changes gradually from oligotrophic to hypereutrophic. These successional changes eventually lead to fen, lowland forest, or other peatland in which the lake is gone and trees, shrubs, and herbs cover the area once occupied by the lake.

The vast majority of lakes in Minnesota are approximately 10,000 to 12,000 years old, having been formed after the recession of the glaciers. Most were probably oligotrophic at the time of their formation, and some, such as Lake Superior and numerous lakes in the northeast, are still oligotrophic. Because numerous factors, such as the characteristics of the watershed and a lake's size, depth, and nutrient content, influence the rate at which successional changes occur, Minnesota has lakes in all stages of succession. Succession proceeds most rapidly in lakes with high productivity and with a high input of sediments in runoff from the watershed basin. When nutrient enrichment and sedimentation are increased because of human activities, the process is often accelerated and is called cultural eutrophication.

Although the natural aging of lakes is essentially irreversible, succession can be slowed by reducing the flow of nutrients and sediments into a lake. Usually, however, the process cannot be reversed without some type of human manipulation, which is often referred to as lake restoration.

Lake restoration may include in-lake measures such as aerating and circulating the water to increase dissolved oxygen, drawing down the water level, dredging to remove sediment, which can be a major source of phosphorus, removing emergent plants by harvesting, using chemical poisons to control algae and aquatic plants, and manipulating fish populations. Changes in the watershed, often more important than in-lake measures, include reducing the input of nutrients and sediments from croplands, feedlots, urban roads, parking lots, and lawns. In addition, restored wetlands and constructed sediment basins can be used to intercept pollutants between their origin and the lake. An important part of lake restoration is education of individuals living in the watershed to encourage practices that enhance, preserve, and protect the lake.

## Animals and community interactions

### Invertebrates

Invertebrates are by far the most abundant animals in Minnesota lakes. Many belong to the taxonomic group crustaceans, but insects, aquatic earthworms, leeches, amphipods, and numerous other kinds of invertebrates are also present in most lakes. These animals are very important in the flow of energy and the cycling of nutrients in lake ecosystems because many of them feed on algae, an important component of primary production. In turn, these invertebrates are eaten by small fish and other animals higher in the food web. Common species of invertebrates were discussed earlier, in the section about lake communities.

### Fish

The fishes that anglers seek in Minnesota are classified into five families and are distributed in a pattern reflecting the productivity of the lakes

**Figure 8.15** *Walleye, members of the perch family, are eagerly sought by Minnesota anglers. (Photo by Bill Schmid.)*

(Waters, personal communication). Lake trout, whitefish, and stream trout of the family Salmonidae occupy oligotrophic lakes in the northeast. Walleye, sauger, and yellow perch, members of the family Percidae, are common in mesotrophic lakes in northern and central Minnesota. Sunfish, crappies, and largemouth bass, in the family Centrarchidae, are favorites of many anglers in central and southern lakes. Most of these lakes are eutrophic and produce large numbers of these game fish. Some eutrophic and hypereutrophic lakes in southern Minnesota also contain catfish and bullheads, in the family Ictaluridae. In some years, bullheads are very abundant and provide excellent fishing. Northern pike and muskellunge, often called muskie, in the family Esocidae, are widespread throughout Minnesota (Waters, personal communication), but they occur most commonly in mesotrophic lakes. These species are excellent predators and put up a tremendous fight when hooked. Over 95 hours of fishing are required, on average, to catch a legal-sized muskie, which must be greater than 36 inches in length (Breining 1989).

Because walleyes (figure 8.15) and members of the Centrarchidae are so highly sought after by anglers, their life histories are described here briefly. Walleyes occur in a wide variety of lakes, especially in the Northern Lakes and Forests and the North Central Hardwood Forests ecoregions. Their pearlescent eyes have a reflective layer, which aids vision at night or in cloudy water (Breining 1989). With this advantage, they often feed at night on yellow perch. Perch cannot see as well as their predator, and thus are easy prey for walleyes.

Natural reproduction occurs in lakes with sand or gravel bottoms and well-oxygenated water. Eggs, which are laid on gravel bars and rocky bottoms in April and May, are fertilized during laying. Adult walleyes provide no parental care for the eggs or young. The eggs hatch in 10 to 30 days. Fry, as the young fish are called, form schools or groups, which are moved about by wind and waves. They feed on large diatoms, rotifers, and other small invertebrates. As the fry grow larger, they feed on

phantom midges and other large zooplankton.

By midsummer, the young walleye are large enough to begin feeding on minnows and other small fish. Most males reach maturity and begin to breed when they are four years old. Most females first spawn when they are five or six years old (Hassinger 1987). In Minnesota lakes, a four-year-old walleye weighs about 1 pound and is about 15 inches long. Walleyes as old as 29 years have been found in Saganaga Lake and the Seagull River in Cook County, where a fish weighing 17.5 pounds, the state record, was caught (Henry Drews, Section of Fisheries, Minnesota Department of Natural Resources, personal communication).

The sunfish family, which includes sunfish, crappies, and bass, is one of the most important families of game fish in the North Central Hardwood Forests ecoregion. Most of these species prefer warmer water with abundant emergent vegetation. Males build nests in shallow water by fanning out a depression in the bottom sediments with their tail fin. Females are enticed to the nest, and courtship activity induces her to lay her eggs. As the eggs are laid, the male releases a cloud of sperm to fertilize them. After egg laying is completed, the male chases the female away from the nest and guards it against all other intruders. The eggs hatch in from three to seven days, depending on water temperature. Once the young fry have moved away from the nest, the male abandons them and the nest as well. The fry feed on zooplankton and are prey for bass, perch, walleye, and northern pike.

*Amphibians and reptiles*

Most of Minnesota's adult amphibians and reptiles prefer terrestrial habitats during much of the year. A few, such as the green frog, mudpuppy, northern water snake, and map turtle, can be observed in and along the shores of lakes. One species, however, that appears to be found almost exclusively in lake habitats is the mink frog. Most mink frogs spend their entire life in the lake. Breeding begins in late May or June, when the males begin calling from

215

their territories along the edge of the lake or in the masses of vegetation along the edge of floating bogs. Females are attracted to the calling males, and the eggs are laid in masses attached to stems of aquatic plants. Young mink frogs spend the winter as tadpoles, and some may even spend two years of their lives in the tadpole stage (Breckenridge 1944; Heeden 1970).

This frog is relatively secretive in its habits and is not commonly seen. Under certain circumstances, however, they can be readily observed and captured. As a field exercise to study population dynamics, students in my vertebrate ecology class at the University of Minnesota's Lake Itasca Forestry and Biological Station captured and marked hundreds of mink frogs on Lake Itasca. During the breeding period a crew of three students in a canoe would approach the floating vegetation along a sedge mat at night, searching with a flashlight for the yellow throats of mink frogs. When a frog was spotted, the paddler would quietly position the canoe so the person in the front could grab the frog and pass it back to the student in the center, who then would measure, mark, and release the frog for future study. As many as 50 to 75 mink frogs could be handled in a few hours by an efficient crew if the breeding population was high. The musky odor of the mink frog stayed on our hands for a long time, but the memorable experience was worth this slight annoyance.

Over several summers we learned that male frogs stayed in nearly the same location for many days, but that females were present in the breeding area for only a few days. Some adults returned to the same breeding area each year. We estimated that the population breeding in one small bay of Lake Itasca was about 286 individuals. Unfortunately, we were not able to gather enough data to estimate the number of mink frogs living in the entire lake.

Two species of reptiles, the snapping turtle and the painted turtle, are likely to be found in many of Minnesota's lakes. Snapping turtles prefer lakes with abundant vegetation. They are sometimes observed moving about in weedy bays with their

backs partly out of the water. They lay their eggs on land in a nest cavity scooped out with their hind legs. Incubation requires several months, and hatching usually occurs in late summer. In some locations, however, the eggs may not hatch until the following spring.

Snapping turtles (figure 8.16) spend most of their life in the water, often crawling about on the bottom or lying partially submerged in the mud. Snappers are omnivorous, taking both animal and plant food. They eat a variety of pondweeds, water-lily leaves, and other plants, and crayfish, frogs, salamanders, fish, insects, and occasionally ducklings and mammals. They may actively search for food or may ambush their prey by lying concealed in the mud (Breckenridge 1944). Snapping turtles spend the winter buried in the sediments or sometimes in a muskrat burrow or under a log. Often they aggregate in small groups, but occasionally individuals can be seen moving about under the ice. Large snapping turtles may weigh as much as 50 or 60 pounds and are highly prized for their meat. W. J. Breckenridge (1944) reported that several market hunters each shipped 50,000 pounds or more of turtle meat in a single season. Recent commercial harvests have been much smaller. In 1992, the total commercial harvest in the state was 42,445 pounds. Sixty trappers reported taking 3,152 turtles larger than 10 inches in carapace width. In 1993, 61 trappers reported harvesting 3,740 turtles weighing 21,671 pounds (Roy Johannes, Section of Fisheries, Minnesota Department of Natural Resouces, personal communication).

Painted turtles, with bright yellow and red markings on the underside of the shell, are found in lakes and ponds throughout Minnesota. Small groups can often be seen basking in the sun on logs, rocks, or muskrat houses. Like the snapping turtle, they lay their eggs in ground nests near the water. Incubation takes place with heat from the sun. Young hatch in late summer two to three months after the eggs are laid. Temperature during incubation determines the sex of newly hatched painted turtles. All are males when incubation

**Figure 8.16** *Snapping turtles spend most of their life in water, feeding on both plants and animals. They are effective predators on fish, salamanders, small mammals, and ducklings. (Photo by Barney Oldfield.)*

temperatures are cooler than 80.6°F, and all are females when incubation temperatures are warmer than 80.6°F (Ewert and Nelson 1991).

Painted turtles are omnivorous, taking both plant and animal food. In addition, they are scavengers and feed on dead fish, dead clams, and other carrion (Breckenridge 1944).

### Mammals

Many kinds of mammals can be observed drinking and foraging along lakeshores and traveling over frozen lakes. Raccoons and minks are good swimmers and often search for food in emergent vegetation in the littoral zone. Wolf and coyote tracks can sometimes be seen on sandy shores and on snow-covered ice. Deer and moose often stand in water up to their bellies, feeding on water-lilies and other aquatic plants. Although many species of mammals have some association with lakes, relatively few are actually dependent on an aquatic environment.

Three mammals closely associated with Minnesota's lakes are beaver, muskrat, and otter. Muskrats and beavers are discussed in other chapters (see chapters 7 and 9). Both species are associated with the littoral community in lakes and rivers, and both make extensive use of the larger aquatic plants.

Otters (figure 8.17), more appropriately called river otters, are found along the shores of northern lakes and many of Minnesota's rivers. They live in dens and are extremely curious and even playful. Their long facial whiskers aid in locating prey in turbid water and at night (Melquist and Hornocker 1983). They are good hunters and are at the

**Figure 8.17** *River otters live along the shores of many lakes and rivers, and are especially common in northeastern Minnesota. Their presence is usually indicated by mud slides leading into the water. (Photo by Thomas D. Mangelsen/Images of Nature.)*

top of the aquatic food web. They eat fish, frogs, clams, snails, and crayfish. Otters swim extremely well and are able to catch fish with ease. Their presence is usually advertised by mud slides leading into the water and by deposits of feces containing the visible remains of prey.

### Birds

A large variety of birds is associated with lakes. Swallows and purple martins can often be seen feeding on insects over water. Other birds catch fish close to the surface. These include eagles, ospreys, and terns, which dive to capture their prey, and white pelicans, which capture fish by scooping them up in their large beaks. Groups of pelicans feed in unison, acting almost like a large net passing through the water. A third group of birds can be observed swimming on the lake and diving for their prey. Mergansers, lesser scaup or bluebills, several species of grebes, and loons capture most of their food in this manner. Herons and egrets wade or stand in shallow water awaiting the telltale movement of a small fish, frog, or crayfish. Their long necks contain powerful muscles that drive their heads forward sharply to capture prey.

Loons (figure 8.18) occur on many lakes in Minnesota during spring and fall migration and during the breeding season. Loons are most common, however, on the more remote northern lakes, where they make an exciting, highly visible addition to the character of the wilderness. We humans expect wilderness to be primitive, and we expect its inhabitants to be wild and somewhat mysterious. The loon fulfills all of these expectations and is truly a symbol of wilderness.

The origin of the loon's common name is not clear. It may be derived from the old English word "lumme," which means lummox or clumsy one. With their legs situated far back on the body, loons have difficulty walking on land. But in the water, they are among the best swimmers and divers of the birds. They weigh from 8 to 12 pounds, are about 3 feet long, and have a 5-foot wingspan. Most of their food, which includes fish, salamanders, frogs, leeches, crayfish, amphipods, and insects, is captured under water.

Loons have a variety of calls. The tremolo, or laughing call, is given as a warning or to signal aggression (McIntyre 1988). When members of the family have become separated, the call is a long, mournful wail that sounds somewhat like the howling of a wolf. During the spring breeding season, the mate advertises his presence and his territory by the yodel. Yodels say "this is my territory" and warn other loons to keep away. A final call, the hoot, is given among family members and appears to signify that all is well.

One to two eggs, and occasionally three, are laid in a slight depression on shore or along a sedge mat. Islands are especially preferred (Olson and Marshall 1952), and the same nest sites may be used year after year but not necessarily by the same individuals. Vegetation is added to the nest during the 29-day incubation period. Both the male and female help with incubation.

Most young hatch in June, and both parents participate in taking care of them. Young may often be seen riding on the back of one of the parents,

which provides them with transportation as well as protection from predators such as northern pike. The young loons eat whole fish, which the parents catch and toss into the water. As the young feed, they are also learning to dive. In about 10 weeks, the young have learned to catch their own fish, and at this time they begin to fly.

Young and adult loons begin to gather in groups on large lakes in August and September in preparation for their fall migration. In October or early November, Minnesota loons begin their migration to either the east coast of the United States or to the Gulf of Mexico, where they spend the winter.

Loons are sensitive to human disturbance, as shown by a study on lakes in the Boundary Waters Canoe Area Wilderness (Titus and VanDruff 1981). Pairs in lakes with low recreational use raised more young than pairs in lakes with high

**Figure 8.18** *Common loons may be seen throughout Minnesota during spring and fall migration, but breeding occurs mostly in northern lakes. In 1989, the summer population in Minnesota was estimated to be between 10,359 and 12,893 loons (Strong and Baker 1991). (Photo by Thomas D. Mangelsen/Images of Nature.)*

recreational use. Humans also affect loons through contamination of their environment. All of 221 loons examined between 1984 and 1990 had mercury residues in their livers, feathers, or both. Livers of 20 loons contained mercury concentrations sufficient to impair reproduction, and several contained mercury in excess of concentrations found to be lethal to other birds (Ensor, Helwig, and Wemmer 1992).

The loon was designated as the official state bird by the Minnesota legislature in 1961. An estimate of the state population was obtained in 1989. Volunteers surveyed 723 lakes on July 15, 1989. Loons were observed on 358 of these lakes, and the population for the entire state was estimated to be between 10,359 and 12,893 loons (Strong and Baker 1991). The survey, which will be repeated at five-year intervals, may prove to be a good index of the quality of our lakes.

## Minnesota's lakes

Nearly all of the discussion in this chapter has dealt with lakes in a general sense. But every lake is unique in many respects. To provide a broader perspective, let us now make a tour of Minnesota from northeast to southwest, looking at a selection of lakes in each of the ecoregions and comparing them in a variety of ways. We shall begin with Lake Superior and then proceed southwestward.

### Lake Superior

Lake Superior, 1,333 feet deep and 31,280 square miles in surface area, is the largest freshwater lake in the world (figure 8.19). It is 360 miles long and 160 miles wide at its widest point. The shoreline extends 1,026 miles (Bell 1989).

Lake Superior was formed primarily through the actions of fire and ice. About a billion years ago, molten rock erupted from volcanic activity in a great rift in the earth's surface extending from the northeast corner of Minnesota to Kansas. A thick layer of basalt formed from the flow of lava, reaching a thickness of 5 miles. The weight of this rock caused the surface of the earth to sink and form a broad, shallow basin. Volcanic activity eventually stopped, and the basin remained dry and level until glacial ice began to change its shape about one million years ago.

About 10,000 years ago, the last glacier began retreating from the area and blocked the drainage path to the northeast. Glacial Lake Duluth was created in the basin. When the ice front was farther north, the present shape and drainage pattern of Lake Superior were established (Bell 1989).

The water supply to Lake Superior greatly exceeds the loss of water through evaporation. About 2.5 feet of water enters the lake as precipitation each year, and about 2.0 feet enters through the 20 rivers and streams flowing into the lake. About 1.6 feet of this 4.5 feet of water input evaporates, and the remainder flows to the ocean through the St. Lawrence Seaway. Although the lake is very large, tidal action changes water levels less than 1 inch. Winds, however, produce significant seiches, sometimes changing water levels by as much as 1 foot (Bell 1989).

Lake Superior is cold and well oxygenated, and its shoreline is rocky. The lake is oligotrophic and is relatively free of aquatic plants. The water is extremely clear, and visibility extends to 65 to 75 feet in many areas. Low nutrients and cold temperatures result in low diversity and slow growth of organisms, which are in turn reflected in the lake's low productivity. Annual fish production was estimated to be about 0.8 pound per surface acre. In comparison, production in nutrient-rich Lake Erie was 7.3 pounds per surface acre (Lydecker 1976). These values are very small in comparison to the thousands of pounds produced annually in the inland lakes in Minnesota.

Because of the slow growth of organisms and the low productivity of the lake, pollutants are not quickly broken down and thus are of special concern. PCBs, dieldrin, toxaphene, and DDT, which enter the lake primarily in precipitation or as dry fallout, are toxic to humans. Their presence in the lake has led to the recommendation that people limit their consumption of fish caught in Lake Superior (Bell 1989).

Before the days of intensive commercial netting and the introduction of non-native species, lake trout were abundant and were at the top of Lake Superior's food chain. Sculpin and burbot, a freshwater cod resembling a catfish, were common near the bottom sediments. Sturgeon, whitefish, and lake herring lived in the pelagic waters. Brook trout lived near the shoreline and migrated upstream in the tributaries to spawn.

In the 1940s and 1950s, the numbers of lake trout and other large predatory fish declined as a result of invasion by the sea lamprey (figure 8.20), which was first observed in Lake Superior in 1938. The eel-like adult lamprey attaches its suckerlike mouth to fish such as lake trout and salmon. The wound caused by the horny teeth and rasping tongue may be lethal. The adult lampreys spawn and then die in tributary streams. Larvae live in the streambeds for four to seven years before moving to Lake Superior as adults. A single female may lay as many as 95,000 eggs (Eddy and Underhill 1974).

In the 1950s, a selective poison that killed the larvae but did not affect other fish was developed to control lampreys in the Great Lakes. Streams are treated each year, but not all of the larvae are killed. This effort at control has resulted in an increase in the population of lake trout, and sportfishing for this trophy is again popular.

The invasion of the sea lamprey resulted in other changes in the lake's fish populations. As numbers of large predatory fish declined, the population of rainbow smelt, first found in Minnesota waters in 1946, increased rapidly (Eddy and Underhill 1974). The fish population in Lake Superior had too many smelt and too few game fish to serve as predators. During the past 25 years, fish populations in Lake Superior have again changed markedly because of stocking, natural reproduction, and lamprey control. Today the fish population in the lake includes numerous predatory species that feed on the smelt and herring. Lake trout are again common, and the stocking of

species such as steelhead, coho, and chinook salmon from the Pacific coast and Atlantic salmon has been quite successful.

Some concern has been expressed, however, about the number of species that have been deliberately introduced by fisheries managers and about those species that have been accidentally introduced, such as the ruffe. Little is known about the long-term effects of these introductions on the overall ecology of the lake. For example, ruffe, native to Europe and Asia, are now the most abundant fish in the estuary of the St. Louis River at Duluth. The perchlike ruffe are too small to be sought by anglers, are too spiny to serve as a favorite food of other fish, and are able to outcompete other fish for food and habitat (Larson 1989).

Recent voyages to the bottom of Lake Superior in a research submarine have led to a number of surprising and important discoveries concerning the ecology of the lake. Zooplankton, sculpin, and burbot were observed in abundance. The large numbers of organisms living in the bottom sediments and the movement of deep currents appear to keep the sediments suspended. Deep currents eventually mix the suspended material throughout the entire body of water, making nutrients available to the food web. J. E. Baker, S. J. Eisenreich, and B. J. Eadie (1991) described this movement of materials from the cold, dark bottom as a nutrient powerhouse, fueling biological productivity throughout the lake.

Mysis shrimp about half an inch long were observed migrating vertically several hundred feet through the water, rising to the surface waters in the evening to feed on plankton, and then returning to the bottom in the morning. Vertical migrations through 950 feet of water were observed (Bell 1986). The submarine voyage to the bottom produced many interesting findings such as these observations about mysis shrimp but also raised

**Figure 8.19** *(overleaf) Lake Superior is the largest freshwater lake in the world. It is 1,333 feet deep, and its shoreline is 1,026 miles in length. (Photo by Blacklock Nature Photography.)*

**Figure 8.20** *Eel-like sea lampreys, which live in Lake Superior, attach their suckerlike mouths to lake trout, salmon, and other fish. The wounds caused by the teeth and tongue are sometimes lethal. (Photo by Jim LaVigne.)*

many questions that need to be answered to provide a better understanding of the ecology of Lake Superior.

### Lakes in the Northern Lakes and Forests ecoregion

In addition to Lake Superior, about 46% of Minnesota's lakes lie in the Northern Lakes and Forests ecoregion. The region is predominantly forested, but small areas have been cleared for agriculture and mining. The ecoregion has little urban development, but homes, resorts, and cabins line the shorelines of many lakes. With the exception of the steep topography along the North Shore of Lake Superior, most of the landscape is gently sloping. Soils are very thin on the rock substrate of the Canadian Shield.

The lakes are relatively deep, having maximum depths ranging from 25 to 60 feet and an average

maximum depth of about 30 feet (Heiskary and Wilson 1990; Minnesota Pollution Control Agency 1992). Most are less than 100 acres in size. Because the lakes are small and deep and precipitation is adequate to balance the loss of water due to evaporation, water levels are relatively stable. Characteristics of a sampling of lakes from five ecoregions are given in table 8.2.

The productivity of lakes in the Northern Lakes and Forests ecoregion is low. In general, these lakes range from oligotrophic to mildly eutrophic. Their low productivity probably results from the low levels of phosphorus in their watersheds and their steep, rocky shorelines, which limit the amount of littoral vegetation present.

The transparency of most of the region's lakes ranges from 8 to 14 feet (Heiskary and Wilson 1990). Only about 6% of the lakes are considered impaired for swimming because of low transparency related to summer algal blooms (Minnesota Pollution Control Agency 1992). Some lakes, however, have low transparency because of the brown color of the water. This brown color comes from dissolved organic chemicals and sediment in the water that has drained through organic soils in the watershed.

Because of the low amounts of limestone in the watersheds of many of the lakes in this ecoregion, they have waters that are low in alkalinity, which means they have a low capacity to neutralize acids. Thus, the lakes have a low pH value.

These low-alkalinity lakes, often called soft-water lakes, are susceptible to acidification by acid rain (Twaroski et al. 1989). A pH of 4.7 is the threshold below which lakes are considered acidified. Because the precipitation in the northeast has an average pH of 4.7, the potential for acidification is great. About 95% of the susceptible lakes in Minnesota are in Cook, Lake, St. Louis, and Itasca Counties (Twaroski et al. 1989). The Minnesota Pollution Control Agency (1992) estimates that as many as 2,200 lakes in Minnesota could be damaged if exposed for several years to precipitation with a pH lower than 4.7. At the present time, however, no lakes in Minnesota are considered to be acidified solely because of acid rain.

Phosphorus levels are comparatively low in lakes in the Northern Lakes and Forests ecoregion, typically ranging from 15 to 30 parts per billion, because of the low concentration of phosphorus in streams entering these lakes. Lakes in the ecoregion do, however, receive pollutants from nonpoint sources such as stormwater runoff from logging operations, urban and shoreland development, mining, and septic systems from homes and cabins. Because of their low alkalinity, and hence their

**Table 8.2** *Characteristics of selected lakes in Minnesota ecoregions*

| Lake | County | Area (acres) | Average depth (ft) | Watershed area (sq mi) | Land use (%) | | | | | Phosphorus parts/ billion | Alkalinity parts/ million | Secchi disk transparency (ft) |
| | | | | | Forest | Wetland and water | Urban | Pasture | Cultivated | | | |
| *Northern Lakes and Forests ecoregion* | | | | | | | | | | | | |
| Bemidji | Beltrami | 6,420 | 31 | 609 | 73 | 7 | 2 | 6 | 12 | 34 | 172 | 8.2 |
| Ten Mile | Cass | 4,640 | 53 | 21 | 51 | 42 | 4 | 3 | | 16 | 120 | 17.7 |
| Greenwood | Cook | 2,047 | 35 | 11 | 69 | 30 | 1 | | | 14 | 6 | 21.3 |
| Musquash | Cook | 141 | | | | | | | | 19 | 8 | |
| Trout | Cook | 259 | 33 | 2 | 80 | 20 | | | | 10 | 12 | 26.2 |
| Whitefish | Crow Wing | 7,370 | | | 75 | 15 | 10 | | | 14 | 123 | 13.8 |
| Trout | Itasca | 1,890 | 50 | 16 | 48 | 46 | 6 | | | 47 | 127 | 10.5 |
| Snowbank | Lake | 4,595 | 50 | 58 | 68 | 32 | | | | 10 | 17 | 16.4 |
| Grindstone | Pine | 529 | 71 | 19 | 29 | 45 | 1 | 7 | 18 | 16 | 46 | 11.5 |
| Shagawa | St. Louis | 2,370 | 19 | 104 | 75 | 15 | 2 | 7 | 1 | 35 | 34 | 6.9 |
| *North Central Hardwood Forests ecoregion* | | | | | | | | | | | | |
| Sallie | Becker | 1,267 | 17 | 91 | 30 | 16 | 9 | 25 | 20 | 43 | 193 | 6.9 |
| Waconia | Carver | 2,607 | 16 | 9 | 4 | 25 | 1 | 9 | 61 | 39 | 160 | 4.3 |
| Miltona | Douglas | 5,950 | 24 | 48 | 21 | 34 | 2 | 3 | 40 | 15 | 184 | 10.8 |
| Calhoun | Hennepin | 421 | 35 | 10 | | 17 | 82 | 1 | | 32 | 99 | 6.6 |
| Harriet | Hennepin | 353 | 29 | 2 | | 17 | 79 | 4 | | 22 | 98 | 10.8 |
| Medicine | Hennepin | 940 | 16 | 18 | 6 | 18 | 45 | 18 | 13 | 40 | 110 | 3.9 |
| Minnetonka | Hennepin | 14,473 | 23 | 95 | 9 | 26 | 34 | 15 | 16 | 37-170 | 125 | 2.0-10.2 |
| Rebecca | Hennepin | 261 | 15 | 2 | 14 | 24 | 2 | 28 | 33 | 86 | 142 | 4.9 |
| Round | Hennepin | 33 | 11 | | | | | | | 66 | | 7.9 |
| Otter Tail | Otter Tail | 14,753 | 23 | 675 | 18 | 22 | 4 | 8 | 49 | 15 | | 10.2 |
| Minnewaska | Pope | 7,110 | 16 | 89 | 4 | 18 | 5 | 15 | 57 | 25 | | 6.6 |
| Como | Ramsey | 70 | 6 | 3 | | | 100 | | | 230 | 53 | 3.6 |
| Big Birch | Todd | 1,980 | 28 | 14 | 22 | 22 | 9 | 17 | 30 | 38 | 161 | 8.2 |
| *Western Corn Belt Plains ecoregion* | | | | | | | | | | | | |
| Crystal | Blue Earth | 396 | 7 | 26 | 1 | 10 | 7 | 7 | 75 | 314 | 118 | 1.0 |
| Bear | Freeborn | 1,560 | 4 | 38 | | 9 | 1 | 10 | 80 | | | |
| Clear | Jackson | 451 | 6 | 2 | | 26 | | | 74 | 97 | 216 | 3.0 |
| Fairmont Chain | Martin | 1,179 | 10 | | | | 10 | 10 | 80 | 150 | 180 | 3.0 |
| Clear | Waseca | 652 | 15 | 6 | 1 | 22 | 38 | 11 | 28 | 82 | 163 | 4.9 |
| *Northern Glaciated Plains ecoregion* | | | | | | | | | | | | |
| Big Stone | Big Stone | 12,610 | 8 | 1,160 | 1 | 5 | 1 | 11 | 82 | 150 | 169 | 3.9 |
| *Paleozoic Plateau ecoregion* | | | | | | | | | | | | |
| Pepin | Goodhue | 25,408 | 18 | 48,634 | | | | | | 226 | 177 | 2.0 |

*Sources:* Heiskary and Wilson (1990); S. Heiskary, personal communication.

low capacity to buffer or neutralize the input of chemicals, these lakes can experience declines in water quality due to pollution. Furthermore, the overall high water quality of the lakes in this region has led users to expect higher transparency and lower algal populations than users in central or southwestern Minnesota expect. Lakes in the northeast were considered to be impaired for swimming when the transparency was less than 6 feet. In central Minnesota, lakes were considered impaired when transparency ranged from 2 to 3 feet, and in the southwest, lakes were considered impaired if transparency was less than 1 to 3 feet.

Many lakes in the region have benefited from measures to preserve or improve water quality, which have been driven in part by the interest of cabin owners and others seeking lake recreation. Ten Mile Lake, a large, deep lake in Cass County, is characterized as being "crystal clear" and "beautiful" during most summers (Heiskary and Lindbloom 1993). In the 1970s transparency values ranged from 12.5 to 16.5 feet. Inspections and upgrading of septic systems and the enforcement of shoreland regulations in the 1980s have contributed to an improvement in the quality of the lake. Transparency values ranged from 16.0 to 19.5 feet in the early 1990s. Ten Mile Lake Association serves as a watchdog for the lake, and its efforts are likely to maintain the high quality of the lake (Heiskary and Lindbloom 1993).

Whitefish Lake in Crow Wing County, which has a transparency of about 14 feet, also has very high water quality. The lake covers 7,370 acres and has a maximum depth of 138 feet. It is borderline between oligotrophic and mesotrophic. The phosphorus concentration is 14 parts per billion and chlorophyll *a* is 3.8 parts per billion. In a recent study, recreational users referred to the lake as being "clear with low algal populations" and "having minor aesthetic problems" (Heiskary and Lindbloom 1993). Earlier studies, however, indicated that transparency was in the 8 to 9 foot range and that algal blooms occurred. This reduced water quality may have been the result of wastewater

entering the lake through the Pine River wastewater treatment facility constructed in 1960. The facility was replaced in 1984, probably reducing the amount of phosphorus entering Whitefish Lake. This reduction, in turn, has contributed to the high quality of the lake in the 1990s (Heiskary and Lindbloom 1993).

Shagawa Lake in St. Louis County, one of the most studied lakes in the state, has a higher level of phosphorus than many lakes in the ecoregion. For many years, Shagawa Lake received effluent containing phosphorus from the wastewater treatment plant at Ely. Beginning in 1973, treatment of Ely's wastewater effluent reduced phosphorus input by 80%. Although the quality of the lake's water has improved, the changes have not been as dramatic or as rapid as hoped because of recirculation of phosphorus from the lake's sediments (Heiskary and Wilson 1990).

Lake Bemidji in Beltrami County has been the focus of studies since about 1970. The lake has a very large watershed—the watershed to lake ratio is 60:1—which is predominantly forest and wetlands. Nonpoint source pollution from urban stormwater and point source pollution from municipal wastewater discharge to the lake have threatened water quality, however. The wastewater treatment facility was upgraded in the mid-1980s to reduce phosphorus input to the lake. In the early 1990s, work was under way to minimize phosphorus input from nonpoint sources in the city and elsewhere in the lake's watershed.

With respect to fish populations, several categories of lakes have been identified in the ecoregion (Heiskary and Wilson 1990). Bass-panfish-walleye lakes are most common, followed by soft-water walleye lakes, and lake trout lakes. Suitable habitat for fish species is influenced by many factors, including the size and depth of the lake, the amount of littoral zone, fertility, spawning substrate, and water temperature (figure 8.21).

Trout Lake and Greenwood Lake in Cook County and Snowbank Lake in Lake County are natural lake trout lakes. Lake trout lakes are usually

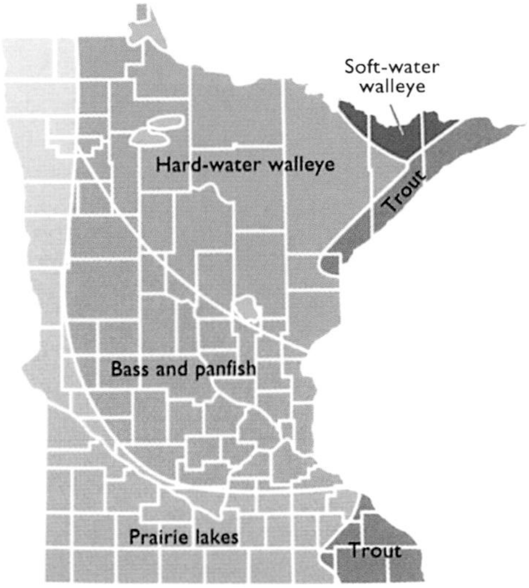

**Figure 8.21** *The Minnesota Department of Natural Resources has classified Minnesota lakes and streams on the basis of the kinds of game fish that are most common. Trout, which require cool waters, are found mainly in the northeast and southeast. Walleye, bass, and panfish are found in northern and central Minnesota lakes. Prairie lakes may contain game fish, but they often winterkill because of the lakes' shallow depth. (Breining 1989, by permission.)*

deep, often exceeding 100 feet, and have low productivity. Phosphorus concentrations are generally less than 15 parts per billion. The hypolimnion retains oxygen during the warm summer months, thus providing the needed combination of cool water and adequate oxygen, more than 5 to 7 parts per million, needed by trout.

In contrast, other deep but eutrophic lakes, such as Trout Lake in Itasca County, do not provide the necessary habitat for lake trout because they lack adequate oxygen in the hypolimnion. As happened in Trout Lake, conditions in a given lake may change over time, leading to a change in the fish species present. For example, if nutrients are added to an oligotrophic lake through pollution, the productivity of plants, algae, and zooplankton may increase, leading to an increase in trout or whitefish. As eutrophication continues, however, these species may decline in abundance, and bass, perch, walleyes, and other mesotrophic fish may increase in abundance. These changes are brought about by changes in oxygen concentrations and transparency,

changes in the abundance of algae and prey organisms, and by parasites and diseases. The total biomass of fish in the lake may increase, but less desirable species may dominate. Trout Lake contained lake trout until the 1940s. The quality of the lake declined as wastewater discharges from Coleraine and Bovey resulted in cultural eutrophication, with associated blooms of blue-green algae. With the increase in phosphorus present in the lake, oxygen was likely reduced in the hypolimnion because of decomposition. Forced to move to a shallower depth to obtain oxygen, the trout were then stressed by the warmer water temperature.

Because of the popularity of recreational fishing, the DNR's Section of Fisheries selected a group of approximately 180 lakes for a special management program designed to establish trout populations. Examples are Musquash Lake in Cook County and Grindstone Lake in Pine County. The existing fisheries in these lakes were considered to be unsatisfactory because not enough game fish were present. These special lakes have no inlets or outlets. Poisons were used to remove all fish from the lakes. In most cases, the fish were stunted sunfish and northern pike. The lakes were then restocked with rainbow, brook, or brown trout, or with splake, a hybrid from a female lake trout and a male brook trout. Because reproduction does not occur naturally in these lakes, they must be restocked periodically with the appropriate trout species.

Such total manipulation of fish populations in a lake is somewhat parallel to logging a forest and replanting the site with different kinds of trees, or to breaking a prairie to grow corn. Trout anglers encourage this type of lake management, and the DNR has been successful in reaching the goal of establishing trout populations for recreational fishing.

The DNR's Section of Fisheries also manages a few larger lakes for both trout and warm-water species, such as walleye and bass. These two-story lakes, as they are called, are managed for trout in the deep-water areas and for the warm-water species in the shallow portions. The fish popula-

227

tions are maintained primarily by stocking and by restrictions on the harvest.

### Lakes in the North Central Hardwood Forests ecoregion

Forty percent of the state's lakes lie in the North Central Hardwood Forests ecoregion. These lakes are similar in size and depth to lakes in the Northern Lakes and Forests ecoregion. The pattern of land use, however, is very different (table 8.2). Only 16% of the land is forested. The majority of the land is used for agriculture; 49% is cultivated and 21% is pasture and open. Five percent of the land is developed, and some of the most heavily used lakes in the state occur in these areas. The trophic status of the lakes ranges from oligotrophic to hypereutrophic. The increasing amounts of land used for agriculture and urban development have contributed to cultural eutrophication of some lakes.

Most lakes in this ecoregion are situated in calcareous glacial drift. The soils are somewhat thicker than in the Northern Lakes and Forests ecoregion and tend to be more fertile and rich in calcium carbonate. As a result, the lakes are primarily eutrophic, hard-water lakes with alkalinity ranging from 40 to 250 parts per million. pH ranges from 8.0 to 8.8, sulfates from 5 to 40 parts per million, and phosphorus from 23 to 50 parts per billion (Eddy 1966; Minnesota Pollution Control Agency 1992).

About 44% of the lakes in this ecoregion are considered impaired for recreational use because of frequent algal blooms and low transparency, which ranges from about 4 to 10 feet (Minnesota Pollution Control Agency 1992). Water levels fluctuate more than in the northeast because evaporation typically exceeds precipitation over much of this ecoregion. Stratification occurs in many lakes. Oxygen depletion is often characteristic of the hypolimnion, and the bottom sediments frequently support populations of midges and oligochaete worms, which do not require oxygen.

Pollutants commonly wash into these lakes from agricultural and urban areas. Productivity is higher than in lakes in the northeast. Chlorophyll *a* typically ranges from 5 to 22 parts per billion in summer (Heiskary and Wilson 1989). In many lakes, the littoral area covers more than half of the total lake area.

Large, deep lakes in western Minnesota, such as Otter Tail Lake in Otter Tail County and Lake Miltona in Douglas County, have very good water quality. They are borderline oligo-mesotrophic despite the fact that large percentages of their watersheds are in agriculture. Their high water quality is a function of their large volume of water and the presence of other lakes and numerous wetlands in their watersheds, which trap sediments and nutrients.

Lakes with lesser volumes of water and higher phosphorus loads from croplands and urban areas in their watersheds, such as Minnewaska in Pope County, Big Birch in Todd County, Waconia in Carver County, and Medicine and Rebecca in Hennepin County, are eutrophic and have algal blooms and reduced transparency. Lake Rebecca is 261 acres and about 15 feet deep. Transparency in the early 1980s ranged from 6.5 to 9.5 feet, but from 1988 to 1991, transparency was only 2.0 to 3.0 feet. Trophic conditions shifted from mesotrophic in the early 1980s to highly eutrophic around 1990. Comments from users in the late 1980s indicated that "algal scums were evident" and that "swimming was impaired" much of the time (Heiskary and Lindbloom 1993).

Phosphorus concentrations in Lake Rebecca ranging from 60 to 90 parts per billion from 1983 to 1985 are indicative of eutrophic to hypereutrophic conditions. The transparency from 6.5 to 9.0 feet for the same years, however, suggests meso-eutrophic conditions. Biological interactions in the lake provide a possible explanation for this discrepancy. If the lake contained a high population of large cladocerans grazing on the algae, higher transparencies might exist than would be expected based only on phosphorus concentrations (Heiskary and Lindbloom 1993).

Studies of such phenomena in lakes has led to

development of new methods of management. A technique called biomanipulation takes advantage of the flow of nutrients and energy through an ecosystem and its effect on certain groups of organisms (Carpenter 1989). Imagine a system from the top down, beginning with the predatory fish that feed on zooplankton. This feeding decreases the abundance of the zooplankton. In turn, populations of many species of algae will respond to the decreased feeding pressure from the zooplankton. In some cases, the algal population may explode. Thus, the result of an increase in zooplankton-eating fish may be an algal bloom in the lake. The cascade of effects from high trophic levels to lower ones, or vice versa, may take place on a timescale of years to decades.

Biomanipulation involves increasing or decreasing the populations of certain organisms by using a variety of techniques. An algal bloom might be controlled by introducing or favoring predatory fish that feed on zooplankton-eating fish. If zooplankton that feed on algae become more common as a result, algal populations may decline.

Round Lake in Hennepin County was managed in this way by restructuring the fish community through poisoning plankton and bottom-feeding fish (Shapiro and Wright 1984). The lake was then restocked with bass, walleye, bluegills, and channel catfish. Catfish were introduced to control bullheads. The results of this management were essentially as predicted. Larger species of zooplankton herbivores became abundant, and algal populations declined because of their grazing. With a reduction in algae, transparency increased, and the clearer water was considered by the public to be more suitable for swimming and other recreation. Biomanipulation is only a temporary solution; it does not solve the problem of pollution, which is often the cause of dense blooms of algae.

Urbanization in the watershed can have as great an impact on a lake as agricultural activity. As cities expand, the percentage of land covered with asphalt and concrete increases, subsequently increasing runoff and transport of pollutants to the lake. Vir-

tually all lakes in the Twin Cities area are affected by urban runoff to some degree.

Como Lake in St. Paul is a prime example of a shallow lake in a completely urbanized watershed. The 70-acre lake has an average depth of 6 feet, transparency of 3.6 feet, and a phophorus concentration of 230 parts per billion. Numerous storm sewers drain its watershed. Efforts were made in the early 1980s to improve water quality by aeration, biomanipulation, and modification of stormwater inflows. Biomanipulation, which involved killing the rough fish and restocking with bass and panfish, resulted in a resurgence in large cladocerans. The cladocerans fed extensively on algae, which led to a temporary improvement in transparency and the establishment of a community of rooted aquatic plants. Although the lake remains eutrophic, its water quality has improved as a result of biomanipulation. Como Lake will probably revert to its prebiomanipulation condition, however, unless nutrient loading is reduced (T. Noonan, personal communication).

Lakes Calhoun and Harriet in Minneapolis also show the effects of increased phosphorus loading from their watersheds, which are approximately 80% urban. As in many urban areas, storm sewers route runoff from streets, parking lots, and lawns to these lakes. Characteristics of particular lakes such as size and depth can affect the impact of the pollution. Lake Calhoun is 421 acres and has an average depth of 35 feet. Lake Harriet is 353 acres and has an average depth of 29 feet. Thus, their water quality is markedly better than the 103-acre and 9-foot-deep Lake of the Isles, which drains to Lake Calhoun. The quality of Lake Harriet is somewhat better than Calhoun because it has a smaller immediate watershed, 1.8 square miles, than Calhoun and because runoff from the upper watershed first drains through Calhoun.

Lake Minnetonka in Hennepin County is another example of a lake that has benefited from reduction of pollution. Forty percent of the lake is classed as littoral. The watershed is 34% urban and 16% cultivated, and the major source of pollution is runoff from fields, lawns, roads, and parking lots. Before 1981, however, seven wastewater treatment plants discharged effluent to the lake. Discharge from six of the seven plants had been diverted by 1981, and discharge from the last treatment plant was diverted in 1986. Overall water quality has improved, largely as a result of this diversion of wastewater. Transparency, which has been monitored continuously in lower Lake Minnetonka since 1977, shows a significant increase from 5.5 to 7.0 feet in 1970 to 8.5 to 10.5 feet from 1986 to 1991, suggesting a shift from eutrophic to mesotrophic conditions (Heiskary and Lindbloom 1993). In the early 1990s, phosphorus concentrations ranged from 30 to 45 parts per billion, and chlorophyll *a* ranged from 7 to 10 parts per billion. Lake users have characterized the recent condition as "clear" to "some algae" and "beautiful" to "minor esthetic impairment" (Heiskary and Lindbloom 1993).

Lake users are now greatly concerned about the rapid increase in non-native Eurasian water milfoil in the lake. Eurasian water milfoil (figure 8.22), introduced from Europe, is very similar in appearance to native water milfoil. It is a rooted perennial plant that was first discovered in Minnesota in Lake Minnetonka in 1987. By 1991 it had spread to many other lakes and was also present in the Mississippi River and Minnehaha Creek (Minnesota Department of Natural Resources, Section of Fisheries, unpublished data). Eurasian water milfoil grows rapidly during spring and summer, as much as 2 inches a day, producing dense mats of vegetation from the shoreline to a depth of about 15 feet. The vegetation may be so dense that it prevents sunlight from entering the water, thereby preventing the growth of other plant species. Dense beds interfere with swimming, boating, and fishing, and the plants block water intakes and may lead to increased populations of mosquitoes (Smith and Barko 1990). In winter, the large mass of decaying plants consumes a great deal of oxygen.

The plant grows on a wide variety of bottom types and can spread rapidly. Small fragments of the plant broken off by boats, waterskiing, and even wave action drift with wind and currents and eventually settle to the bottom. These fragments produce roots, which develop into new plants (Smith and Barko 1990). Because of its ability to produce new plants from small fragments, boaters are required by law to remove all Eurasian water milfoil from boats and trailers when leaving a lake.

In addition to enforcing the law prohibiting transportation of Eurasian water milfoil, the DNR is attempting to eradicate existing infestations. Methods used range from removing plants by hand to applying herbicides. Research is also in progress to determine the feasibility of raising and releasing a species of beetle that is believed to feed exclusively on Eurasian water milfoil. Another control method being studied involves application of a native fungus that selectively feeds on and destroys the plants.

Most of the lakes in the North Central Hardwood Forests ecoregion are classified as bass-

**Figure 8.22** *Eurasian water milfoil was first discovered in Minnesota in Lake Minnetonka in 1987. It has spread rapidly to many lakes and to some streams. Its dense mats prevent growth of other aquatic plants and interfere with swimming, fishing, and boating. (Photo by Jim LaVigne.)*

panfish lakes. Northern pike, white suckers, and yellow perch are also common. Although only 10% of the lakes in this ecoregion are classified as walleye lakes, this species is perhaps the fish most sought after by anglers.

Changes in the fish population as a result of cultural eutrophication are well documented for some lakes in this ecoregion. Lake Sallie in Becker County received sewage effluent from the city of Detroit Lakes until 1972. In the early 1950s, blooms of algae and excessive growths of vegetation were reported (Colby et al. 1987). Phosphorus concentration was in the range of 200 to 450 parts per billion in the early 1970s. Following divergence of the sewage effluent in 1972, the phosphorus concentration decreased to about 100 parts per billion during the 1980s. Fish populations responded to these changes in lake characteristics. In 1949, lake surveys indicated high populations of walleye, perch, northern pike, crappie, and largemouth bass, but by 1975 the abundance of walleye, perch, and suckers was significantly lower. By the 1980s, following divergence of the sewage, walleye and bluegill populations had tripled, and perch had nearly doubled. In the early surveys, yellow bullheads, which are not tolerant of eutrophic conditions, were present in low numbers. In 1968, bullheads were abundant, but nearly all were black bullheads, a species tolerant of eutrophic conditions. In the 1980s, bullhead populations had declined, and yellow bullheads were more abundant than black (Colby et al. 1987).

### Lakes in the Western Corn Belt Plains and Northern Glaciated Plains ecoregions

Agriculture is the primary land use in these two southern Minnesota ecoregions. Over 80% of the land is under cultivation, and less than 4% of the land is forested. The major crops are corn and soybeans. Along the western border of the state, about 11% is pasture and open land. Soils tend to be silty and very fertile.

Approximately 12% of Minnesota's lakes occur in these ecoregions. Most are large and shallow, ranging from 200 to 700 acres in size and from 5 to 15 feet deep and having an average depth of 10 feet (table 8.2). The combination of row crops, rich soils, and shallow lake basins leads to eutrophic or hypereutrophic lakes. About 18% of the lakes are eutrophic, and 80% are hypereutrophic (Minnesota Pollution Control Agency 1992). Over 90% of the lakes are assessed as impaired for swimming (Minnesota Pollution Control Agency 1992).

Phosphorus concentrations range from 65 to 250 parts per billion, and chlorophyll *a* ranges from 30 to 140 parts per billion, values which are much higher than in the other ecoregions. The high nutrient levels cause abundant growth of aquatic plants and algae. Water from melting snow and from rains washes sediments and fertilizers containing high levels of phosphorus from farm fields into streams and lakes. Drainage from feedlots is also a significant source of nutrients and organic matter when wastes are not contained. Alkalinity ranges from 45 to 195 parts per million. Transparency is very low in about 90% of the lakes, and Secchi disk readings of 1 to 3 feet are common. The water in these lakes often appears to be brown because of its high sediment content.

These large, shallow lakes do not stratify and often have low oxygen content because of large amounts of decaying organic matter. As a result, both summer and winter fish kills may occur.

Bear Lake in Freeborn County has many of the characteristics typical of lakes in these ecoregions. It covers about 1,600 acres and averages less than 5 feet in depth. Nearly all of the wetlands in Bear Lake's 38-square-mile watershed have been drained by tile and open ditches. Corn and soybeans are grown on nearly all of the watershed, and runoff from the fields carries soil, fertilizers, and pesticides into the lake. The sediment covers the lake bottom and forms deltas of silt at the places where streams and ditches flow into the lake. Eventually the sediment will fill the lake, and the mixture of open water and aquatic vegetation so important for wildlife will disappear.

This lake provides valuable recreation in the

form of waterfowl hunting, winter fishing, trapping, and opportunities to observe wildlife in a natural setting (Fandrei, Mockovak, and Reetz 1990). The DNR is managing Bear Lake for waterfowl production by manipulating the water depth to provide good growing conditions for aquatic plants, which provide essential food and cover for waterfowl. Control of the water level, however, does not curb the polluted runoff entering the lake. If Bear Lake is to survive, management practices must be implemented to control the pollution from the surrounding agricultural lands (Fandrei, Mockovak, and Reetz 1990).

Crystal Lake in Blue Earth County is an extreme example of a lake affected by cultural eutrophication. It is only 7 feet deep and has a large, highly agricultural watershed with numerous feedlots. The lake has an extremely high phosphorus level of approximately 314 parts per billion. Thus, algal levels are high, and blooms are common. Transparency is usually 1 foot or less. Improving water quality in such lakes is difficult because of the many sources of nonpoint pollution.

Cultural eutrophication is common in many other lakes in the southwest. In the 1970s, dense growths of blue-green algae were reported in the Fairmont chain of lakes in Martin County (Stefan and Hanson 1981). Algae limited swimming and boating activities, created taste and odor problems for the municipal water supply, and poisoned a number of domestic animals. Oxygen concentration in the bottom waters was very low, and it was believed that there was a significant time during the growing season when no oxygen was present along the interface between the water and the bottom sediment. The high concentration of phosphorus in the Fairmont chain of lakes, ranging from 30 to 150 parts per billion, was thought to be in large part a result of phosphorus contained in the lake sediments being mixed into the water (Heiskary and Wilson 1990). These problems have not been solved, and the lakes are still under investigation.

The water quality of Big Stone Lake in the Northern Glaciated Plains ecoregion on the Min-

nesota–South Dakota border has declined during the past 20 to 50 years, leading to a significant decline in sportfishing. Frequent blooms of blue-green algae limit recreational use of the lake in summer. The lake covers 12,610 acres and has an average depth of about 8 feet. In 1939, the lake became a reservoir when a concrete dam replaced the natural outlet. Construction of this dam doubled the watershed to over 700,000 acres. Most of the surrounding land is used for farming, and the increase in surface runoff containing sediment, fertilizers, and pesticides resulted in the decline in water quality. The lake is characterized as hypereutrophic with a phosphorus concentration of 150 parts per billion and transparency ranging from 1 to 4 feet.

Watershed management through restoration of wetlands and control of runoff from agricultural land and feedlots is being implemented to reduce pollution. Water quality in Big Stone Lake is apparently improving, and tourism to the area is increasing (S. Heiskary, personal communication).

Not all lakes in these ecoregions are declining in quality. Clear Lake, a 451-acre, 10-foot-deep lake in Jackson County, has shown a significant increase in transparency over the past 20 years. In the late 1970s, transparency ranged from 1.5 to 2.0 feet. During the late 1980s, it ranged from 2 to 4 feet, suggesting a shift from hypereutrophic to eutrophic conditions. But, in spite of the improvement in transparency, users surveyed said that the lake was "algal green" and "swimming was impaired" (Heiskary and Lindbloom 1993).

Another Clear Lake, in Waseca County, has been restored. Before restoration, Clear Lake, which has a maximum depth of 30 feet, had a phosphorus concentration of about 130 parts per billion and experienced severe nuisance blooms of blue-green algae. Restoration included use of sedimentation basins and wetlands to treat stormwater and agricultural runoff and an alum treatment of the lake sediments to reduce internal phosphorus loading. In the early 1990s, phosphorus concentration was approximately 80 to 100 parts per billion, and nuisance algal blooms occurred much less frequently than in the 1980s (Heiskary and Wilson 1990).

Some of the lakes in the Northern Glaciated Plains ecoregion have high concentrations of sulfate salts and are designated as alkaline or saline lakes. Their pH ranges from 8.4 to 9.0. During periods of drought, the concentration of dissolved salts may increase as much as 50 times as the lake water evaporates (Eddy 1966). Heavy blooms of blue-green algae often occur in these lakes, and the bottom fauna is usually sparse, consisting mostly of chironomids and annelids.

Lakes in these ecoregions with the highest water quality are classified as bass-panfish lakes. Some of these are also stocked with walleyes. Bullheads are the dominant fish in some lakes, and the poorest quality lakes are classified as winterkill-roughfish lakes (Heiskary and Wilson 1990).

## Present status of Minnesota's lakes

Minnesota is a state rich in lake resources, which have a positive impact on our economy and quality of life. Much progress has been made in improving the quality of these resources over the past few decades. Our lakes are cleaner than they were 20 years ago, largely because of federal, state, and local programs to control pollution (Minnesota Pollution Control Agency 1992). Much improvement still must be made, however.

Most of the lakes in Minnesota are very heavily used for a variety of kinds of recreation, which is a very important industry in Minnesota. But many of our lakes are also used as dumping grounds for sewage and industrial wastes, and surface runoff entering lakes from farmland, cities, and roads carries many kinds of pollutants. An assessment of the trophic status of over 15% of the lakes and over 51% of the lake acres in the state showed that 68% of our lakes fully support swimming, 2% fully support fish consumption, and 98% support fish consumption with some restrictions. Nutrients in agricultural runoff are the primary source of pollution in the south and central ecoregions, whereas atmospheric deposition of mercury is a primary concern in the Northern Lakes and Forests ecoregion (Minnesota Pollution Control Agency 1992).

Control of pollution is a primary concern of several state agencies in Minnesota, as well as of lakeshore property owners and recreationists. Wastewater treatment facilities in many of Minnesota's cities, management of manure from feedlots, and the use of minimum or no-till agriculture are reducing nutrient loading. Improvement in water quality has largely been the result of the control of point source discharges of pollutants. Present efforts now include controlling stormwater and other nonpoint source pollution, protecting groundwater, and preventing accumulation of toxic substances. Further improvement of the quality of Minnesota lakes is dependent on continuation of all of these measures.

Unfortunately, not all sources of pollution can be controlled by local efforts. The high level of mercury in fish from some Minnesota lakes, first noted in 1969 (Moyle 1972), is a potential health hazard for both humans and wildlife. Levels increased significantly from 1930 to the 1980s in many northeastern lakes, suggesting that the mercury comes from atmospheric pollution. Water chemistry and watershed geology influence the uptake of mercury by fish. Fish from soft-water lakes and colored lakes are likely to have higher levels than fish from hard-water lakes (Swain and Helwig 1989). Because the process of accumulation of mercury in fish is not well understood, little can be done to reduce the contamination. Therefore, the number of meals of fish from contaminated lakes must be limited.

We have considered many ecological aspects of lakes in this chapter, but in the process we may have lost sight of one of their most significant qualities, namely, the tranquil beauty that lakes contribute to the landscape. Lakes are like sparkling jewels in their effect on humans and in their contribution to Minnesota's environment. We must make sure that our stewardship guarantees the proper future for all of our 12,000 lakes.

# STREAMS AND RIVERS  9

A fisher in waders is standing beside the cool waters of a Minnesota trout stream (figure 9.2). The angler is casting an artificial fly in hope of luring a brown trout from beneath the overhanging bank. The gurgling of the water and the occasional calls of birds living in the adjacent forest are the only sounds. If the angler is skillful, the splashing of the hooked trout will add to the gentle sounds of the stream environment.

If we could examine the food in this trout's stomach, we would likely find a collection of nymphs of mayflies and stone flies. If we could then examine with a microscope the food items in the gut of the mayflies and stone flies, we would be able to identify algae, small bits of vegetation and animals, and perhaps some small invertebrates such as copepods. If we were fortunate enough to have access to a very high-powered electron microscope, we might also be able to identify a number of different kinds of bacteria and fungi attached to the bits of organic matter in the gut.

The food chain described is just one aspect of the functioning of stream ecosystems. In this chapter, I will explain the complex structure and functioning of stream ecosystems in terms of their physical and biological aspects, and I will describe some of the major streams and rivers in Minnesota.

235

**Figure 9.1** *Beauty and power are obvious at this spot on Tait Creek, a tributary of the Poplar River, in Cook County. The large tree trunks were carried and deposited by floodwaters. (Photo by Bob Firth, Firth PhotoBank, Minneapolis.)*

**Figure 9.2** *Fly-fishing for trout is a popular recreation in Minnesota streams. (Photo by Barb LaVigne.)*

## Origins of streams and rivers

Streams and rivers in Minnesota, like wetlands and lakes, originated as a result of the activity of glaciers about 10,000 years ago. In geologic terms, our rivers and streams are relatively young compared to ancient rivers such as the Nile and the Amazon. Some rivers follow the paths of drainage ways that developed under the glacial ice or that carried meltwater from the front edge of the glacier. Some, such as the Glacial River Warren (now the Minnesota River), drained immense glacial lakes. Other rivers began in wetlands or lakes, and their path followed the lowest parts of the landscape. In some locations, water originating from springs and seepages created a stream.

Rivers receive their water from their drainage basin, or watershed area, which extends all of the way from the source of the river to its mouth. The drainage pattern that develops is often shaped like the branches of a tree, with small streams feeding into larger streams, which eventually empty into the main stem of the river. Eight major river basins are present in the state (figure 9.3). The Rainy and Red Rivers, which drain the Canadian Shield and the Lake Agassiz Basin, flow northward toward

Hudson Bay. Water from the St. Louis River and the many swift-flowing streams emptying into Lake Superior along its North Shore reaches the Atlantic Ocean via the Great Lakes and the St. Lawrence River. The Des Moines River, which drains a portion of southwestern Minnesota, enters the Missouri River, which eventually joins the Mississippi River. The St. Croix and Minnesota Rivers also enter the Mississippi, which eventually enters the Gulf of Mexico.

The Mississippi is, of course, the largest river flowing out of Minnesota. The drainage basin of this great river nearly spans the state from west to east and from north to south. At the river's beginning at Lake Itasca, it is inches deep and only a few feet wide (figure 9.4). In the southeast where the Mississippi leaves Minnesota, it is wide and deep and flows in a gorge nearly 4 miles from bluff to bluff and 500 feet deep (figure 9.5).

Water flowing in our streams and rivers comes primarily from subsurface flow, with contributions from rainfall and snowmelt. Of course, the subsurface water from seepages and springs also originates from rain and snow. Snowmelt in March and April increases the flow in all of our rivers. Spring rains add to this flow, and in some years, floods

**Figure 9.3** *(above) Eight major drainage basins are present in Minnesota. Two drain north to Hudson Bay, one drains east through the St. Lawrence River, and the rest flow into the Mississippi River. (Modified from Schwartz and Thiel 1954, by permission.)*

**Figure 9.4** *(below) The headwater of the Mississippi River, flowing out of Lake Itasca, is only inches deep and a few feet wide. (Photo by Alfred K. Peterson, Firth PhotoBank, Minneapolis.)*

develop on certain drainages. Flow throughout the remainder of the year is dependent to a large extent on rainfall, which in Minnesota can be highly variable (see chapter 2).

## Physical and chemical characteristics

### Physical characteristics

Physical characteristics of streams and rivers are often considered to be more important than chemical and biological characteristics in terms of control of the ecosystem (Hynes 1970). Physical characteristics include the characteristics of the stream channel, such as its shape, length, gradient or slope, and composition of the streambed; the characteristics of the moving water, such as its discharge rate and current velocity; and the characteristics of the water, such as turbidity and temperature.

Many of the physical characteristics of a stream are interrelated and are correlated with the stream's order, which is its position in the hierarchy of

streams in its drainage basin. Stream-order categories can be visualized by thinking of a tree branch. First-order streams are the smallest, youngest segments with no tributaries. When two first-order streams join, they create a second-order stream. At least two streams of any given order are required to form the next order. Stream-order numbers are also related to the length of the channel and the watershed area. Stream orders higher then 10 are rare. The Mississippi River is probably the only example in North America.

Runoff from the watershed influences the entire stream. In the headwaters, as the upper segments of streams are called, erosion may be important. Water concentrates in rills and then in furrows, which erode to form gullies. When the gullies keep eroding at their head, they serve to extend the drainage area of the stream. Spring floods may make the bottom unstable in the headwaters area.

Sediments carried into the stream by erosion are moved downstream and are eventually deposited in the lower reaches, where they have a marked impact on bottom characteristics. The moving water erodes the shore at the outside of bends and deposits the eroded material on the inside of bends. This process causes the river to meander and may result in the formation of oxbows. Thus, the bed and shores of a stream frequently change in character from the headwaters to the mouth.

Current velocity, or flow rate, is one of the most important characteristics of streams and rivers because it influences many of their other characteristics. For example, fast-flowing water (figure 9.1) has the power to move large particles, such as grains of sand, gravel, and stones, thus changing the composition of the streambed. Fast-flowing water is usually cooler and has a higher oxygen content than slow-flowing water. Fast-flowing streams also have better ability to recover from

Figure 9.5 *In southeastern Minnesota, the Mississippi River is wide and deep. The river carries much barge traffic and is also heavily used for recreation. (Photo by Bob Firth, Firth PhotoBank, Minneapolis.)*

pollution then do slow-flowing streams. All these factors influence the kinds of plants and animals present in the stream.

Current velocity and the shape of the riverbed determine the volume of water passing a given location over a given period of time. This volume is called the discharge rate. The St. Croix River discharges 5,000 cubic feet of water per second as it empties into the Mississippi River. In contrast, the Otter Tail River discharges 300 cubic feet per second at its mouth in the Red River of the North (Waters 1977).

A given stream or river may have a high current velocity in its headwater reaches and a much slower rate farther downstream. Lowland, downstream reaches usually contain deeper water and higher loads of sediments, which reduce light and therefore the growth of algae. These lowland rivers often meander over the landscape, sometimes forming oxbows and shallow wetlands. In some, the current velocity may be so slow that even with low light, algae develop in much the same manner as in wetlands and lakes.

Current velocity may exhibit marked seasonal changes. In spring, at the time of snowmelt or when heavy rains occur, the stream may be in a flood stage. This high volume of rapidly moving water may carry away organisms from the streambed and may even shape the stones and cobble making up the bed. Ice often scours the bottoms of streams during spring breakup.

When the river overflows its banks, the floodplain receives a plentiful supply of moisture and deposits of rich sediments. These sediments result in rich soil, valued for agricultural crops. The annual supply of moisture frequently enables trees to grow along rivers far out into the prairies. Floodplains are not actually part of the river channel but seem to be part of the river during floods. In spite of the dangers of flooding, human settlements are common on floodplains.

Dry weather during July and August reduces stream flow and may cause some streams to dry out completely. Such intermittent streams are more common farther west but are also typical of Minnesota prairies and agricultural areas.

In contrast to lakes, rivers seldom stratify, because of the mixing action of the moving water. Occasionally the flow may be so slow that deep pools stratify for short periods of time. Pools on the margins of large rivers isolated from the main flow pattern may also stratify.

Flowing water transports a variety of materials. Elements can be transported as dissolved ions,

**Figure 9.6** *Streams originating from peatlands may be brownish because of the large amount of organic matter carried by the water. The Upper Manitou River on the North Shore of Lake Superior flows through peatlands, as shown by the brown water plunging over the falls. (Photo by Bob Firth, Firth PhotoBank, Minneapolis.)*

invisible to the observer. Tiny particles of organic material, soil, and rock are carried in suspension, making the water appear dirty or cloudy. A decrease in transparency due to suspended particles carried in by erosion may result in low productivity of algae and other aquatic plants in the lower reaches of a river.

Large objects such as tree trunks and branches are often carried by floodwaters. Stones and gravel may be rolled and forced along the streambed by the moving water. Flowing water removes over 60 million tons of topsoil from Minnesota uplands each year (Minnesota Environmental Quality Board 1988), and much of this is carried away by streams and rivers.

Temperature is variable in most rivers. In upstream portions, cool temperatures may prevail because of the cool temperature of the groundwater and because of dense shade if the site is forested. Rivers usually warm downstream as the flow slows, the bed widens, and shade decreases.

Many species of fish are sensitive to temperature. For example, brook trout prefer temperatures from 55°F to 66°F, whereas one strain of rainbow trout can live in water as warm as 85°F (Eddy and Underhill 1974). An increase in temperature reduces the amount of oxygen the water can hold, which in turn influences the kinds of fish that can live in the river. Species such as carp can survive much lower oxygen levels than many game fish can.

Most streams freeze over in winter, but water usually continues to flow under the ice. During warm periods in midwinter, the ice may disappear, only to form again as the temperature drops. During a typical winter in the Twin Cities, ice cover on the Mississippi River appears and disappears many times.

### Chemical characteristics

The chemical parameters of streams that are most important to their ecology are hardness, pH, alkalinity, nitrogen, and oxygen. Just as the chemistry of runoff that enters a lake or wetland is influenced by the area over which it runs, the water entering a river is influenced by the rock and soil in its watershed. Streams that flow through watersheds containing limestone rocks are hard-water streams and have a relatively high pH (between 7 and 9). These streams support many kinds and high numbers of organisms. Soft-water, more acidic streams often originate from peatlands. The brownish water attests to their peatland origin (figure 9.6). As is the case in soft-water lakes, these streams have fewer kinds and lower numbers of organisms.

Runoff entering streams along the North Shore of Lake Superior flows over rock outcrops and thus carries little in the way of sediments or nutrients. In contrast, streams in the southeastern corner of Minnesota receive much of their runoff from agricultural lands. As a result, they carry heavy loads of sediment and nutrients.

Most of the nitrogen in streams comes from sources in the watershed such as agricultural fields and livestock feedlots. In addition, significant amounts of nitrogen enter the stream from the atmosphere through precipitation, and some nitrogen is produced by the decomposition of organic detritus in the watershed and in the stream itself. Streams with a high nitrogen content have a higher productivity of plants and animals than streams low in nitrogen.

The oxygen content in streams is highly variable. Some oxygen is carried into the stream with runoff, and some enters in the flow of groundwater. In addition, some enters the stream directly from the atmosphere. Agitation of the water molecules as they flow results in oxygen entering from the air and being mixed from top to bottom in the water column. In headwater reaches, little oxygen is produced by photosynthesis because of the scarcity of algae and aquatic plants. In lower reaches, however, photosynthesis by algae and aquatic plants may contribute a significant amount of oxygen to the stream. Oxygen is used in the process of respiration, which is carried out by all organisms throughout the entire length of the stream. Decomposition of organic matter introduced into the stream can severely deplete oxygen.

## Stream communities

### Flowing water and stream organisms

Stream ecologists agree that the structure or diversity of stream communities is determined mainly by physical and chemical factors and secondarily by food availability, predation, and competition. For organisms living in streams, hydraulics, the mechanics of moving water, are an extremely important characteristic of the stream habitat. For example, a rock has a top and a bottom, an upstream front surface and a downstream back surface. Water may flow smoothly over the top surface, and the bottom of the rock may be embedded in the streambed. The full force of the water strikes the front surface, whereas the flow past the protected back surface is much gentler. Because of these differences in hydraulics, each surface is occupied by a slightly different group of plants and animals (Usinger 1967).

The sediments carried in the water also play a role in hydraulics. Water containing bits of sand is more likely to disturb plants and animals attached to the rock surfaces than is water containing finer particles of silt or clay. These hydraulic stresses may have a strong influence on the structure of the community of organisms occupying the rocks on a stream bottom. In contrast, in pools where the hydraulic stress is low, biotic factors such as grazing by aquatic insects and other invertebrates, predation, and competition may be more important influences on community structure.

### Adaptations to flowing water

To live in the environment of flowing water, plants and animals have adapted in numerous ways. All organisms must face the constant downstream pull of the water currents. Basic activities, such as feeding, mating, laying eggs, or even staying in one place present formidable challenges for animals. In some cases, the adaptations to flowing water are so specialized that animals that live well in fast water are not able to live and reproduce in slow-moving reaches.

Some animals have developed behavioral adaptations to the swift currents (figure 9.7). Crayfish may hide beneath stones or in crevasses in the stream bottom. Caddis fly larvae construct tubelike cases with ballast stones attached to the sides to keep them from moving downstream.

Other animals have developed physiological adaptations to the stream environment. Blackfly larvae spin a pad of silk on a rock and attach hooks on the rear of their body to this pad. The silk acts as a lifeline when the larva breaks loose from the rock. "Hand over hand," the larva pulls itself back and reattaches to its pad (Moss 1988). Other organisms have developed special feeding mechanisms that allow them to filter or strain their food directly from the flowing water.

Many animals have developed morphological or anatomical adaptations for the stream environment. Perhaps the most obvious is the streamlined shape of the trout's body, which offers little resistance to the current. The sculpin, a small fish common in many streams, is not streamlined but has oversized front fins, which it braces against stones to hold itself in place. Larvae of net-winged midges have suction disks on their underside, and snails have suction-pad feet that they attach to the streambed. Stone fly and mayfly larvae are generally flat and somewhat streamlined. Some cling to the bottom of rocks and others build burrows in the sediments.

### Community structure

One way to examine community structure in a stream is by grouping organisms according to their relationship to flowing water. In any reach of a stream, organisms can be classified into four groups: the fixed or sedentary organisms living on the bottom, the drifting organisms, the organisms floating on the surface of the stream, and the active swimmers that move upstream and downstream at will.

Because the velocity of flow changes from place to place and time to time, a wide range of substrates exists on the stream bottom for colonization and community establishment. Plants growing here are mainly algae and mosses attached to rocks and pebbles. Filamentous algae growing in long tufts attached to rocks may be especially abundant. Aquatic mosses with a similar growth form are also common. Insect larvae live among the stones and pebbles on the bottom, some attached by adhesive devices and others clinging with their sharp claws. Several kinds of worms are found in the sediments with one end or the other buried in the mud. Freshwater sponges may be found attached to rocks. A variety of microscopic animals, such as copepods and cladocerans, live in the pore spaces of the sponges or amid the tufts of the attached algae and mosses. Here the rush of the water is slowed, and the organisms are safe from the sweep of stronger currents passing just above them.

The drifting component of the stream ecosystem includes detritus from a variety of sources, plankton, and larger invertebrates. Organic detritus, such as branches, leaves, needles, small bits of plants, dead insects, and other animals, falls into the stream or is carried in by runoff or by wind.

Stonefly larva

Net-spinning caddisfly larva

Blackfly larva

Decomposition and physical breakdown of dead plants and animals in the stream also contribute particles to the drift. These particles are further broken down by the physical action of the stream, by a group of organisms called shredders, by other consumers, and by decomposers. The biomass of bacteria and fungi increases as we move downstream. These decomposers use the carbohydrates in the coarse particles of organic matter and take up dissolved inorganic nitrogen from the stream in their processes of metabolism and reproduction.

Smaller particles of organic matter are referred to as fine particulate organic matter. Some of this enters the stream in the detritus, and the remainder comes from the breakdown of coarse particles by physical action and by decomposers. The smallest particles are referred to as dissolved organic matter. These are the molecules resulting from the breakdown of the fine particulate organic matter, or from leaching directly from the detritus on the ground

or in the stream. In turbulent water, dissolved organic particles may flocculate, or adhere, to each other. These conglomerations of organic molecules, which look like foam on the stream, are recycled by fungi and other organisms, which use the molecules as nutrients.

Plankton make up another component of the drifting category. Algae and zooplankton are relatively rare in the headwater reaches because they have no adaptations for maintaining a population in the headwaters—they drift downstream and have no way of getting back upstream. A copepod, for example, requires about 15 days to develop from egg to maturity (Coker 1954). If the developing organism is in a river flowing at the rate of 1 mile per hour, the adult stage will be reached 360 miles downstream, probably in a totally different environment from the point where the egg was laid.

A large number of invertebrates occur in the drift. Many of these are animals pulled free of

**Figure 9.7** *Invertebrates that live in streams have developed different adaptations to the swift current.*

attachment or pulled out of the sediments by the flowing water. Immature insects and amphipods (figure 9.8) are particularly common in the drift. Higher drift rates have been observed at night than during the day for most of these organisms (Waters 1972). The typical daily pattern is low numbers in the drift during daylight and a great increase just after sunset. During the night, numbers gradually decline. A secondary peak in abundance may occur just before sunrise. The increase in drift at night may be related to the possibility that predators may have difficulty in seizing prey at night.

If all these organisms drift downstream, how are populations in the headwaters replenished? Some organisms actually crawl upstream against the flow, and some adult insects fly upstream to deposit their eggs to begin the drift process anew. In this way the young lost to the drift are replaced. And these young are not really lost, because they are important food for trout, smallmouth bass, and other species of fish, as well as for predatory insects, who position themselves to take advantage of the abundant food in the drift.

The drift may also be a means of dispersal. T. F. Waters and J. C. Hokenstrom (1980) studied the production and drift of a species of amphipod in a stream in southeastern Minnesota and concluded that the high productivity of this amphipod may exceed the carrying capacity of the stream. The excess amphipods were carried downstream in the drift, thereby decreasing crowding.

A few organisms can be observed floating on the stream surface. Duckweeds, which are free-floating plants, may be carried on the water surface but undoubtedly originated from a quiet pool or backwater. Water striders and water beetles can be seen moving on the surface of pools and in calm downstream reaches. Birds such as ducks and grebes are also users of the surface layer, although their mobility hardly qualifies them as exclusive members of the river ecosystem.

A number of animals are active swimmers and have the ability to move upstream or downstream at will. These include fish, crayfish, tadpoles, snakes, and turtles.

Many kinds of lifestyles occur among stream animals, and many different kinds of animals are present. These animals are partitioned in the stream by both time and space. Interactions of many kinds, including foraging, competition, and predation, occur throughout the stream. Thus, the structure of the community, although it changes spatially, dictates the functioning of the stream ecosystem.

## Succession

If one accepts the idea that succession occurs in communities because of disturbance, then one would conclude that succession must be common in streams. Disturbances such as floods can remove nearly all of the organisms from a reach and alter

**Figure 9.8** *Fairy shrimp, one of the many kinds of amphipods, are common in the drift of streams. Most are less than 0.5 inch long. They eat algae and are eaten by trout and other fishes. (Photo by Don Rubbelke.)*

the nature of the substrate. The structure then reestablishes itself based on the rearranged substrate and the availability of colonizing organisms. If flooding occurs frequently, populations of stream organisms may be held to such low levels that competition and predation are relatively unimportant. A given stream reach may gain and lose species because of disturbances such as floods and channel developments, and this process could be viewed as succession. Little is known about this aspect of succession in streams, however, because the high frequency of disturbance does not allow adequate time for measurable changes to occur in the stream community.

In spite of the difficulty in measuring succession in streams, stream ecologists have obtained some data about succession and have found it to be highly variable (Fisher 1983). Succession is influenced by both the timing and severity of disturbance; frequent disturbance results in simpler systems with reduced diversity. If colonizers are present nearby, the stream may be quickly repopulated after a disturbance.

Attached algae and mosses on the stream bottom may survive major disturbances. This community may therefore change gradually over time. In the absence of disturbance, succession in the classic time-related sense may also occur in streams.

Because adjacent upland communities are the source of the organic materials carried into the stream by runoff, succession in these communities has a direct effect on the stream ecosystem. For example, the nutrient content of aspen leaves is lower than that of sugar maple leaves. Therefore, a stream flowing through a maple forest is likely to have a higher nutrient level than a stream flowing through an aspen forest.

## Ecosystem function

### Nutrient cycling and energy flow

To further our understanding of stream communities, we will next examine nutrient cycling and energy flow. Primary production in streams is defined differently than it is for the other ecosystems we have examined: primary production is the production in the stream plus plant parts, such as leaves and twigs, that fall or are carried into the stream. The proportion of organic matter washed in from the land within the watershed to primary production from the stream itself (algae and vascular plants) has a strong influence on community structure. These differences in sources of energy will, to a large extent, determine the dominance of the different groups of invertebrates (Cummins and Klug 1979).

Because vegetation growing in typical streams is sparse, the major input of plant material is the plant detritus washed into the stream from the watershed (figure 9.9). Thus, the vegetation along the banks of a stream may be much more important to the ecology of the stream than the vegetation growing in the stream. Leaves and twigs provide food and hiding places for animals, and as they decompose, they release nutrients into the stream. The availability of nutrients controls the processes of productivity and decomposition and serves to link the watershed and the stream. For example, dissolved organic carbon, which originated in watershed detritus, is used by the bacteria and fungi in the detritus food web. Particles released during decomposition then move downstream, carrying the nutrients with them for use in other communities.

The large input of plant material from external sources has been reported in numerous localities (Hynes 1963) and has been studied in detail in a hardwood forest stream in New Hampshire by S. G. Fisher and G. E. Likens (1973). They analyzed the annual energy budget in terms of caloric input and output for one year. In lower reaches of the stream, direct litter fall contributed 37.4% of the input of organic matter, litter blown in contributed 6.3%, water flowing down the trunks and stems of plants contributed 0.5%, surface seepage water contributed 24.8%, transport of matter from upstream contributed 30.7%, and photosynthesis by the plants in the stream contributed only 0.2%. Thus, nearly all of the input is from the stream watershed. These findings are probably applicable to similar Minnesota streams. Much of this input occurs in autumn when leaves fall from the deciduous trees. In these areas much of the decomposition of the coarse particulate organic matter, the leaves, occurs in winter under quite cold temperatures. This makes streams unique when compared to other Minnesota ecosystems, in that an important part of productivity occurs during the cold winter (Cummins 1974).

The growth of aquatic plants in streams is influenced mainly by the shape of the stream and by the availability of light and nutrients. In many streams, mosses are common in headwater reaches, and aquatic plants are common in downstream reaches. Plants are not often abundant in shady woodland streams. In a prairie stream, however, J. K. Neel (1985) reported finding numerous species of algae attached to rocks. Filamentous green algae and diatoms were most abundant.

The animals that consume plants can be grouped in terms of their feeding habits. Herbivores, which eat living and dead plant material, can be divided into four groups: shredders, collectors, scrapers, and piercers (figure 9.9).

Shredders, which are mostly invertebrates, eat leaves, needles, and other bits of vegetation and also obtain energy from the bacteria and fungi attached to the vegetation (Cummins 1974). The shredders actually select those bits of vegetation with high numbers of microbes, especially fungi. These fungal tissues contain rich protein formed from carbohydrates taken from the vegetation and nitrogen taken from the water. Stone fly, caddis fly, and crane fly larvae and scuds are all shredders.

Invertebrates in the collector group filter or strain food particles out of the water using hairs, bristles, or nets. Collectors were the most important secondary producer group in three Minnesota streams with different water qualities (Kreuger and Waters 1983). The collectors consume particles that are left over from shredders and scrapers. Bac-

teria and fungi on these particles are also important food for the collectors.

Two groups of collectors can be found in a typical stream. Filterers capture detritus particles in suspension. Larvae of some species of caddis flies (Hydropsychidae) spin small nets (figure 9.7) to catch food carried by the current. These nets capture fine particles of plant and animal material, fecal pellets, and even small insects. Blackfly larvae have comblike hairs on their antennas that filter fine particles from the current. Gatherers, the second group, forage on fine particles on the bottom and in sediments. Large numbers of individuals of many species belong in this group. Mayfly nymphs and midge larvae are the most common in Minnesota streams.

The third group of herbivores, scrapers, includes snails and some species of mayfly and caddis fly larvae. They use their mouth parts to scrape algae and moss from rock surfaces. They also feed on aquatic plants and occasionally on detritus. These scrapers or grazers are less important then the shredders or collectors in terms of secondary production.

Piercers are the final group of herbivores. They include the corixid bugs and caddis fly larvae that suck juices from plant cells.

All of these herbivores are eaten by a wide variety of predators or carnivores that capture their prey alive. Some predatory insects are piercers, which feed by seizing their prey and sucking juices from its tissues and organs. Often piercing predators capture other insects, but some, like the giant water bug (Belastomatidae), may capture small fish and tadpoles. Other predators such as fish, turtles, and some immature insects can be classified as engulfers. They bite off chunks or swallow their prey whole.

In earlier chapters I pointed out that the detritus food web is usually far more important than the grazing food web in terrestrial and lake ecosystems. The detritus food web is similarly significant to the functioning of the stream ecosystem. Detritivores can ultimately consume a major part of the biomass in the stream. Measuring nutrient cycling and

246

energy flow in the detritus food web, however, is difficult. Bacteria and fungi are probably present on all dead organic matter and are consumed along with the plant and animal remains. Bacterial and fungal tissues, as mentioned earlier, are a significant energy source for many species of consumers.

Nutrient cycling from the processing of organic matter is critical if the stream is to neutralize the effects of pollution. The high nitrogen content of wastes dumped into a stream, for example, can be reduced by stream organisms to levels at which the nitrogen is no longer detrimental. Such processing

of nutrients in a stream is influenced by the streambed itself and by the cover provided by vegetation and structures such as logs, large rocks, and gravel bars. These determine how fast materials move along the stream and thus how long waste may remain in a given part of the stream to be consumed or decomposed.

### Productivity

In the same way that alkalinity and other aspects of water chemistry influence the productivity of lakes in Minnesota (Moyle 1956), water chemistry also

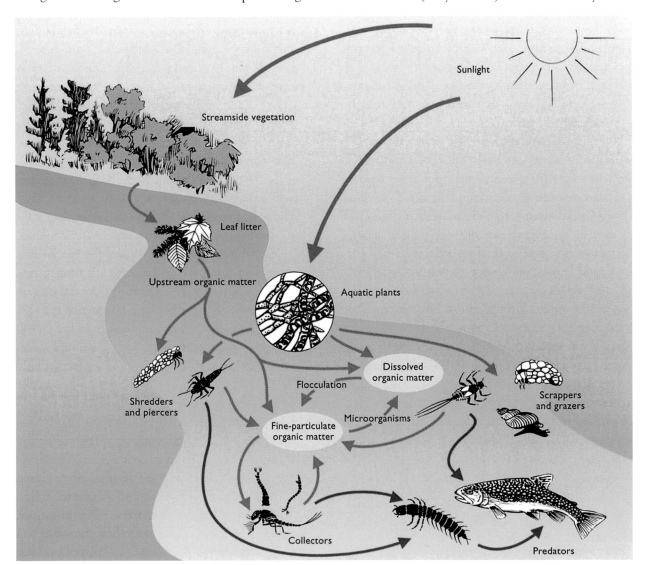

**Figure 9.9** *Movements of nutrients and energy in a stream ecosystem are shown by the arrows in this food web. The green arrows represent movement of vegetative material through the grazing and detritus food web. The red arrows represent predation.*

influences the productivity of streams. The same northeast to southwest pattern that exists for lakes can be observed in streams; alkalinity increases from 15 or 20 parts per million to 175 parts per million. High values in the southwest are probably related to the increased nutrients carried in runoff from agricultural fields. The carrying capacity of the streams for fish increases from about 20 pounds per acre in the northeast to 400 pounds per acre in the southwest.

In a recent analysis of streams throughout Minnesota, T. F. Waters (unpublished manuscript) used alkalinity to develop a productivity classification. His designators of oligotrophic, mesotrophic, and eutrophic follow the lake classification described in chapter 8. In general, oligotrophic streams occurred in the northeast, mesotrophic in central Minnesota, and eutrophic streams in western and southwestern regions.

Productivity and energy flow in Minnesota streams were examined with reference to four trophic levels: net primary production, production of invertebrates, production of small fishes, and production of fish-eating fish in the top carnivore category. Net primary production included plants growing in streams plus detritus carried in from the watershed. In the northeast, the production of plants in the streams was about equal to the production of plants in lakes (table 9.1). The addition of plant detritus made the streams two to five times more productive than lakes. In the southwest, where the flow rate was slower and where sunlight was more available, plant growth in the streams was greater in spite of the high turbidity of the silt-laden water. Input from the watershed resulted in net primary productivity being about twice as high in streams as in lakes.

Waters (unpublished manuscript) calculated net primary production in Minnesota streams in terms of pounds per acre per year of plant material. Net primary production ranged from 15,076 pounds per acre per year in the northeast to 47,500 pounds per acre per year in the southwest (table 9.1).

Table 9.1 *Annual production in selected Minnesota streams (wet weight in pounds/acre/year)*

| County | Net primary production | Bottom organisms | Small fish | Fish carnivores |
|---|---|---|---|---|
| Cook | 15,076 | 475 | 48 | 1 |
| Kanabec | 23,104 | 1,005 | 102 | 10 |
| Jackson | 47,500 | 2,690 | 230 | 35 |

*Source:* Modified from Waters (unpublished manuscript).

Primary production is used as food by insects, crustaceans, mussels, worms, and other consumers. Because a high percentage of the primary production actually comes from the land surrounding the stream, most of these consumers are detritus feeders. Even where rooted plants are present in the stream, they appear to be little used by grazers. In some locations, however, algae may be heavily grazed (Gregory 1983).

In his analysis of productivity in Minnesota streams, Waters (unpublished manuscript) included invertebrate carnivores and detritivores in a category labeled bottom organisms. Their annual production ranged from 475 pounds per acre per year in the northeast to 2,690 pounds per acre per year in the southwest (table 9.1).

The third trophic level included small fishes, like minnows and darters, as well as stream trout, smallmouth bass in northern streams, and channel catfish, suckers, and carp in southern streams. The productivity of this group of fishes ranged from 48 pounds per acre per year in the northeast to 230 pounds per acre per year in the southwest (table 9.1).

The abundance of invertebrates observed in the streams did not seem adequate to support the abundance of fish. This apparent paradox was examined in some detail by C. C. Kreuger and Waters (1983), who concluded that the fish were feeding on insects from the surrounding land, and suckers, minnows, and carp were probably obtaining a large amount of their food from the detritus washed in from the land.

The fourth trophic level consisted of the top carnivores, the fish-eating fish such as large trout, northern pike, and walleye. As explained in the discussion of ecological principles, productivity declines by about 90% from one trophic level to another. Thus it is not surprising that Waters (unpublished manuscript) reports that the production of top carnivores varied from 1 pound per acre per year in the northeast to 35 pounds per acre per year in the southwest.

If we could observe the movement of materials in a stream or river until the mouth is reached, we would find that consumption by higher trophic levels and by decomposition removes about 34% of the organic input. Thus 66% is transported to the ocean or lake at the mouth of the river. Although these estimates were obtained in an analysis of the energy budget of Bear Brook in New Hampshire (Fisher and Likens 1973), they are probably typical of streams flowing through deciduous forests.

## River continuum concept

All of this productivity that we have just discussed occurs even though the detritus and organisms are continually moving downstream with the current. Energy flow and nutrient uptake and release are occurring along with this movement. Because the nutrients are displaced downstream as they move through the trophic levels, the term resource spiraling is used to describe this process.

In recent years scientists conducting research on rivers have begun to view the spiraling process in the river ecosystem in terms of a continuum of patches or communities. They think that nutrient spiraling and the dynamics of populations are determined largely by the physical state of the river at a given location, the form of the channel being the key factor (Naiman et al. 1987; Vannote et al. 1980). The continuum concept provides a frame of reference for examining rivers and enables us to examine processes such as succession and productivity in this unique ecosystem.

The continuum concept divides a river into three sections: headwater streams (orders 1 through 3), the middle of the river (orders 4 through 6),

and the lower reaches (orders greater than 6) (figure 9.10). In any given river, the continuum is not necessarily stable. Disturbances such as fire or logging may shift the continuum upstream or downstream because of changes in light, temperature, and input of organic material in runoff.

The headwater portion of the continuum of a typical river receives high input of organic matter from the watershed. Shoreline vegetation shades the entire stream, and because of the shade and the current, algae and aquatic plants are sparse. Shredders are more abundant here than in the middle or lower reaches. Primary production in the stream is low, and fish, such as trout, sculpin, and dace, feed mainly on insects.

In the middle of the continuum, streams are wider, and only a portion is shaded. Algae and aquatic plants grow here, and the stream is supported by energy produced within the system. Collectors and grazers are the common herbivores in this reach. Because of the availability of organisms from the headwaters, drift is an important source of food for organisms living in this part of the river. Insect-eating and predatory fish such as smallmouth bass, rock bass, shiners, and darters live here.

The lower portion of the continuum receives much fine particulate organic matter from upstream and little coarse particulate organic matter from its surroundings. Thus, particle size has decreased through the continuum. The water is now considered to be aged, and its high turbidity limits light and consequently plant growth. Here productivity is less than respiration because the major energy source is from organic matter produced upstream and along the river edge. Because the water is moving slowly, some algae develop in spite of the turbidity, and rotifers and copepods, which feed on the algae, are present. The collector group of herbivores is most abundant in this reach because of their ability to consume fine particles.

Plankton-eating fish such as carp and buffalo are common in the lower reaches. The organisms living in the bottom sediments here are similar to

249

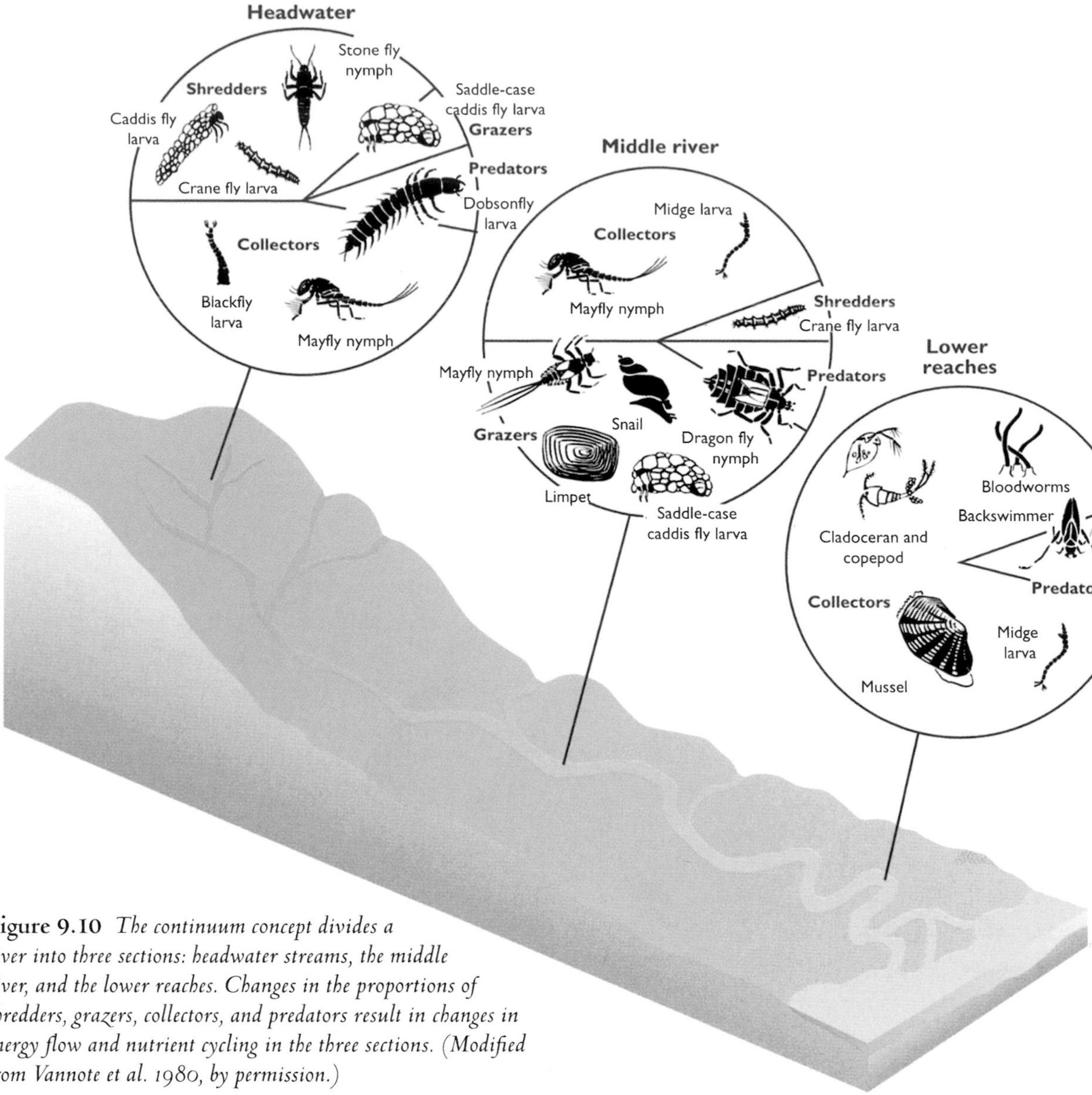

**Figure 9.10** *The continuum concept divides a river into three sections: headwater streams, the middle river, and the lower reaches. Changes in the proportions of shredders, grazers, collectors, and predators result in changes in energy flow and nutrient cycling in the three sections. (Modified from Vannote et al. 1980, by permission.)*

those in many lakes. Insect larvae, tubifex worms, and mussels are likely to be especially common.

Some biologists think that competition may not be important in stream communities because of the harsh physical environment and the spatial and temporal variation that occurs (Hart 1983). D. D. Hart, however, presented evidence from studies to show that competition sometimes occurs. He indicated that competition for nutrients has been observed and that competition within and among species influences spacing, dispersal, and drift. He concluded that the role of competition in streams

has been underestimated and that it sometimes plays a critical role in habitat selection, foraging behavior, survivorship, and growth.

## Animals and community interactions

### Invertebrates

Most animals in a typical Minnesota stream are aquatic insects. Insects can be found clinging to objects in the stream with their legs and claws, attached to objects with suctionlike devices, or burrowed in the streambed. In North Shore streams, three groups of immature insects make up the majority of the stream community, stone flies, mayflies, and caddis flies (figure 9.7) (Waters 1986). In some streams, blackfly larvae, which use tiny hooks to anchor themselves to rocks, are very abundant. Fanlike combs on the head of the larva catch food particles from the water. Sometimes the blackfly larvae are so abundant that they look like a coating of moss on the rock surface. Midge larvae, similar to those found in the sediments of wetlands and lakes, also occur in stream and river sediments. These are likely to be common in the lower reaches of a river, whereas the blackfly larvae are usually found in headwater reaches.

Among the other invertebrates common in streams are tubifex worms, which are related to the common earthworm. The worms live on the stream bottom, with their heads buried in the mud. As the mud passes through their body, the organic matter is filtered out and used as food. Leeches and mussels are also common in Minnesota streams.

### Fish

Fish populations are highly variable depending on the characteristics of the stream or river. Several species of trout and smallmouth bass are the common game fishes in cool, clear streams. They require high levels of oxygen, cool temperatures, and a sand or gravel bottom without silt in which to spawn. Trout lay their eggs in depressions in the gravel and cover them to prevent washing away and predation.

The lower reaches of streams and rivers may contain many of the same species of fish present in our lakes. Several species such as catfish, paddlefish, and sturgeon are more common in river environments than in lakes. Smaller fish such as minnows, darters, and sculpins are also common in many Minnesota streams.

### Amphibians and reptiles

The largest Minnesota amphibian, the bullfrog, can be found in southeastern Minnesota, particularly in wetlands and backwaters along the Mississippi River. Bullfrogs, which may exceed 12 inches in total length, prefer slow-moving, permanent water with abundant vegetation. Tadpoles do not metamorphose until their third summer. Bullfrogs are rare in Minnesota, perhaps because they are adapted to a warmer climate, but also because they are easy to catch and are used for bait and food.

Pickerel frogs, somewhat similar in appearance to northern leopard frogs, can be found in some of the clear, cool springs and spring-fed streams in southeastern Minnesota. Unlike the leopard frog, which moves into fields and meadows, pickerel frogs remain near the stream, where vegetation affords protection. Shade provided by the deciduous forest may also be beneficial to this amphibian. Secretions from the skin of pickerel frogs are so distasteful that many snakes and birds refuse to eat them. R. C. Vogt (1981) reported that other kinds of frogs placed in the same containers with pickerel frogs may be killed by the secretions.

Pickerel frogs are apparently sensitive to water quality and therefore to pollution. Too much agricultural runoff containing fertilizers or pesticides may eliminate pickerel frogs from a stream (Vogt 1981). Thus, their presence is a good indicator of high-quality water.

Almost any species of Minnesota amphibian may be found associated with streams and rivers. In addition to those just mentioned, green frogs and leopard frogs are the most common.

Two groups of turtles, map turtles and softshell turtles, are especially adapted to streams and rivers. Three species of map turtles, Ouachita, false, and common, occur in Minnesota. They often live together in slow-moving waters, such as the St. Croix and Mississippi Rivers. Stream reaches with dense aquatic vegetation seem to be especially preferred. Eggs are laid in early summer, often on a sandbar or near the river. Some hatchlings emerge in September, but if a clutch of eggs is laid late in the summer, the young will not emerge until the following spring. Map turtles feed primarily on invertebrates, including mussels, crayfish, and

251

insect larvae (Vogt 1981). False map turtles eat more vegetation than common map turtles do but also consume a variety of invertebrates and fish. All three species have been found overwintering together, often behind rock piles or wing dams but sometimes in muskrat or beaver burrows.

Two kinds of softshell turtles, smooth and spiny (figure 9.11), live in Minnesota rivers. They have long necks and a leathery carapace, which gives them the appearance of a large pancake. These species are most common in large rivers with muddy bottoms. Nests are located on sandbars or beaches, but the females are so wary that they are rarely observed laying their eggs. They are carnivorous, eating crayfish, mussels, fish, amphibians, and aquatic insects, which they often capture by ambush while lying on the bottom partly covered with sediment. Softshell turtles have the ability to spend

long periods, up to five hours, underwater. Consequently, they are susceptible to low levels of oxygen and may be one of the first species to disappear from a polluted river.

### Birds

Many species of birds can be observed along streams and rivers, taking full advantage of the available food, cover, and water. A number of the species that actually live on water, described in the chapters on wetlands and lakes, also use stream and river environments.

A spectacular member of the stream ecosystem is the belted kingfisher (figure 9.12), with its rattling call, large head, heronlike bill, and iridescent blue plumage. Kingfishers feed largely on small fish but also eat insects, crayfish, mussels, frogs, lizards, and small snakes. At times they may even consume

**Figure 9.11** *Spiny softshelled turtles are excellent swimmers and are found in many Minnesota rivers. Often they lie covered with sand or mud with only their neck and head showing. Unwary crayfish and aquatic insects are quickly snapped up by their powerful jaws. (Photo by Barney Oldfield.)*

**Figure 9.12** *Belted kingfishers can easily be observed as they dive into the river to capture minnows and other prey. They nest underground in tunnels in the riverbank. (Photo by Thomas D. Mangelsen/Images of Nature.)*

berries and fruit (Roberts 1932). Kingfishers observe the stream from a strategic perch and then quickly dive to capture prey with their bill. At other times kingfishers may hover above the water, watching for fish in much the same manner that a kestrel searches for mice in a meadow.

Kingfishers nest in tunnels dug 5 to 10 feet into a bank. Both members of the pair dig the tunnel with their large bills. Nests consist of only a few sticks of grass and leaves and usually contain an abundance of fish bones and scales. Young kingfishers remain in the nest until they are ready to fly.

### Mammals

Mammals associated with lakes or wetlands are also likely to be found in streams and rivers. Beavers (figure 9.13), mink, otters, muskrats, and raccoons are common.

Water is an essential part of beaver habitat, and Minnesota's many streams and rivers held large numbers of beavers before European exploration and settlement. Demand for beaver pelts, used to make coats and top hats, drew trappers to Minnesota. The North West Company established several fur trading posts in the area in the late 1700s, and unrestricted trapping continued until populations of beavers were eliminated or greatly reduced. Like the loggers, the trappers then moved farther west to the mountains.

Populations have recovered, largely because of the growth of hardwoods in the succession following logging and because of restrictions on trapping seasons and limits. Today, beavers are present throughout the state. Detailed studies on the impact of beavers on the northern coniferous forest landscape in Voyageurs National Park found that beaver ponds and impoundments increased from 1% to 13% of the total area between 1940 and 1986 (Johnston and Naiman 1990). The

beaver population increased from near extirpation to nearly three colonies per square mile. By 1986, beavers had built dams on 87% of the fourth-order streams in a major drainage basin in the park.

On lakes, beavers may live in bank dens or in lodges that they build. On streams, dams are constructed to provide a deep pond for the lodge and food cache, which is also underwater (figure 9.14). Dams are constructed of branches, twigs, and mud, with a gradual slope on the upstream side and a steeper slope downstream. On streams with steep banks, beaver dams may be 6 feet or more high. On streams with gentle banks, dams are lower but may extend for hundreds of yards. Lodges are constructed of branches and mud as islands in a pond or are dug into the bank. The living chamber in the

**Figure 9.13** *Beavers sometimes cut trees as far as 400 feet from water and then carry or drag the sticks and branches back to their food caches. (Photo by Blacklock Nature Photography.)*

254

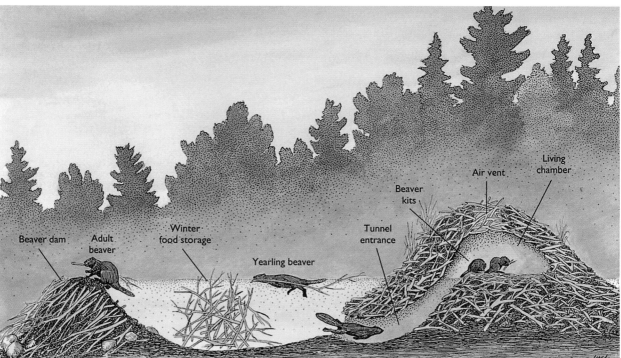

**Figure 9.14** *The living chamber in a beaver lodge is above water and is used year-round. Deep entrances enable beavers to obtain food from underwater caches in winter. (Illustration by Don Luce.)*

lodge is above water (figure 9.14) and is used year-round. Because entrances to the chamber are below water and below the ice in winter, beavers in the lodge are safe from most predators.

A typical beaver colony consists of a female and her mate, yearlings from the previous breeding season, and the current litter. They all live in the same lodge or den. In spring, when the new litter is born, the two-year-old young are chased out of the lodge by the adults and must establish themselves in new colonies.

In summer, beavers feed mainly on aquatic plants and bark from nearby trees and shrubs. In addition to feeding and repairing their dam and lodge, the members of the colony must store food for the winter. The lake or pond must be deep enough so that the winter food supply can be stored beneath the ice. Food stored in the cache, which is anchored in the mud at the bottom of the pond, consists mostly of branches from trees and shrubs that the beaver have cut. Aspen, alder, and willow are favored winter foods in Minnesota. The cambium, or growth layer, under the outer bark, is probably the most important part of their winter diet.

Although beavers will cut trees as large as 15 inches in diameter, they cannot transport "sticks" much larger than 4 inches in diameter. The trunk and larger branches are wasted, although they sometimes feed on the bark of these parts of the tree. Small branches are cut off in short lengths, and beavers have been known to transport these sticks as far as 400 feet.

Activities of beavers may have both positive and negative effects. On the one hand, the ponds provide habitat for ducks, muskrats, mink, and other wildlife and play a role in retaining sediments and organic matter, conserving runoff, and maintaining the water table. On the other hand, dams may flood valuable forest or farmland by blocking drainage ditches and culverts, and the slow-moving water may become too warm and silted for trout.

Beavers are frequently compared to loggers, and the comparison is apt. A colony living near a stand of aspen trees averaging 2 inches in diameter can harvest over 2 acres per year. In view of the flooding caused by their dams and the large acreage of trees required for food and construction materials each year, beavers can obviously have a major impact on their habitat.

Another mammal commonly associated with streams is the water shrew (figure 9.15). Not much is known about this shrew because of its secretive

**Figure 9.15** *A streamlined shape and large hind feet fringed with long, stiff hairs enable water shrews to capture crayfish, minnows, and insects in streams. They also eat earthworms and insects that they capture on land. (Photo by Don Rubbelke.)*

nature. It is especially adapted to an aquatic environment. Its large hind feet fringed with long stiff hairs, its short, dense fur, and its streamlined shape enable it to move easily underwater as well as on the surface of the water. The water shrew has even been observed running across the surface of a small pond, supported by the surface tension of the water (Jackson 1961).

Water shrews feed on a wide range of animal foods, taking slugs, earthworms, and spiders from upland areas (Whitaker and Schmeltz 1973) and crayfish and insects, such as mayflies, stone flies, and caddis flies, from the stream. Their movements are extremely quick, and I have observed them capturing minnows in an aquarium with little difficulty. Presumably they feed on small fish in the natural environment as well. Larger fish such as trout and northern pike may prey on water shrews, however.

## Minnesota's streams and rivers

Streams and rivers can be classified by watershed, by size, by order, or by any one of several other characteristics. Much of the information on individual streams and rivers in the remainder of this chapter comes from the Minnesota Department of Natural Resources (DNR) Section of Fisheries and is organized in the same manner as the section's files. These files, which contain reports of every survey of every stream and river in the state, are classified into two categories: cold-water streams supporting trout and warm-water streams supporting warm-water game fish.

### Cold-water streams

Cold-water trout streams in southeastern Minnesota (figure 9.2) lie in the unglaciated area, and many are spring fed. They frequently have high alkalinity from the limestone bedrock in which they originate, and some have very high levels of nitrogen, probably originating from runoff from cropland. Aquatic insects and amphipods thrive in these streams and serve as major foods for trout.

**Table 9.2** *Comparison of trout streams on the North Shore and in southeastern Minnesota*

|  | North Shore | Southeast |
|---|---|---|
| Water source | lakes and runoff | springs and runoff |
| Flow | variable | stable |
| Gradient | up to 100 ft/mile | about 25 ft/mile |
| Geology | volcanic and glacial | limestone bedrock |
| Temperature range | 32° to 80°F | 32° to 75°F |
| Oxygen | 0 to 14 ppm | 5 to 14 ppm |
| Alkalinity (dissolved calcium carbonate) | 15 to 50 ppm | 200 to 300 ppm |
| pH | 5.0 to 8.5 | 6.0 to 8.5 |
| Productivity (fish/acre/yr) | 25 to 40 lbs | up to 500 lbs |

*Source:* Adapted from Breining (1989).

The South Branch of the Whitewater River in Olmsted and Winona Counties is a typical stream. It is 30.5 miles long, about 25 feet wide in its middle reaches, and 37 feet wide in its lower reaches. Most upland in the watershed is used for crop production. Some pasture and row crops occur in the valley bottom, which is narrow and wooded. Flow is fast because the streambed drops about 22 feet per mile. The main sources of sustained flow are groundwater seepage and input from tile lines draining agricultural fields. Floods from snowmelt and heavy rains are common and cause bank erosion and degradation of trout habitat.

Brown and rainbow trout, as well as white suckers and dace are common. Fishing pressure is high; 2,277 hours of fishing per mile were recorded in 1987 (Minnesota Department of Natural Resources, Section of Fisheries, unpublished data).

Trout production, mostly of brown trout, in southeastern streams may be as high as 200 pounds per acre per year, which greatly exceeds the productivity of trout streams along the North Shore of Lake Superior (table 9.2). If one adds the production of sculpin and suckers, production in these southeastern streams may reach from 400 to 500 pounds per acre per year (Waters, unpublished manuscript).

**Figure 9.16** *The Beaver River in Lake County and most other streams along the North Shore of Lake Superior support populations of brook and rainbow trout. (Photo by Blacklock Nature Photography.)*

The valleys of the Whitewater and other streams cut through the unglaciated area on their way to the Mississippi River. The DNR lists 75 streams or reaches in this area that are managed for trout. Cold springs originating near the bases of bluffs are ideal for brook trout, which are native to the area. After European settlement, many of the streams deteriorated because of pollution and sedimentation. Introduced brown and rainbow trout, which are more tolerant of such conditions, have replaced brook trout in many of these streams.

The DNR lists over 150 streams or reaches managed for trout along the North Shore of Lake Superior (figure 9.16). Most of these streams are short, less than 25 miles, and have gradients of 50 to 100 feet per mile (Waters 1977). The Temperance River in Cook County is a typical North Shore stream. It begins in a complex of lakes and wetlands about 26 miles from its mouth in Lake Superior. The many lakes and wetlands in the headwaters ensure a somewhat stable flow in the Temperance, but in much of the river the flow is turbulent because of the steep gradient with its waterfalls and rapids. Most of the watershed was logged or burned in the early 1900s. The resulting loss of shade led to an increase in midsummer water temperature, possibly limiting use of the river by trout. The present forest provides shade for much of the river, but trout populations are very low.

Brook trout were first stocked in the Temperance River in 1910, smallmouth bass in 1924, brown trout in 1926, rainbow trout in 1928, steelhead in 1971, and chinook salmon in 1980 (Minnesota Department of Natural Resources, Section of Fisheries, unpublished data). A small population of wild brook trout is believed to be present in the middle reaches, but stream censuses by the DNR have collected few individuals. Repeated stockings of other species were made but were not considered to be successful in establishing wild breeding populations.

In most North Shore streams, brook trout were found originally in the lower reaches below the first waterfall barrier. Now these lower reaches are used as nursery areas by rainbow or steelhead trout, which were introduced to Lake Superior in the late 1800s. The rainbows can outcompete the brook trout and essentially eliminate them from a stream reach.

In spring, male and female rainbows move upstream where the female selects a location on a bar or gravel bed, builds her nest by fanning a depression in the gravel, and lays her eggs. The eggs are immediately fertilized by the male and then covered with gravel. The adult rainbows then return to Lake Superior. The young must fend for themselves when they hatch and squirm out of their gravel nest.

But what about the brook trout that were native in these lower reaches and were eliminated by the rainbow trout? Through the fisheries management program of the DNR, brook trout have now been stocked in the upper reaches of many North Shore streams. The rainbows cannot travel to these upstream areas because of the waterfall barriers.

Cold-water streams are fascinating to tourists, naturalists, and anglers. Their inner workings are hidden from view, but the waterfalls, rapids, pools, and eddies reveal some of their dynamics. Clear,

cool, fast-moving streams provide enjoyment to humans as well as homes for trout and many other organisms.

*Warm-water streams*

Most of the remaining flowing waters in Minnesota are classified as warm-water streams. Game fish present in warm-water streams include northern pike, walleye, crappie, largemouth bass, and smallmouth bass. As the adjective warm-water implies, water temperature is higher than in cold-water streams, but these streams differ from cold-water streams in many other characteristics as well. In general, alkalinity is higher, sedimentation is more common, transparency is lower, the gradient is lower, pollution is greater, and productivity is higher. To provide a broad picture, I have chosen the Kettle, Des Moines, Minnesota, Otter Tail, and Mississippi Rivers as examples.

The Kettle is one of Minnesota's outstanding recreational rivers, with excellent opportunities for canoeing, fishing, camping, and hiking. It begins in an area of shrub wetlands in Carlton County and continues eastward for about 80 miles, where it empties into the St. Croix River. The drainage area is 1,060 square miles, and the river has a total drop of 500 feet. This drop is not uniform, however. In a 2-mile stretch in the Kettle River gorge, the gradient is about 40 feet per mile. In other reaches, the gradient is as low as 3 feet per mile. The fast reaches provide exciting canoeing and kayaking as well as excellent smallmouth bass fishing.

Before 1900, the landscape through which the Kettle River flowed was predominantly white pine forest. Following logging, the landscape changed to open pasture and second-growth forests of aspen, birch, and other hardwoods. Remnants of an old logging dam are still present near Sandstone. Dissolved organic chemicals from the bogs and other wetlands in the headwaters color the water brown. In the 1930s, these reaches were ditched and channeled to speed the flow of water off the land.

Today, ruffed grouse, woodcock, and deer are abundant in the aspen forests bordering the Kettle, but rare species also live here. Threatened species include the wood turtle and Blanding's turtle. Lake sturgeon, listed for special concern, spawn in the Kettle, and the largest sturgeon ever legally caught in Minnesota, weighing 94 pounds, 4 ounces, was taken from the river in 1994 (Henry Drews, Minnesota Department of Natural Resources, Section of Fisheries, personal communication).

In recognition of its special values, the Kettle was selected as the first river to be designated under the Wild and Scenic Rivers Act passed by the Minnesota legislature in 1973. This designation provides protection of the river resources and, I hope, ensures that the natural values of the Kettle River will be preserved.

Moving to Murray County in southwestern Minnesota, to the top of the Coteau des Prairies, we find the origin of the Des Moines River. It begins in Lake Shetek, winds 94 miles southward to the Iowa border, and empties into the Mississippi after flowing across the entire state of Iowa. With an average gradient of 2.5 feet per mile, the river drops about 235 feet from its headwater lake to the Iowa border. It averages about 60 feet in width. The East and West Forks of the Des Moines River drain 1,550 square miles. The land adjacent to the stream is composed of about equal amounts of wooded pastures, open pastures and croplands, and deciduous forests. Surface runoff from the rich loam soils used for growing corn and soybeans carries sediments, fertilizers, and pesticides into the river, which is classified as eutrophic by the DNR's Section of Fisheries (unpublished data).

Many lakes and wetlands occur in the Des Moines River watershed, and some, such as Talcott Lake and Heron Lake, are especially important for waterfowl and other aquatic wildlife. Stream surveys show large populations of carp, bigmouth buffalo, river carpsucker, and black bullhead. Walleye, northern pike, and channel catfish are also present and are highly sought after by anglers. Walleye and northern pike probably reproduce in lakes and wetlands through which the river flows, and channel catfish reproduce naturally in the river (Minnesota

Department of Natural Resources, Section of Fisheries, unpublished data). As one would expect in a shallow, eutrophic river, fish kills due to low oxygen content of the water occur occasionally.

Rivers flowing north and east off the Coteau des Prairies empty into the Minnesota River. This river begins at the border between Minnesota and South Dakota in Big Stone Lake in Big Stone County and flows 333 miles through the riverbed of Glacial River Warren to join the Mississippi River at St. Paul. It drops 280 feet with an average gradient of 0.8 foot per mile, has 201 tributaries and 6 dams, and drains nearly 17,000 square miles, 14,751 in Minnesota and the rest in South Dakota and Iowa (Kirsch et al. 1985). Surface water flow to the river comes from 1,208 minor watersheds. Tributary streams generally have low gradients until they reach the steep walls of the river valley. Here they drop about 200 feet to the floodplain in a relatively short distance.

The Ortonville dam controls the level of Big Stone Lake. Dams at Big Stone National Wildlife Refuge, Marsh Lake, and Lac qui Parle are used for flood control. The dams at the latter two sites are located where the deltas of tributary streams impeded the natural flow of the river. These three dams create extensive headwater lakes, which are important wildlife management areas and public hunting grounds. Near Granite Falls in Chippewa County, two natural waterfalls occurred. These have both been altered by construction of dams to produce hydroelectric power, but only one of the dams is still used for power generation.

Today, most of the watershed of the Minnesota River is farmland carved from the prairie sod and clearing of forests. Row crops, primarily corn and soybeans, cover approximately 82% of the watershed (figure 9.17). Eight percent is pastured and 3% is forested. Cultivation up to the river edge causes slumping and erosion of the banks. The river is muddy and polluted to the extent that swimming is not recommended, and anglers are warned to limit consumption of fish taken from the river.

259

**Figure 9.17** *This soybean field bordering the Minnesota River was recently subjected to heavy rainfall. Erosion of the soil is obvious, as is the path of runoff carrying soil particles to the river. (Photo by Bob Firth, Firth PhotoBank, Minneapolis.)*

Productivity is impeded by the low transparency caused by flooding, bottom scouring, disturbance of bottom sediments by carp and suckers, silt deposition, bank erosion, and runoff from adjacent cropland. Low populations of aquatic plants and invertebrates and low oxygen concentrations limit production of desirable fish, such as walleye, sauger, northern pike, and catfish (Minnesota Department of Natural Resources, Section of Fisheries, unpublished data).

The Minnesota River is considered one of the state's most polluted waterways. In 1989, more than 30 federal, state, and local agencies began a comprehensive study, the Minnesota River Assessment Project, to evaluate the sources and effects of pollution. Four categories of pollutants—bacteria, sediments, nutrients, and substances that reduce the oxygen level in the water—were found to be

of greatest concern (Minnesota Pollution Control Agency 1994).

Runoff from livestock feedlots and inadequately treated sewage from septic systems and cities often contain disease-associated bacteria that pose health risks for people swimming and waterskiing. Sediments, mostly nonpoint source fine particles of silt and clay from croplands, are carried into the river during rainstorms and snowmelt via ditches and tributaries. Tile and open-ditch drainage of croplands and wetlands within the watershed has speeded removal of water and sediments from the land. In the river, these sediments destroy fish spawning sites and eggs and limit the oxygen available to aquatic life.

Phosphorus and nitrogen are the two most important nutrients causing problems. Both result in increased production of aquatic plants and algae, which cause nuisance blooms during times of heavy runoff. Nitrogen in the form of ammonia may accumulate in sediment and become toxic to aquatic organisms. In addition, high nitrate levels in drinking water taken from the river are dangerous to human health.

Oxygen is essential for all aquatic life. Levels in the river may reach critically low concentrations because oxygen is used in the breakdown and decomposition of raw manure, feedlot runoff, domestic sewage, industrial wastes, stormwater runoff, and sediment (Minnesota Pollution Control Agency 1994).

The study's report contains many recommendations for improving the Minnesota River. Practices to control soil erosion on land used for agriculture, transportation, construction, and urban development are recommended, and use of minimum or reduced tillage methods on cropland is especially encouraged. To reduce nutrient pollution, communities must properly manage municipal sewage and waste, including stormwater runoff, and farms must control inputs from septic systems and feedlots. Vegetated buffers should be established on streambanks, and wetlands should be restored in selected locations to settle sediments, remove nutrients, and reduce peak water flows. Channel straightening and clearing should be reduced. If these recommended actions are successful, restrictions on swimming and fish consumption will be lifted, a result that would be highly visible to and appreciated by users of the river. With this comprehensive study and a detailed plan for improvement, the Minnesota River can be restored.

Next let us move to the north and explore the Otter Tail River, which drains 1,922 square miles in west-central Minnesota. It begins in Becker County and flows 190 miles from its source in the northern coniferous forest biome through the deciduous forest biome and then across the tallgrass prairie biome to its mouth at the Red River of the North. Its upper reaches pass through forested glacial moraines, outwash plains, and 18 lakes. The lower reaches flow through the flat, fertile Red River valley, the lake bed of Glacial Lake Agassiz. A total of 45 tributaries enter the Otter Tail River. Overall, the river drops 550 feet, with an average gradient of 2.9 feet per mile (Hanson et al. 1984).

Flow is highly controlled by the presence of 21 functional dams and numerous beaver dams. Eleven dams control lake levels, and three control the river level. Two are for water supply and flood control, one diverts water to a hydroelectric station, and four are used to produce hydroelectric power. Water use for power production takes precedence over other uses and poses problems for management of the flow in downstream reaches (Minnesota Department of Natural Resources, Section of Fisheries, unpublished data).

Although the Otter Tail River begins as a clear stream and flows through many clear lakes, its downstream reaches have high turbidity and fertility. Algal blooms are common in downstream reservoirs. Secchi disk readings ranged from 8.0 feet near the source to 1.4 feet near the mouth (Hanson et al. 1984). Municipal sewage, waste from several thousand resident Canada geese, and warm water from power plants enter the river near Fergus Falls. Farther downstream, runoff from livestock feedlots and croplands, eroding streambanks, and municipal wastewater inputs cause further increases in turbidity and pollution.

River water is used by the hydroelectric plants, by several cities as their source of municipal water, and by about 15 irrigation projects. Fishing, especially below the dams and in the many lakes the river flows through, hunting, canoeing, snowmobiling, cross-country skiing, and wild rice harvesting are important recreational uses of the river and its corridor. Walleye, northern pike, largemouth bass, bullhead, and channel catfish are the common game fish. Other fish species in the river include a large number of minnows and other small fish, white sucker, redhorse, and carp.

How a Japanese industry might affect the Otter Tail and other Minnesota rivers provides an interesting example of the interrelationships between humans and natural ecosystems. Production of

261

commercial pearls by implanting small "seeds" into oysters has developed into a major industry in Japan. Seeds are made by cutting and polishing small pieces of mussel shell. The high quality of shells from Minnesota rivers makes them highly desirable as a source of seeds, and strong interest in commercial harvesting of mussels developed in the late 1980s. In 1990, 18 permits were issued for harvesting mussels by hand from the Otter Tail River. Two operators harvested 32,930 pounds, with an estimated market value of about $30,000 to $45,000 (Roy Johannes, Minnesota Department of Natural Resources, Section of Fisheries, personal communication). Because of the lack of data on the impact of such harvests, the entire state except the lower Mississippi River below Red Wing was closed to commercial harvesting of mussels in 1991. In 1993, 200,000 pounds of mussels were harvested from this portion of the Mississippi River (Roy Johannes, personal communication).

About 80 years earlier, another industry based on mussels developed along the lower reaches of the Mississippi River. In the early 1900s, mussel shells were used in the manufacture of buttons (Carlander 1954). Factories were located in Iowa, and the shells of many species of mussels were deemed suitable for buttons. The commercial harvest was so large in some areas that mussel populations declined and propagation attempts were made. Not even propagation was adequate, however, to maintain the harvest necessary to sustain the factories, and restrictions were established too late to allow recovery of the populations. Pollution in the river further reduced mussel populations, and eventually the industry shut down.

Last, let us consider Minnesota's great river, the Mississippi, which begins in Lake Itasca in Clearwater County (figure 9.4) and flows 2,350 miles, 660 in Minnesota, to the Gulf of Mexico. It is the largest river in North America in terms of volume, the seventh largest in the world, and drains over 1.25 million square miles (Waters 1977). In Minnesota the river drains 45,000 square miles, about half the state, drops 833 feet, has an average gradi-

ent of 1.26 feet per mile, and flows through one of the state's largest lakes, Lake Pepin (table 8.2).

The Mississippi River begins in the springs and wetlands of the northern coniferous forest. It first flows northward through many wetlands and forests characteristic of a wilderness environment, then eastward through many lakes ringed with cabins and resorts, and through dense forests and wild rice beds before beginning its flow south to St. Anthony Falls. Forests become less common along the banks, and agricultural activities become more common closer to the Twin Cities. Numerous dams have been constructed to control lake levels, to establish flood-control reservoirs, and to provide power for hydroelectric stations. In the northern reaches, water quality is good, but as the river passes through more and more cities and farms, it gathers sediments from erosion of corn and soybean fields and becomes polluted by municipal and industrial wastes and runoff from feedlots and croplands. Flow is sometimes greatly reduced by the heavy demand for irrigation water and by diversion of water through electric power plants. Fluctuating water levels may have a detrimental effect on fish spawning and on the survival of newly hatched fish (Minnesota Department of Natural Resources, Section of Fisheries, unpublished data).

St. Anthony Falls is perhaps the most significant feature of the river in Minnesota. It divides the river into two parts and has affected both the river and the communities around it. Until the construction of the two locks in 1963, the falls served as a barrier for many aquatic species that could not move upstream over the falls. For example, only 9 species of unionid clams occurred above the falls, whereas 38 species were estimated to occur below the falls (van der Schalie and van der Schalie 1950). The falls also served as a barrier to barges and other boats. St. Anthony Falls was important as a source of power for the lumber industry, flour milling, and the first hydroelectric station in the United States. Minneapolis and St. Paul grew around the falls, and the first bridge to span the river was constructed here in 1854 (Waters 1977).

Downstream from St. Anthony Falls, the river changes dramatically. This reach, about 135 miles long to the Iowa border, flows through a broad, deep valley carved by meltwaters from glacial lakes. The valley is partially filled with sediment, and the river channel meanders gently through the broad floodplain (figure 9.5).

Large boats and commercial barges seem to dominate the main channel, which is dredged to maintain a minimum depth of 9 feet and a minimum width of 400 feet. Use of the Mississippi, and of other rivers and streams, for transportation has been important throughout Minnesota's history. American Indians and fur traders made extensive use of waterways to move supplies and furs by canoe, and early settlers traveled by paddle-wheel steamboats. Today most of the transportation on the Mississippi River consists of barges hauling grain, coal, gravel, and other bulky commodities.

Ten locks and dams have been built between Minneapolis and the Iowa border, including the two completed in 1963 that allow boat traffic to move 4.5 miles above St. Anthony Falls (Waters 1977). Each dam creates a navigation pool, which contains the 9-foot channel but also covers the former floodplain. Although the deep channel may not have appreciable negative effects on the river environment, the dredge spoil must be disposed of somewhere. Usually it is deposited near the riverbank, burying whatever is growing there. Spoil deposits seem to be relatively sterile and are slow to be colonized by new communities of plants and animals.

Many islands, wing dams, and rock piles can be found between the 9-foot channel and the river-bank. Some have been created to protect the main channel and some to improve the river habitat for game fish. Walleye, sauger, and catfish are common game fish, and carp is the most abundant rough fish in this habitat. Carp and other rough fish are harvested commercially in both the Minnesota and Wisconsin portions of the river.

Backwater lakes and wetlands occur in oxbow loops of old channels and in areas flooded by dams.

Here the water has little or no current, and the bottom typically contains several feet of silt. Aquatic plants are common, and some backwaters contain standing trunks of flooded trees. These areas often hold large numbers of ducks and geese, and some provide excellent fishing for bass, crappie, and northern pike.

Because the Mississippi affects such a large portion of the state, the river could be considered a barometer measuring Minnesota's environmental status. People damage the river directly by depositing harmful materials into the water and indirectly by their use of the lands within the watershed. Nonpoint inputs from the watershed to the river include sediments from erosion and fertilizers and pesticides from cropland. Straightening and deepening of the channel and drainage of wetlands in its watershed increase the speed at which water is removed from the landscape, creating the possibility of flooding in some places and shortages of water in other places.

How are we dealing with these environmental problems? Individuals and society, through governmental and nongovernmental organizations, know and care about the Mississippi River, but few seem to know what to do to improve the river and its environment (Craig and Anderson 1991). Conflicting interests, such as commercial shipping and recreational boating, and lack of clear legal jurisdiction over Minnesota-Wisconsin boundary waters present difficult matters for resolution. Interest in making improvements is high, and I believe that we should be optimistic and anticipate that improvements in the environment of the Mississippi River are imminent.

## Present status of Minnesota's streams and rivers

Minnesota's streams and rivers have changed in many ways during the past century. The logging of forests and the clearing of land and breaking of sod for agriculture have caused more water to run off the land into streams and rivers. Without the

263

shade from the trees, the water in streams running through logged forests warms, possibly to the extent that trout can no longer exist in these streams. Silt carried in by erosion may alter a stream channel and reduce transparency. Thus, the complete character of many streams and rivers is altered by what happens in their watersheds.

Our society seems to have a penchant for physical alteration of streams and rivers. Nearly 22,000 miles of Minnesota streams had been channelized by 1971 (Waters 1977). Many attractive meandering streams with pools, riffles, and bars were deepened, widened, and straightened into ditches. True, the water now moves downstream more quickly, but this movement only serves to intensify problems in the downstream reaches. A channelized stream is basically unable to function by itself. It requires maintenance and thus often proves to be a burden on the society that created it (Waters 1977).

Another physical alteration, the construction of dams and the subsequent formation of pools, has had a major effect on the Mississippi River. The action of the current and of waves caused by wind, boats, and barges has reduced the number and size of islands in the pools (figure 9.18) and has eroded shorelines in many places. These changes are detrimental to fish and wildlife and make the river less desirable for recreational uses.

Streams and rivers are often used as sewers, sometimes by design and sometimes not. Some pollution may act as fertilizer in a stream, increasing the productivity of plants and animals. Decomposition of these organisms and of organic wastes in the pollution may then lead to depletion of the oxygen supply. Furthermore, accumulation of pollutants may create a habitat for disease organisms.

Another type of pollution, thermal, occurs when electric-generating plants dump their cooling water back into a river. This thermal plume extends downstream and may have a significant impact on the kinds of fish and other organisms that can survive (Ross and Winter 1981).

Acid rain, which I have discussed in reference to lakes in the previous chapter, is another source

of pollution. Fortunately no Minnesota streams have yet become acidified. Several interesting sets of data exist to substantiate the stability of the chemistry of North Shore streams, those considered to be most susceptible to acidification. Data collected from seven streams from 1988 through 1990 (Swain et al. 1993) indicated that the pH was in the same range, around 7.4, as found in previous studies by L. L. Smith Jr. and J. B. Moyle (1944) and by T. F. Waters (1986). During periods of heavy runoff from snowmelt, pH decreased to 6.9, which is sufficiently acid to be detrimental to fish and other aquatic organisms, but quickly rose again to about 7.4 when runoff returned to normal. E. B. Swain and co-workers (1993) concluded that the capacity of the streams to buffer input of acidic meltwater was sufficiently high to mitigate the potential effects of acid rain.

Because of the strong interest in fishing for trout and other game fish, considerable effort has been devoted to the management of stream habitat. These efforts can be grouped into two major categories, changing the physical characteristics of the stream and changing the watershed.

The most common management practice has been the addition of structure to the streambed by constructing wing dams or placing large logs and boulders to deflect the current. Such structures serve to hold leaf litter and woody debris for longer periods, leading to an increase in productivity. Change in flow pattern caused by the structures serves to remove silt from the streambed and to create deep pools and overhanging banks, which provide excellent habitat for fish. Because cool water is so important for trout, trees and shrubs are sometimes planted along steambanks to create shade, and beaver dams are removed to release the water before it becomes too warm for trout.

Management of an entire watershed can also be highly beneficial to sportfishing. As discussed earlier, as much as 99% of the organic matter in the middle reaches of a stream comes from the watershed. Therefore, increases or decreases in the amount of runoff will have a marked impact on

**Figure 9.18** *The loss of islands in lower Pool 8 in the Mississippi River from 1939 to 1989 is believed to be related to the completion of Lock and Dam 8 near Reno in Houston County in 1937. Nearly 80% of the island acreage was eliminated by erosion caused by waves and currents in the pool. (Modified from unpublished maps, T. Owens and B. Drazkowski, Environmental Management Technical Center, National Biological Survey.)*

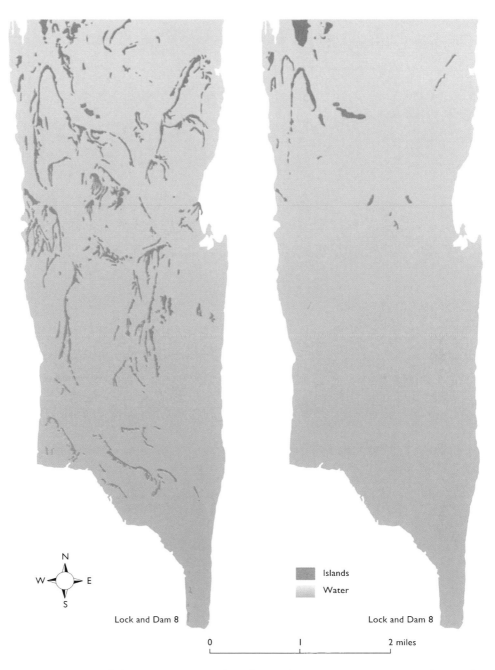

Islands
Water

N
W · E
S

Lock and Dam **8**                    Lock and Dam **8**

0          1          2 miles

the nutrients and detritus in the stream and thus on its productivity. For example, increases in plant detritus are likely to lead to increases in the population of shredder invertebrates. An increase in the amount of fine particles of soil carried into the stream will decrease transparency and reduce populations of algae and mosses. Changes in all or a part of a watershed may result from the kinds of agricultural crops being grown or may be the result of major planning by the governmental agency responsible for the entire watershed. Management plans for many stream watersheds that are currently in preparation will, I hope, lead to higher quality water and the return of our streams and rivers to more pristine conditions.

# THE FUTURE

In a clearing in a northern Minnesota forest, I stood with a small group of people watching three beautiful bald eagles soaring in a sky dotted with puffs of clouds. A soft voice from the group said, "This is a special place, and we will have a good summer. The eagles are an omen." The person making this statement obviously felt in close contact with the environment and had complete confidence in nature's ability to convey a message.

This experience caused me to reflect on the way we live in relation to our environment. Consider the impacts that settlement, modern industry, and agriculture have had on Minnesota. We have a high quality of life, a sound economy, and abundant natural resources. Yet, more natural resources have probably been destroyed or used in the past 100 years than in all preceding time. Most of the prairie, many wetlands, much of our iron ore, and nearly all of the great pine forests are gone. Land and other resources have been used for housing, cities, railroads, and many tangible products. Our society has readily adapted to changes and indeed has thrived. The question that we must now ponder is how long this intensive use of natural resources can or should continue and what it might mean to our future.

Some American Indian cultures believed that the land, its plants, and wildlife would provide all their needs. Therefore, they respected nature and were protective of the land. They viewed land as

267

**Figure 10.1** *The future, like Minnesota weather, is difficult to predict. (Photo by Blacklock Nature Photography.)*

something that was to be used and then passed on to future generations, not owned as a commodity and used as one pleased. It is this trust in nature and respect for the land that I would like to keep in mind as we consider the future of Minnesota from an ecological perspective. If we hope to live a good life in this state without compromising the quality of life of future generations, we must live in harmony with our environment.

Many ecological issues are of concern to our present society, and many unknowns face future generations (figure 10.1). Most of these issues are related to human activities, and as a result, solutions to problems are also related to human values and activities. An understanding and knowledge of the impact of human activities on Minnesota's ecosystems are essential to informed decision making. Let us consider some of the major issues that may affect the ecosystems and the environment of Minnesota.

## Human impacts on Minnesota ecosystems

Human activities have had direct and well-known impacts on plants and animals, as well as on entire ecosystems. As the world's population has increased and has developed more and more land and resources, habitats have been lost and fragmented, resulting in the loss of species and in less diverse communities. Certain species have become rare or endangered, and non-native species have in some cases displaced native species. Commerical uses of natural resources have often resulted in significant changes in habitats, and air and water pollution have degraded habitats. The impacts discussed in this section illustrate some of the most significant ways that human activities have affected Minnesota's ecosystems and some of the steps that are being taken to minimize these impacts.

### Human population growth

I have no doubt that the most important issue regarding the future of Minnesota's ecosystems is the growth of the human population in the world.

A larger population means that less space is available for each individual and that the demand for land, material goods, and services increases. This demand drives the economy, of course, but also increases the use of natural resources on which humans depend. In some parts of the world, forests that took hundreds of years to develop and soils that formed over thousands of years are being used up in one human generation.

Every ecosystem has a limit to the number of individuals of each species that it can support. We know that in some species of animals numbers are limited by food, in some by predators or disease, and in some by the amount of habitat or space that is available. Starvation, sickness, and death are consequences of exceeding such limits. We do not know, however, how many people planet Earth can actually support, nor do we know what factors will actually operate to limit the human population. We do know, however, that a limit certainly exists.

The population density in Minnesota is low compared to many other parts of the United States and the world. In some parts of the state, however, our population continues to grow (figure 10.2). From 1980 to 1992, the population in the 10-county metropolitan area centered around Minneapolis and St. Paul increased 18.6%, to 2,482,329 people (Minnesota Office of the State Demographer and the U.S. Bureau of Census). As the Twin Cities expands, we lose croplands, wetlands, and forests to suburban development and highways.

Large increases in the world population will also affect Minnesota. These increases will lead to increased demand for food, fuel, and other resources, and some of these will come from Minnesota fields and forests. Furthermore, population increases will likely lead to more crowding, more slums, more clearing of forests, and more health-related problems. Considering the anticipated growth in the world population, reductions in the capacity of the planet to produce food, fuel, and fiber are likely to have drastic consequences. We must be aware of these inevitable consequences of growth in the world population.

**Figure 10.2** *Is this lake too crowded? Everyone has his or her own response to space and crowding. This issue is perhaps more sociological than biological in terms of human perception. There is no question, however, that intensive boat traffic has a negative effect on many physical and biological features of a lake. (Photo by Bob Firth, Firth PhotoBank, Minneapolis.)*

Use of the land for disposal of the waste produced by our society also has implications for the future. Minnesotans produce approximately 10 to 12 million cubic yards of solid waste each year, about 1 ton per person. Waste materials cannot simply be dumped without having considerable social and ecological effects. In addition to removing the land from other uses, drainage from landfills has contaminated groundwater supplies and nearby wells. No community wants a landfill nearby, and proposing a site for the disposal of radioactive and other kinds of toxic material always generates public actions and legal battles.

The best approach to the problem of disposal is to reduce the generation of waste in as many ways as possible. Furthermore, disposal of waste materials can be facilitated by composting, recovery, and recycling. Over 1,500 dumps have been used to dispose of solid waste in the state (Minnesota Environmental Quality Board 1988). By 1988, this number had been reduced to 87 licensed landfills.

These landfills are tightly regulated and supervised. Although they are large, they probably pose less threat to the environment than many small, uncontrolled sites. Ultimately the quantity of waste produced depends on our population size, on a commitment to minimizing waste, and on innovative practices for doing so. These efforts must receive maximum encouragement from both a moral and an economic viewpoint. Landfills must be considered to be a last resort (figure 10.3).

Nearly all aspects of the ecology of Minnesota have connections to the issue of world population size and the impacts of that population on the environment. Now let us look at some of these impacts.

### Habitat fragmentation

Minnesota's natural ecosystems developed over many hundreds of years through interactions between climate, topography, soil, drought, fire, and other forces. In the last 150 years, however, European settlement and related human activities have

**Figure 10.3** *Space for landfills is very difficult to find. The best approach to disposal is to reduce the generation of waste in as many ways as possible. (Photo by Richard Hamilton Smith.)*

affected the air, water, land, plants, and animals in these natural ecosystems in numerous ways, as I have discussed in other chapters. In some cases, only small fragments of natural ecosystems remain, the rest having been developed for human uses.

The loss of prairies, woodlands, and wetlands (figure 10.4) to agriculture, logging, and urbanization is probably the most important cause of the loss of species. Tracts of native prairie and prairie wetlands exist in small, isolated fragments. These habitat fragments may be too small for animals with large home ranges and too far apart for animals to move freely from patch to patch. Large, wide-ranging species, such as antelope and prairie chickens, which require extensive areas of prairie, are absent or rare.

The Big Woods, Minnesota's largest tract of deciduous forest, was nearly all cleared for agriculture. Only a few small remnants of the Big Woods remain, none large enough to support black bears.

Nearly all of the great forests of white and red pine that covered much of central and northern Minnesota are gone. In their place are abandoned dams, railroad grades, tote roads, and vast stands of aspen and birch. And now these aspen and birch forests are being harvested at a faster rate than they are being replaced.

The impacts of fragmentation of our forests on most animals are not well known. Ecologists believe, however, that fragmentation of forests in Mexico and Central America has led to reduced populations of migratory songbirds. These same species nest in Minnesota forests, and fragmentation of their nesting habitat may cause further reductions in abundance.

Recognizing the impact of habitat fragmentation on species, government agencies such as the Minnesota Department of Natural Resources (DNR) and the U.S. Fish and Wildlife Service and private organizations such as The Nature Conservancy are now working toward the protection of tracts large enough to support wide-ranging species. Where large tracts cannot be protected, they are trying to provide linear corridors to link patches of habitat. The DNR is attempting to develop such a network of patches and corridors of native tallgrass prairie in Clay County and is planning to link two Scientific and Natural Areas in Norman and Polk Counties, Agassiz Dunes and Prairie Smoke Dunes, through acquisition of a quarter-mile-wide corridor about 1.5 miles in length (R. Djupstrom, Scientific and Natural Areas Program, Minnesota Department of Natural Resources, personal communication).

Considerable attention has been and is currently being given to the status of wetlands in Minnesota. Drainage for agriculture and urban development has eliminated nearly all of our prairie wetlands. As a result of the recognition of the many values of wetlands, however, some are being restored by breaking tile lines or blocking drainage ditches. Both state and federal agencies recognize the importance of preserving wetlands. In 1991, the state legislature passed the Wetland Conservation Act to strengthen laws protecting wetlands, and legislative

policies supporting no net loss of wetlands are being discussed. Under such policies, wetlands destroyed for whatever reason must be replaced by new wetlands. The future status of wetlands is dependent on the wishes of an informed public carried out through their voice in local land-use decisions and the larger legislative process.

## Rare and endangered species and the concept of biodiversity

In spite of the many changes in the landscape because of human development, the terrestrial and aquatic environments in Minnesota contain many species of organisms. Some species, such as robins and maple trees, are widely distributed, but a few species, such as buffalo and whooping cranes, no longer live in the state, and passenger pigeons, once abundant, are extinct. Other species, such as Minnesota trout lilies, certain mussels, and peregrine falcons, are very rare. A total of 57 species in the state are listed as endangered, 49 as threatened, and 181 of special concern (Pfannmuller and Coffin 1989). Of the total of 287 species, 96 are animals and 191 are plants. About one-third of these are found in prairie habitats.

**Figure 10.4** *Draining a wetland may provide an economic benefit. This must be evaluated, however, in terms of the loss of the numerous benefits that wetlands contribute. This type of cost-benefit analysis provides essential information for decision makers. (Photo by Matt Kerschbaum, U.S. Fish and Wildlife Service.)*

271

What has caused certain species to become so rare? In Minnesota, the primary causes are loss and degradation of habitat (Pfannmuller and Coffin 1989). Widespread pollution degrades habitats and sometimes affects species directly. Recent studies in Minnesota have revealed that Higgins eye and fat pocketbook mussels have become extremely rare as a result of pollution, overharvesting, and habitat loss from dam construction and dredging in the Mississippi River. Live specimens of Higgins eye mussels have been collected at only six sites in Minnesota since 1965, and no fat pocketbook mussels have been collected in the Mississippi River since 1947 (Coffin and Pfannmuller 1988). These species are classified as both state and federally endangered.

The remaining factors causing plants or animals to be rare or endangered have affected comparatively few species. Intensive collection of ginseng for profit, described in chapter 4, nearly eliminated this plant in Minnesota. Today trade in rare species is regulated by an international convention (Pfannmuller and Coffin 1989). Unregulated hunting of passenger pigeons throughout North America led to their extinction in the early 1900s. Present-day programs of natural resource agencies are directed to the control of harvests and to giving complete protection to many native species.

Some species have always been rare because they live in habitats that are rare. For example, Leedy's roseroot grows only on limestone cliffs, and is found in only four locations in the United States, three in southeastern Minnesota and one in New York (Pfannmuller and Coffin 1989). The very limited distribution of Minnesota trout lilies, which occur only in Minnesota, is due to their unique biology and habitat requirements. Only recently, a new species of prairie moonwort, *Botrychium gallicomontanum*, was discovered on Frenchman's Bluff, a dry prairie in Norman County (Farrar and Johnson-Groh 1991). This plant had never been listed or described by botanists and may be endemic to the state; that is, this location may be the only place on earth where it grows.

Several species of snakes and bats have become rare because of misunderstanding or poorly guided human activities. Bounties were paid for rattlesnakes in several counties, although the incidence of rattlesnake bites in Minnesota is very low. Bats consume thousands of insects and pose little threat to humans. Nevertheless, they are often killed by poorly informed people.

Why should we worry about rare plants or animals that are of little apparent value? The answer is complex and has both practical and esoteric aspects. Native species are the direct source of many valuable medicines, insecticides, and other products. The genetic basis for crops such as corn and wheat came from wild plants. Pollination of some crops, such as seed clover, depends on the presence of insects. Without the insects, no crop would be produced.

These practical reasons are easily understood. Harder to understand perhaps are the more esoteric reasons related to values. All species have evolved over time as part of the natural world. This world is extremely complex, and the ecology of most species is not well understood. Therefore, we cannot predict the impact of the loss of a particular species on an ecosystem, because we do not know its role. Although the loss of species has occurred throughout the history of the planet, the rate of loss has increased dramatically in recent years because of human activities, and continues to increase. Do we not have a responsibility and obligation to help preserve the diversity of life that makes up our natural world?

The loss of species may also affect the functioning of ecosystems because the parts of every ecosystem are interconnected in many ways through the processes of energy flow, nutrient cycling, competition, and predation. The importance of having high species diversity in natural communities has long been assumed by ecologists, who believe that the presence of many species allows for multiple pathways for energy flow and nutrient cycling and for resistance to perturbations. Recent studies of species diversity in grasslands, for example, at the

Cedar Creek Natural History Area in Anoka County revealed that areas with only a few species were highly stressed by drought and had low productivity. In contrast, areas with higher species diversity had higher productivity and recovered more quickly when the drought ended (Tilman and Downing 1994).

What does the future hold for rare and endangered species and the biodiversity of Minnesota ecosystems? Fortunately, most species will probably survive, largely because the organisms are remarkably adaptable to change. In fact, the rate of evolution of many organisms may be higher now than in the past in response to recent and pronounced changes in their environment induced by humans. Biodiversity will probably remain relatively stable, although populations of some species will increase and populations of others will decrease as the environment changes. Furthermore, the survival of species will be aided by the strong interest in conservation by citizens and government. As mentioned in earlier chapters, many groups including the U.S. Fish and Wildlife Service, The Nature Conservancy, and the DNR, through its Natural Heritage, Nongame Wildlife, and Scientific and Natural Areas programs, have assumed responsibility for the stewardship of rare species and habitats.

## Introduction of non-native species

The addition of non-native species to ecosystems, although admittedly adding to biodiversity, has often been detrimental because of the negative impact these species have on native species. Examples of species inadvertently introduced include Norway rats, Eurasian water milfoil, Dutch elm disease, sea lampreys, and zebra mussels. Management of selected ecosystems, although well intended, has also resulted in the introduction of undesirable species such as carp and purple loosestrife. Non-native organisms, such as Eurasian water milfoil, outcompete native species. Some non-native species, such as sea lamprey, prey on native species but have no native predators of their own. Others, such as carp, alter habitat, making it

less favorable for native species, and still others, such as purple loosestrife, outcompete native plants and are not suitable as food or cover for native animals.

Unfortunately we do not know all of the impacts of these manipulations or of the additions and losses of species. We do know, however, that once non-native species are present, they are often extremely difficult to control or eliminate. To reduce the spread of undesirable non-native species, the Minnesota legislature passed the Harmful Exotic Species Management Act in 1992. One measure of this act imposes a severe fine for the transportation of Eurasian water milfoil on boats and trailers. Passage of this act is a significant step in efforts to control non-native species.

## Commercial use of natural resources

The commercial use of some of our natural resources has had and will continue to have a significant impact on Minnesota's ecosystems. Intensive agriculture, with its widespread use of chemical fertilizers and pesticides and use of groundwater for irrigation, places heavy demands on both soil and water resources. Erosion by wind and water reduces the amount of topsoil, and the eroded sediment finds its way into wetlands, lakes, and streams. Agriculture, however, is a major economic force in Minnesota. Ecologically sound methods must be found to minimize detrimental aspects such as erosion and contamination of drinking water, while at the same time maximizing production of crops and livestock.

Harvest of Minnesota's forests for production of wood products also has the potential for significant ecological impact. Expansion in the number of wood-processing plants in northern Minnesota (figure 10.5) is anticipated to lead to heavy use of the available resources of aspen and other species. Intensive harvest will have the effect of removing nutrients from the soil and setting back succession. Such changes will benefit those species associated with early stages of succession and will be detrimental to those species associated with mature

273

forests. These ecological changes must be considered along with the economic benefits related to industrial expansion.

No coal, oil, or gas deposits suitable for commercial use have been found in Minnesota, but biological sources of energy exist in the state. The extensive peat deposits in northern Minnesota have been considered for use as an energy source, but the DNR has not issued permits for their use on a commercial scale. Wood provides a renewable source of energy, as does biomass from other types of vegetation. The possibility of using cattails as an energy source has received considerable attention

**Figure 10.5** *An increase in the number of wood-processing plants in Minnesota will lead to heavy use of aspen and other tree species. Cost-benefit analysis, taking into consideration the numerous values of uncut forests, must be used to evaluate this type of industrial expansion. (Photo by Richard Hamilton Smith.)*

in Minnesota (Dubbe, Garver, and Pratt 1988). Cattails may be especially appropriate because of their high productivity and their rapid growth rate. Alfalfa is also being considered as a fuel to produce electricity. Although uses of biological materials as energy sources have only been discussed, implementation of these proposals could have strong impacts on the future of Minnesota's farmlands, forests, peatlands, and wetlands.

### Air and water pollution

The industrial revolution and the associated growth in the use of fossil fuels have caused numerous kinds of air pollution. Burning of fossil fuels to heat homes and to provide electric power and transportation is a major source. Air pollution also originates from industrial sources such as paper manufacturing plants and facilities that use large amounts of fossil fuels. Agriculture contributes to air pollution in the form of dust and chemicals from fertilizers, insecticides, and herbicides blown from the surface of croplands.

Many of the lakes, streams, and rivers in Minnesota would probably not be considered polluted by most standards. Some rivers, however, carry heavy loads of pollution from industrial, municipal, or agricultural sources, and some lakes are not suitable for swimming. For example, wastewaters from industries and cities carry large quantities of waste into the Mississippi River. In addition, during the years from 1982 to 1991, 17 major spills of more than 1,000 gallons or 8,000 pounds, mostly oils and chemicals, also occurred in the river (Craig and Anderson 1991). In agricultural areas, runoff carrying fertilizers and pesticides enters the groundwater and many first- and second-order streams. This same runoff may carry fine particles of silt and clay eroded from the soil surface, leading to high loads of sediment in larger rivers.

Many of the impacts that humans have had on air and water can be easily measured. The chemical contents of both air and water have been monitored and recorded for many years at numerous locations in the state. Analyses of the changes that have occurred over time in levels of toxic substances and other chemicals provide some of our best indicators of the health of our environment.

## Environmental indicators

As we have seen from the previous discussion, changes due to human activities have been affecting plants and animals in Minnesota ecosystems for a long time. Still other changes in the environment, such as global warming and alteration of the ozone layer, are also taking place. Scientists are measuring and recording these changes, but their effects on organisms and ecosystems are not readily observed or even understood. These changes may affect air, water, and soil, and thus they have the potential to affect all plants and animals, including humans. Many scientists believe that these changes are indicators that can warn us of the potential deterioration of our environment. The following discussion features those indicators that have proved most useful in predicting or demonstrating change in the health of our environment.

### Global warming

A growing problem resulting from air pollution is the increase in the carbon dioxide content of the atmosphere. This change is well documented by measurements taken at Mauna Loa Observatory in Hawaii (Keeling et al. 1989). Annual fluctuations recorded at the observatory reflect the use of carbon dioxide by photosynthesis in summer and the production of carbon dioxide by respiration in winter in the northern hemisphere. The data clearly show that the carbon dioxide in our atmosphere is increasing. Several other gases, such as methane, which is produced by decomposition and by certain industries, are also increasing in the atmosphere.

The increase in carbon dioxide, methane, and other gases may cause global warming. Long-wave radiation from the sun is reflected from the earth's surface toward the sky. These gases reduce the amount of this radiation that can escape from the atmosphere, thereby causing an increase in tem-perature (figure 10.6). This warming has been compared to the process that occurs when solar energy warms a greenhouse, and the term greenhouse effect is frequently used when discussing global warming.

What might happen in Minnesota if our average annual temperature were to increase by a few degrees? We know that average temperatures since the time of glaciation have increased by about 9°F (Webb 1989) and that this increase has had an impact on Minnesota's ecosystems (see chapter 1). We also know that the average annual temperature in Minnesota varies from about 38°F in the forested regions of Clearwater and Beltrami Counties to about 41°F in the agricultural prairies in Becker and Mahnomen Counties. Thus, the effects of a small change in climate can be seen today within about 50 miles as one travels from the forest to the prairie in these northwestern counties.

Predictions about climate change, made from general circulation computer models, indicate that central North America is likely to become warmer and drier (Manabe and Wetherald 1987). If global warming increases the average annual temperature in Minnesota by about 4°F, the spruce, fir and pine forests of northern Minnesota will probably be replaced by hardwood forests (Davis 1990). The present hardwood forests will be replaced by farms, and in some places where propagules are present or where we create and plant new prairie reserves, by tallgrass prairie. Present tallgrass prairies may be replaced by mid- or short-grass prairies. Some species now present may disappear, and new species may move into Minnesota if the climate changes. Agriculture will also change in response to the climate change, and farmers will probably grow very different crops, perhaps peanuts and cotton.

If our climate becomes warmer and drier, wetlands and lakes will lose water, and many will become completely dry. Concentrations of chemicals already present in the lakes will increase as the volume of water is reduced. These changes, in turn, will affect the populations of plankton and fish. As the water recedes from lakeshores, cabin owners

275

**Figure 10.6** *Incoming short-wave (ultraviolet) and long-wave (infrared) radiations from the sun, shown in yellow, heat the surface of the earth. The surface reradiates these waves, shown in red. The ultraviolet rays escape through the atmosphere. Some of the reradiated infrared rays are reflected back to earth by clouds, and some are absorbed by carbon dioxide, methane, and other "greenhouse" gases, causing further heating of the earth. Increased concentrations of these greenhouse gases in the atmosphere may lead to global warming.*

and the resort industry will suffer. This change occurred in Minnesota during the drought of the 1930s, but fortunately the drought ended and rainfall filled the wetlands and lakes. Flow in streams and rivers will also decrease, and some may dry up. A reduction in flow reduces the ability of streams to cleanse themselves of pollution. Game fish populations are likely to decline, and rough fish populations are likely to increase.

*Atmospheric pollutants*

Substances causing air pollution also include gases such as sulfur dioxide and oxides of nitrogen. These chemicals may lead to deposition of acid rain and dust because these compounds are converted into sulfuric and nitric acid. The detrimental effects of acidification on lakes and streams in Scandinavia, the northeastern United States, and southeastern Canada are well known (Graedel and

Crutzen 1989). All levels of the food web can be affected, and the size and diversity of fish populations may be reduced. On land, both wet and dry deposition is believed to have damaged conifer forests in the northeastern United States and in parts of central Europe.

The potential impacts of acid rain on Minnesota's lakes and rivers were discussed in chapters 8 and 9. Laws have been passed requiring Minnesota industries to reduce air pollution, but about 90% of the pollutants causing acid rain in Minnesota are generated outside the state (Minnesota Environmental Quality Board 1988).

About 25 years ago, mercury was recognized as another kind of atmospheric pollutant in Minnesota. During the past few decades, mercury levels have been slowly increasing in many northern Minnesota lakes. The Minnesota Department of Health advises that consumption of fish from certain lakes and rivers be limited to prevent intake of mercury and other toxic substances. Most of the mercury comes from atmospheric pollution originating from incineration of household and industrial wastes, from latex paints, and from burning fossil fuels. Deposition of mercury on the landscape is three to four times greater now than in preindustrial times. Some of the mercury in lakes comes directly from the atmosphere, and some enters in runoff from the lake's watershed. The contribution from local geology is believed to be minor (Swain et al. 1992).

Contamination of fish with mercury was first investigated in Minnesota in 1969, and several years later, J. B. Moyle (1972) reported high mercury concentrations in lakes unaffected by point discharges. Once in a lake, mercury in the form of methylmercury is absorbed throughout the food web and accumulates in fish. It is tightly bound to proteins, and no method of cleaning or cooking fish is known to reduce mercury levels.

Fish-eating birds and mammals are especially vulnerable to mercury contamination. In the 1970s and 1980s, mercury concentrations in fish high enough to be detrimental to mink, otter, loons, and perhaps to eagles and osprey were reported (see summary in Swain and Helwig 1989). As discussed in chapter 8, a study in the late 1980s revealed that every loon of 221 examined contained mercury residues (Ensor, Helwig, and Wemmer 1992).

Chlorofluorocarbons, used in refrigerators, air conditioners, industrial solvents, and aerosols, escape into the atmosphere and destroy ozone molecules. Holes in the ozone layer over the polar regions are believed to be caused entirely by changes in the global environment due to human activities. Depletion of ozone in the upper atmosphere allows more ultraviolet radiation to reach the earth, which may cause increased skin cancer, cataracts, and changes in the growth of zooplankton, algae, crops, and other organisms. Even if emission of chlorofluorocarbons ceases very soon, depletion of ozone in the atmosphere will continue for at least a century because the chemical compounds presently causing ozone depletion will not break down for many decades (Graedel and Crutzen 1989).

Congress and state legislatures have provided numerous incentives to improve air quality throughout the country. Further efforts should be made to improve the quality of air by minimizing pollution from smokestacks and from other sources. We, as citizens, must encourage Minnesota and other states to continue their efforts to minimize air pollution in the future. We should hope also that such efforts will be made throughout the world. Movement of air masses is a global phenomenon unimpeded by political decisions or boundaries.

### Water pollution

The basic requirements for animals and plants include nutrients and water. Fresh, clean water in lakes, streams, and underground reservoirs is thus a valuable natural resource. Aquifers, geologic formations below the earth's surface, contain groundwater, which constitutes a large amount of the water used by humans today. Seventy-five percent of Minnesota's population depends on groundwater for daily use (Alexander and Alexander 1989). Much of the groundwater actually entered the ground thousands of years ago, long before civilization as we know it existed. This water is relatively pure in contrast to that which is currently entering the groundwater (Alexander and Alexander 1989), which is contaminated as a result of human activities. Many kinds of pollutants are carried into groundwater from landfills, toxic waste sites, buried storage tanks, septic drain fields, and by runoff from roads and parking lots containing heavy metals, salts, and hydrocarbons.

Today many farm wells contain nitrate, coliform bacteria from animal manure, and a variety of pesticides (figure 10.7). Recent studies suggest that the nitrate common in recent groundwater, but rare in vintage groundwater, is probably derived from some type of human activity such as the heavy use of agricultural nitrate fertilizer (Alexander

277

and Alexander 1989). Areas in the state with the highest susceptibility to contamination are the central and southeastern counties with sand and gravel deposits or with karst topography (Porcher 1988). A Minnesota Department of Health survey reported that atrazine, a widely used agricultural herbicide, has been found in about 30% of Minnesota's municipal wells. All of these substances are detrimental to the health of humans and animals. Unfortunately, the pollutants are likely to be present in the groundwater for many years after pollution ceases. Analysis of groundwater and removal of contaminants are difficult because of its location belowground.

Use of groundwater sometimes exceeds the recharge or replenishment rate. This imbalance is especially likely during drought years, when large amounts of water are used for lawns and crop irri-

gation. If global warming leads to a decrease in the amount of water in Minnesota's landscape, it also will lead to a higher demand for water for irrigation (figure 10.8), industrial, and domestic uses. Much of this increased demand will be placed on the aquifers. With a decrease in the rate of recharge of the aquifers and an increase in use, it would seem only a matter of time until demand exceeds supply. If this occurs, who will make decisions about who gets to use the available water? Water has been viewed as public property, and its use has not been controlled to a large extent in Minnesota. This situation may change, and citizens must be prepared to take steps to accommodate this change.

Certain types of pollution in our lakes and streams are easily observed. Cloudiness or turbidity of water may be related to fine particles of sediment, and frequently outlets from wastewater

**Figure 10.7** *Many kinds of chemical weed and insect killers (herbicides and insecticides) are applied to agricultural fields in Minnesota. Some of these chemicals are carried into streams and rivers by runoff. Some enter the groundwater and show up in wells that provide drinking water. Proper use of agricultural chemicals and control of runoff and erosion are essential. (Photo by Bob Firth, Firth PhotoBank, Minneapolis.)*

278

**Figure 10.8** *Irrigation of agricultural fields uses vast quantities of water. Who will decide how many and whose fields can be irrigated when demand for water from our underground aquifers exceeds supply? (Photo by Bob Firth, Firth PhotoBank, Minneapolis.)*

treatment plants and industries flow directly into a river. Because this pollution is often obvious, it has received attention from both private and government organizations. Some sources of pollution, especially those from industries and cities, have been cleaned up, and efforts are continually being made to improve the quality of our water. In spite of these efforts, over half of the streams and rivers monitored by the Minnesota Pollution Control Agency from 1983 to 1988 were designated as being too polluted for swimming or fishing (Minnesota Environmental Quality Board 1988). Surveys by the same agency in 1987 and 1988 indicated that 25% of Minnesota lakes under 5,000 acres in size and 10% of larger lakes did not support swimming or fishing. Fecal coliform bacteria, nutrients, and turbidity from nonpoint sources such as agricultural runoff, feedlots, and forestry activities were the most common pollutants. Nonpoint source pollution is difficult to monitor and control because it may come from many sources in the landscape and may enter a body of water at many places.

The vast resource of freshwater in Lake Superior is clean and pure compared to waters in the eastern Great Lakes. The lake has high water quality in spite of pollution from numerous sources. Large quantities of tailings, as much as 67,000 tons per day, from the processing of iron ore at the Reserve Mining Company taconite plant were dumped into Lake Superior for many years beginning in 1955 and ending in 1980. Attempts have been made to document the impact of this pollution, but the final analysis has not yet been made. Industrial and municipal wastes also enter the lake via the St. Louis River and other streams. Fortunately pollution from these sources is being reduced, and the maintenance of Lake Superior's clean water appears feasible. Lake Superior provides an example showing that keeping large bodies of water clean is possible even in today's highly industrialized society.

*Soil erosion*

Erosion by water and wind is the greatest single threat to the conservation of Minnesota's soil resources. Estimates indicate that 154 million tons of topsoil are lost per year. Over 96% of this topsoil comes from cropland. Water erosion removes 42%, and wind erosion removes 58%. In many areas, erosion is removing topsoil faster than the rate of soil formation, resulting in lower soil fertility and increased costs of agricultural production (Minnesota Environmental Quality Board 1988). Much of the soil lost to erosion is deposited in wetlands, lakes, and streams, where it causes sedimentation and other problems.

Soil erosion can be reduced to a large extent by conservation practices, which have been advocated for decades. Contour farming (figure 10.9), strip cropping, and establishment of permanent vegetative cover are effective agricultural practices that can be implemented. Erosion can also be reduced by changes in drainage patterns. These changes can be accomplished by building terraces, diversions, and erosion control structures.

As with many ecological problems, economic considerations play a major role in dealing with soil erosion. Maximizing production of agricultural crops over the short term often receives a higher priority than protection of the soil resource for future generations. The increase in the early 1990s in the use of conservation practices, such as reduced tillage or no-tillage, should lead to less erosion and may result in restoration of soil nutrients and organic matter. Economic analyses of these conservation practices, although not complete, suggest that when they are used, profits to farmers in some cases may be equal to or greater than when conventional tillage is used.

## A perspective to live with

A discussion of societal responsibilities and obligations to all organisms is appropriate in a book on ecology, but how do these ideas play in the real world? All people do not see forests, rivers, soils,

**Figure 10.9** *Contour farming reduces soil erosion from runoff and creates an attractive landscape. (Photo by Richard Hamilton Smith.)*

and wetlands in the same way. Many people overlook the threats that industrialization and intensive agriculture pose to the environment and to many species of plants and animals and think that achievement of a high standard of living at the

expense of the environment is acceptable. They believe that economic growth should be the most important driving force in our society.

Thinking in terms of costs and benefits can add some perspective to this issue. Costs of efforts to improve our environment can often be estimated. We can determine the costs of scrubbers to clean smokestack emissions, of treatment plants to purify sewage and industrial wastes, and of pre-

serving a tract of prairie or old-growth forest. The problem with this approach, however, is that the benefits may be very difficult to value in dollars. How does one determine the benefit or value of a duck, a butterfly, a clean river, or clean air? Is a tree or a forest worth more standing than as lumber or paper? In an analysis of costs and benefits of environmental management or improvement, how do we weigh the benefits of improved physical health,

**Figure 10.10** *This view of a familiar sign behind the big bluestem grass on a patch of native prairie conveys, in a way, the status of Minnesota today. We, as informed citizens, have the privilege and obligation to decide the future. (Photo © Jim Brandenburg.)*

improved mental health, possibilities of a longer life, or stated more simply, having a better environment in which to live against the preservation of jobs in the logging industry, the creation of more farmlands in place of wetlands, or the need to designate more places to dump sewage or bury wastes? Ultimately these decisions involve our values and the importance that we place on living in harmony with our environment, but as a start we should insist that all aspects of costs and benefits be considered in making decisions.

The process by which such decisions are made is complex and controversial. In most states and countries, environmental matters and economic matters are handled by separate agencies. But the two areas are inextricably intertwined. Growth is considered to be desirable and beneficial to our economic system, and businesses and communities usually take the position that larger is better. Yet

such growth surely affects our environment because it increases demands for space, natural resources, and waste disposal. The decision-making process should reflect the need to consider environmental and economic matters together.

In accordance with what can be considered to be natural law, the economy must function within the limits of a planet with finite capabilities to produce fresh water, create new topsoil, and absorb pollution (Postel and Flavin 1991). Growth cannot continue forever. Sustainable agriculture and forestry, a stable population size, and a stable economy may be highly desirable.

The future of Minnesota, and of the world, depends on informed citizens making the proper decisions (figure 10.10). I hope that the information and ideas presented in this book will enrich your memories and experiences regarding Minnesota and will be of help in guiding its future.

APPENDIXES

# APPENDIX B

# TREES FOUND IN MINNESOTA

| Common name | Scientific name | Prairie | Deciduous forest | Coniferous forest | Wetlands |
|---|---|---|---|---|---|
| Ash, black | *Fraxinus nigra* | | ▲ | ▲ | ▲ |
| Ash, green | *Fraxinus pennsylvanica* | | ▲ | | |
| Ash, mountain | *Sorbus americana* | | | ▲ | |
| Ash, white | *Fraxinus americana* | | ▲ | | |
| Aspen, large-toothed | *Populus grandidentata* | | ▲ | ▲ | |
| Aspen, trembling | *Populus tremuloides* | ▲ | ▲ | ▲ | |
| Balsam fir | *Abies balsamea* | | | ▲ | |
| Balsam poplar | *Populus balsamifera* | | ▲ | ▲ | |
| Basswood | *Tilia americana* | | ▲ | ▲ | |
| Birch, paper | *Betula papyrifera* | | ▲ | ▲ | |
| Birch, river | *Betula nigra* | | ▲ | | |
| Birch, yellow | *Betula alleghaniensis* | | ▲ | ▲ | |
| Black walnut | *Juglans nigra* | | ▲ | | |
| Box elder | *Acer negundo* | | ▲ | | |
| Butternut | *Juglans cinerea* | | ▲ | | |
| Cedar, eastern red | *Juniperus virginiana* | | | ▲ | |
| Cedar, white | *Thuja occidentalis* | | | ▲ | ▲ |
| Cherry, black | *Prunus serotina* | | ▲ | | |
| Cherry, pin | *Prunus pennsylvanica* | | ▲ | ▲ | |
| Cottonwood | *Populus deltoides* | ▲ | ▲ | | |
| Elm, American | *Ulmus americana* | | ▲ | | |
| Elm, rock | *Ulmus thomasii* | | ▲ | | |
| Elm, slippery | *Ulmus rubra* | | ▲ | | |
| Hackberry | *Celtis occidentalis* | | ▲ | | |
| Hickory, bitternut | *Carya cordiformis* | | ▲ | | |
| Hickory, shellbark | *Carya ovata* | | ▲ | | |

| Common name | Scientific name | Prairie | Deciduous forest | Coniferous forest | Wetlands |
|---|---|---|---|---|---|
| Hornbeam | Carpinus caroliniana | | ▲ | | |
| Ironwood | Ostrya virginiana | | ▲ | ▲ | |
| Kentucky coffee tree | Gymnocladus dioica | | ▲ | | |
| Maple, red | Acer rubrum | | ▲ | ▲ | ▲ |
| Maple, silver | Acer saccharinum | | ▲ | ▲ | |
| Maple, sugar | Acer saccharum | | ▲ | ▲ | |
| Oak, black | Quercus velutina | | ▲ | | |
| Oak, bur | Quercus macrocarpa | ▲ | ▲ | ▲ | |
| Oak, northern pin | Quercus ellipsoidalis | | ▲ | ▲ | |
| Oak, red | Quercus rubra | | ▲ | ▲ | |
| Oak, white | Quercus alba | | ▲ | | |
| Pine, jack | Pinus banksiana | | | ▲ | |
| Pine, red | Pinus resinosa | | ▲ | ▲ | |
| Pine, white | Pinus strobus | | ▲ | ▲ | |
| Spruce, black | Picea mariana | | | ▲ | ▲ |
| Spruce, white | Picea glauca | | | ▲ | |
| Tamarack | Larix laricina | | | ▲ | ▲ |
| Willow, black | Salix nigra | | ▲ | | ▲ |
| Willow, pussy | Salix discolor | | ▲ | ▲ | ▲ |

Sources: Scientific names from Ownbey and Morley (1991); common names from Aaseng et al. (1993) and Ownbey and Morley (1991).

# APPENDIX C

# SHRUBS FOUND IN MINNESOTA

| Common name | Scientific name | Prairie | Deciduous forest | Coniferous forest | Wetlands |
|---|---|---|---|---|---|
| Alder, dwarf | Rhamnus alnifolius | | | ▲ | ▲ |
| Alder, green | Alnus viridis | | | ▲ | |
| Alder, speckled | Alnus rugosa | | ▲ | ▲ | |
| American yew | Taxus canadensis | | ▲ | ▲ | |
| Blackberry | Rubus sp. | | ▲ | ▲ | |
| Blueberry, lowbush | Vaccinium angustifolium | | | ▲ | |
| Blueberry, velvet-leaf | Vaccinium myrtilloides | | | ▲ | ▲ |
| Bog birch | Betula pumila | | | | ▲ |
| Bog laurel | Kalmia polifolia | | | ▲ | ▲ |
| Bog rosemary | Andromeda glaucophylla | | | ▲ | ▲ |
| Buckthorn, European | Rhamnus cathartica | | ▲ | | |
| Bush juniper | Juniperus communis | | ▲ | ▲ | |
| Cherry, choke | Prunus virginiana | ▲ | ▲ | ▲ | |
| Cherry, sand | Prunus pumila | | ▲ | ▲ | |
| Chokeberry | Aronia melanocarpa | | | ▲ | ▲ |
| Creeping snowberry | Gaultheria hispidula | | | ▲ | ▲ |
| Currant | Ribes sp. | | ▲ | ▲ | ▲ |
| Dogwood, gray | Cornus foemina | | ▲ | | |
| Dogwood, pagoda | Cornus alternifolia | | ▲ | ▲ | |
| Dogwood, red osier | Cornus stolonifera | | ▲ | ▲ | |
| Dogwood, round-leaved | Cornus rugosa | | ▲ | ▲ | |
| Downy arrow-wood | Viburnum rafinesquianum | | ▲ | ▲ | |
| Elder, common | Sambucus canadensis | | ▲ | | |
| Elder, red-berried | Sambucus pubens | | ▲ | ▲ | |
| False heather | Hudsonia tomentosa | ▲ | | | |
| Fragrant false-indigo | Amorpha nana | ▲ | | | |

| Common name | Scientific name | Prairie | Deciduous forest | Coniferous forest | Wetlands |
|---|---|:---:|:---:|:---:|:---:|
| Gooseberry | Ribes sp. | | ▲ | ▲ | |
| Hazelnut, American | Corylus americana | | ▲ | ▲ | |
| Hazelnut, beaked | Corylus cornuta | | ▲ | ▲ | |
| High-bush cranberry | Viburnum trilobum | | ▲ | ▲ | |
| Honeysuckle, bush | Diervilla lonicera | | ▲ | ▲ | |
| Honeysuckle, fly | Lonicera canadensis | | | ▲ | |
| Honeysuckle, mountain fly | Lonicera villosa | | | ▲ | |
| Honeysuckle, Tartarian | Lonicera tatarica | | ▲ | | |
| Juneberry | Amelanchier humilis | | ▲ | ▲ | |
| Labrador tea | Ledum groenlandicum | | | ▲ | ▲ |
| Leadplant | Amorpha canescens | ▲ | | | |
| Leather-leaf | Chamaedaphne calyculata | | | ▲ | ▲ |
| Leatherwood | Dirca palustris | | ▲ | | |
| Mountain maple | Acer spicatum | | ▲ | ▲ | |
| New Jersey tea | Ceanothus americanus | | ▲ | | |
| Ninebark | Physocarpus opulifolius | | ▲ | ▲ | |
| Prairie rose | Rosa arkansana | ▲ | | | |
| Prickly ash | Zanthoxylum americanum | | ▲ | | |
| Raspberry | Rubus sp. | | ▲ | ▲ | |
| Red mulberry | Morus rubra | | ▲ | | |
| Russian olive | Eleagnus angustifolia | ▲ | | | |
| Saskatoon | Amelanchier alnifolia | | | ▲ | |
| Shrubby cinquefoil | Potentilla fruticosa | | | ▲ | |
| Spirea | Spiraea alba | | ▲ | ▲ | ▲ |
| Sumac, poison | Rhus vernix | | ▲ | | |
| Sumac, smooth | Rhus glabra | ▲ | ▲ | ▲ | |
| Sumac, staghorn | Rhus typhina | | ▲ | | |
| Wild plum | Prunus americana | ▲ | ▲ | | |
| Wild rose | Rosa blanda | ▲ | ▲ | | |
| Willow | Salix spp. | ▲ | ▲ | ▲ | ▲ |
| Winterberry | Ilex verticillata | | ▲ | ▲ | |
| Wintergreen | Gaultheria procumbens | | | ▲ | |
| Wolfberry | Symphoricarpos occidentalis | ▲ | ▲ | | |

Sources: Scientific names from Ownbey and Morley (1991); common names from Aaseng et al. (1993) and Ownbey and Morley (1991).

289

# APPENDIX D

# COMMON HERBS FOUND IN MINNESOTA

| Common name | Scientific name | State status | Prairie | Deciduous forest | Coniferous forest | Wetland, lake, river |
|---|---|---|---|---|---|---|
| Alexanders, golden | Zizia aurea | | ▲ | | | |
| Alexanders, heart-leaved | Zizia aptera | | ▲ | | | |
| Alkali plantain | Plantago eriopoda | | ▲ | | | ▲ |
| Alum root | Heuchera richardsonii | | ▲ | | | |
| Angelica | Angelica atropurpurea | | | | | ▲ |
| Arrowhead | Sagittaria latifolia | | | | | ▲ |
| Aster, aromatic | Aster oblongifolius | | ▲ | | | |
| Aster, azure | Aster oelentangiensis | | ▲ | | | |
| Aster, big-leaved | Aster macrophyllus | | | ▲ | ▲ | |
| Aster, heath | Aster ericoides | | ▲ | | | |
| Aster, New England | Aster novae-angliae | | ▲ | | | |
| Aster, panicled | Aster lanceolatus | | ▲ | | | ▲ |
| Aster, prairie golden | Heterotheca villosa | | ▲ | | | |
| Aster, silky | Aster sericeus | | ▲ | | | |
| Aster, smooth | Aster laevis | | ▲ | | | |
| Beach pea | Lathyrus japonicus | | | | | ▲ |
| Beard-tongue, small white | Penstemon albidus | | ▲ | | | |
| Beard-tongue, white-flowered | Penstemon grandiflorus | | ▲ | | | |
| Beggar-ticks | Bidens frondosa | | | ▲ | ▲ | ▲ |
| Bellwort | Uvularia grandiflora | | | ▲ | ▲ | |
| Bird's-foot violet | Viola pedatifida | | ▲ | | | |
| Black snakeroot | Sanicula marilandica | | | ▲ | ▲ | |
| Bladderwort | Utricularia vulgaris | | | | | ▲ |
| Blanket flower | Gaillardia aristata | | ▲ | | | |
| Blazing star | Liatris aspera | | ▲ | | | |
| Bloodroot | Sanguinaria canadensis | | | ▲ | ▲ | |

| Common name | Scientific name | State status | Prairie | Deciduous forest | Coniferous forest | Wetland, lake, river |
|---|---|---|---|---|---|---|
| Blue vervain | Verbena hastata | | | | | ▲ |
| Bristly sarsaparilla | Aralia hispida | | | | ▲ | |
| Bugleweed | Lycopus uniflorus | | | | | ▲ |
| Bulrush, hard-stem | Scirpus acutus | | | | | ▲ |
| Bulrush, prairie | Scirpus paludosus | | | | | ▲ |
| Bulrush, soft-stem | Scirpus validus | | | | | ▲ |
| Bulrush, water | Scirpus subterminalis | | | | | ▲ |
| Bunchberry | Cornus canadensis | | | | ▲ | |
| Bur-reed, floating | Sparganium fluctuans | | | | | ▲ |
| Bur-reed, giant | Sparganium eurycarpum | | | | | ▲ |
| Canada mayflower | Maianthemum canadense | | | ▲ | ▲ | |
| Canada waterweed | Elodea canadensis | | | | | ▲ |
| Carrion flower | Smilax lasioneura | | ▲ | ▲ | ▲ | |
| Cattail, common | Typha latifolia | | | | | ▲ |
| Cattail, narrowleaf | Typha angustifolia | | | | | ▲ |
| Chara | Chara sp. | | | | | ▲ |
| Cinquefoil, prairie | Potentilla pennsylvanica | | ▲ | | | |
| Cinquefoil, tall | Potentilla arguta | | ▲ | | | |
| Cinquefoil, white | Potentilla tridentata | | | | ▲ | |
| Clearweed | Pilea pumila | | | ▲ | | ▲ |
| Coast jointweed | Polygonella articulata | | | ▲ | | |
| Columbine | Aquilegia canadensis | | | ▲ | | |
| Compass plant | Silphium laciniatum | | ▲ | | | |
| Coneflower, prairie | Ratibida columnifera | | ▲ | | | |
| Coneflower, purple | Echinacea angustifolia | | ▲ | | | |
| Coontail | Ceratophyllum demersum | | | | | ▲ |
| Cow-wheat | Melampyrum lineare | | | | ▲ | |
| Cranberry | Vaccinium oxycoccos | | | | ▲ | ▲ |
| Cut-leaf ironplant | Haplopappus spinulosus | ● | ▲ | | | |
| Cutgrass | Leersia oryzoides | | | | | ▲ |
| Daisy fleabane | Erigeron strigosus | | ▲ | ▲ | ▲ | |
| Dock | Rumex spp. | | ▲ | ▲ | ▲ | ▲ |
| Duckweed, common | Lemna minor | | | | | ▲ |
| Duckweed, greater | Spirodella polyrhiza | | | | | ▲ |
| Duckweed, star | Lemna trisulca | | | | | ▲ |
| Dutchman's breeches | Dicentra cucullaria | | | ▲ | | |

| Common name | Scientific name | State status | Prairie | Deciduous forest | Coniferous forest | Wetland, lake, river |
|---|---|---|---|---|---|---|
| Enchanter's nightshade | Circaea alpina | | | ▲ | ▲ | |
| False boneset | Kuhnia eupatoriodes | | ▲ | | | |
| False Solomon's seal | Smilacina triflora | | | | ▲ | ▲ |
| Feathermoss | Pleurozium schreberi | | | | ▲ | |
| Feathermoss | Dicranum undulatum | | | | ▲ | ▲ |
| Feathermoss | Polytrichum strictum | | | | ▲ | ▲ |
| Fern, bracken | Pteridium aquilinum | | | ▲ | ▲ | |
| Fern, cinnamon | Osmunda cinnamomea | | | | ▲ | |
| Fern, lady | Athyrium angustum | | | ▲ | ▲ | |
| Fern, marsh | Thelypteris palustris | | | | | ▲ |
| Fern, ostrich | Matteuccia struthiopteris | | | ▲ | | |
| Fern, sensitive | Onoclea sensibilis | | | ▲ | ▲ | |
| Field chickweed | Cerastium arvense | | ▲ | | | |
| Gayfeather | Liatris pycnostachya | | ▲ | | | |
| Gentian, bottle | Gentiana andrewsii | | ▲ | | | |
| Gentian, fringed | Gentiana puberulenta | | ▲ | | | |
| Gentian, northern | Gentiana affinis | ● | ▲ | | | |
| Ginseng | Panax quinquefolium | ● | | ▲ | | |
| Glasswort | Salicornia rubra | | ▲ | | | ▲ |
| Gold-thread | Coptis groenlandicum | | | | ▲ | ▲ |
| Goldenrod, Canada | Solidago canadensis | | ▲ | | | |
| Goldenrod, grass-leaved | Euthamia graminifolia | | ▲ | | | |
| Goldenrod, gray | Solidago nemoralis | | ▲ | | | |
| Goldenrod, Missouri | Solidago missouriensis | | ▲ | | | |
| Goldenrod, Riddell's | Solidago riddellii | | ▲ | | | |
| Goldenrod, soft | Solidago mollis | ● | ▲ | | | |
| Goldenrod, stiff | Solidago rigida | | ▲ | | | |
| Goldenrod, tall | Solidago gigantea | | ▲ | | | |
| Goldenrod, zig-zag | Solidago flexicaulis | | | ▲ | | |
| Goosefoot | Chenopodium spp. | | ▲ | | | |
| Grass, alkali | Puccinellia nuttalliana | | ▲ | | | ▲ |
| Grass, big bluestem | Andropogon gerardii | | ▲ | | | |
| Grass, blue grama | Bouteloua gracilis | | ▲ | | | |
| Grass, blue-eyed | Sisyrinchium mucronatum | | ▲ | | | |
| Grass, blue-joint | Calamagrostis canadensis | | ▲ | | | ▲ |
| Grass, bog reed | Calamagrostis inexpanse | | | | | ▲ |

| Common name | Scientific name | State status | Prairie | Deciduous forest | Coniferous forest | Wetland, lake, river |
|---|---|:---:|:---:|:---:|:---:|:---:|
| Grass, bottlebrush | Elymus hystrix | | | ▲ | | |
| Grass, cotton | Eriophorum spissum | | | | ▲ | ▲ |
| Grass, false melic | Schizachne purpurascens | | | ▲ | ▲ | |
| Grass, fowl meadow | Poa palustris | | | | | ▲ |
| Grass, foxtail | Setaria spp. | | ▲ | | | |
| Grass, hairy grama | Bouteloua hirsuta | | ▲ | | | |
| Grass, Indian | Sorghastrum nutans | | ▲ | | | |
| Grass, June | Koeleria macrantha | | ▲ | | | |
| Grass, Kentucky blue | Poa pratensis | | ▲ | | | |
| Grass, little bluestem | Schizachyrium scoparium | | ▲ | | | |
| Grass, mat muhly | Muhlenbergia richardsonis | | ▲ | | | |
| Grass, mountain rice | Oryzopsis asperifolia | | | ▲ | ▲ | |
| Grass, needle | Stipa comata | | ▲ | | | |
| Grass, northern reed | Calamagrostis inexpansa | | ▲ | | | |
| Grass, oat | Helictotrichon hookeri | | ▲ | | | |
| Grass, panic | Panicum spp. | | ▲ | | | |
| Grass, plains muhly | Muhlenbergia cuspidata | | ▲ | | | |
| Grass, porcupine | Stipa spartea | | ▲ | | | |
| Grass, prairie cord | Spartina pectinata | | ▲ | | | |
| Grass, prairie dropseed | Sporobolus heterolepis | | ▲ | | | |
| Grass, quack | Agropyron repens | | ▲ | | | |
| Grass, red triple-awned | Aristida purpurea | ● | ▲ | | | |
| Grass, redtop | Agrostis stolonifera | | ▲ | | | |
| Grass, reed canary | Phalaris arundinacea | | | | | ▲ |
| Grass, salt | Distichlis stricta | | ▲ | | | ▲ |
| Grass, sand dropseed | Sporobolus cryptandrus | | ▲ | | | |
| Grass, sand reed | Calamovilfa longifolia | | ▲ | | | |
| Grass, sideoats grama | Bouteloua curtipendula | | ▲ | | | |
| Grass, sweet | Hierochloë odorata | | ▲ | | | |
| Grass, wigeon | Ruppia occidentalis | | | | | ▲ |
| Grass-of-Parnassus | Parnassia glauca | | ▲ | | | ▲ |
| Greater water dock | Rumex orbiculatus | | | | | ▲ |
| Ground plum | Astragalus crassicarpus | | ▲ | | | |
| Ground-pine | Lycopodium obscurum | | | | ▲ | |
| Hairgrass | Deschampsia cespitosa | | | | | ▲ |
| Harebell | Campanula rotundifolia | | ▲ | ▲ | ▲ | |

| Common name | Scientific name | State status | Prairie | Deciduous forest | Coniferous forest | Wetland, lake, river |
|---|---|---|---|---|---|---|
| Heartleaf twayblade | Listera cordata | | | | ▲ | ▲ |
| Hog peanut | Amphicarpaea bracteata | | | ▲ | | |
| Horsetail | Equisetum spp. | | | | | ▲ |
| Indian hemp | Apocynum sibiricum | | ▲ | | | |
| Indian pipe | Monotropa uniflora | | | | ▲ | |
| Indian plantain | Cacalia plantaginea | ◗ | ▲ | | | |
| Jack-in-the-pulpit | Arisaema triphyllum | | | ▲ | ▲ | |
| Jeweled shooting-star | Dodecatheon amethystinum | ● | | ▲ | | |
| Jewelweed | Impatiens capensis | | | | | ▲ |
| Joe-pye weed | Eupatorium maculatum | | ▲ | | | ▲ |
| Ladies tresses | Spiranthes cernua | | ▲ | | | |
| Lady-slipper, showy | Cypripedium reginae | | | | ▲ | ▲ |
| Lady-slipper, white | Cypripedium candidum | ● | ▲ | | | |
| Leedy's roseroot | Sedum integrifolium | ○ | | ▲ | | |
| Lily, bluebead | Clintonia borealis | | | | ▲ | |
| Lily, Philadelphia | Lilium philadelphicum | | ▲ | | | |
| Lingonberry | Vaccinium vitis-idaea | | | | ▲ | ▲ |
| Liverwort | Riccia fluitans | | | | | ▲ |
| Locoweed | Oxytropis lambertii | | ▲ | | | |
| Loosestrife, prairie | Lysimachia quadriflora | | ▲ | | | |
| Loosestrife, purple | Lythrum salicaria | | | | | ▲ |
| Lotus | Nelumbo lutea | | | | | ▲ |
| Lousewort | Pedicularis lanceolata | | ▲ | ▲ | | |
| Marsh marigold | Caltha palustris | | | ▲ | | ▲ |
| Marsh vetchling | Laythrus palustris | | ▲ | | | |
| Meadow parsnip | Thaspium barbinode | | ▲ | | | |
| Meadow ragwort | Senecio pseudaureus | | ▲ | | | ▲ |
| Meadow-rue, early | Thalictrum dioicum | | | ▲ | ▲ | |
| Meadow-rue, tall | Thalictrum dasycarpum | | ▲ | | | |
| Milfoil, Eurasian water | Myriophyllum spicatum | | | | | ▲ |
| Milfoil, water | Myriophyllum exalbescens | | | | | ▲ |
| Milkweed | Asclepias spp. | | ▲ | ▲ | ▲ | ▲ |
| Milkweed, oval-leaved | Asclepias ovalifolia | | ▲ | | | |
| Milkweed, showy | Asclepias speciosa | | ▲ | | | |
| Milkweed, swamp | Asclepias incarnata | | | | | ▲ |
| Moccasin flower | Cypripedium acaule | | | | ▲ | |

294

| Common name | Scientific name | State status | Prairie | Deciduous forest | Coniferous forest | Wetland, lake, river |
|---|---|---|---|---|---|---|
| Mountain mint | Pycnanthemum virginianum | | ▲ | | | |
| Naiad, large | Najas marina | | | | | ▲ |
| Naiad, slender | Najas flexilis | | | | | ▲ |
| Nettle, stinging | Urtica dioica | | ▲ | ▲ | ▲ | |
| Nettle, wood | Laportea canadensis | | | ▲ | | |
| Northern bedstraw | Galium boreale | | ▲ | | ▲ | |
| Northern miterwort | Mitella nuda | | | | ▲ | |
| Partridgeberry | Mitchella repens | | | | ▲ | |
| Pasque flower | Pulsatilla nuttalliana | | ▲ | | | |
| Pink corydalis | Corydalis sempervirens | | | | ▲ | |
| Pipewort | Eriocaulon septangulare | | | | | ▲ |
| Pitcher plant | Sarracenia purpurea | | | | | ▲ |
| Plains paintbrush | Castilleja sessiliflora | | ▲ | | | |
| Pliant milk-vetch | Astragalus canadensis | | ▲ | ▲ | | |
| Poison ivy | Rhus radicans | | ▲ | ▲ | ▲ | |
| Pondweed, clasping-leaf | Potamogeton richardsonii | | | | | ▲ |
| Pondweed, fine-leaf | Potamogeton filiformis | | | | | ▲ |
| Pondweed, flat-stem | Potamogeton zosteriformis | | | | | ▲ |
| Pondweed, floating-leaved | Potamogeton natans | | | | | ▲ |
| Pondweed, Illinois | Potamogeton illinoensis | | | | | ▲ |
| Pondweed, large-leaved | Potamogeton amplifolius | | | | | ▲ |
| Pondweed, sago | Potamogeton pectinatus | | | | | ▲ |
| Pondweed, spiral | Potamogeton spirillus | | | | | ▲ |
| Prairie clover, hairy | Petalostemum villosum | | ▲ | ▲ | | |
| Prairie clover, purple | Petalostemum purpureum | | ▲ | | | |
| Prairie clover, white | Petalostemum candidum | | ▲ | | | |
| Prairie larkspur | Delphinium virescens | | ▲ | | | |
| Prairie moonwort | Botrychium gallicomontanum | ◗ | ▲ | | | |
| Prairie phlox | Phlox pilosa | | ▲ | | | |
| Prairie sagewort | Artemisia frigida | | ▲ | | | |
| Prairie smoke | Geum triflorum | | ▲ | | | |
| Prairie turnip | Psoralea esculenta | | ▲ | | | |
| Prairie white fringed orchid | Platanthera praeclara | ○ | ▲ | | | |
| Prickly pear | Opuntia spp. | | ▲ | ▲ | | |
| Puccoon, hoary | Lithospermum canescens | | ▲ | | | |
| Puccoon, narrow-leaved | Lithospermum incisum | | ▲ | | | |

| Common name | Scientific name | State status | Prairie | Deciduous forest | Coniferous forest | Wetland, lake, river |
|---|---|---|---|---|---|---|
| Pursh's plantain | Plantago patagonica | | ▲ | | | |
| Pussytoes | Antennaria parlinii | | ▲ | | | |
| Quillwort | Isoetes echinospora | | | | | ▲ |
| Quillwort | Isoetes macrospora | | | | | ▲ |
| Ragweed | Ambrosia artemisiifolia | | ▲ | ▲ | | |
| Rattlesnake master | Eryngium yuccifolium | ● | ▲ | | | |
| Red baneberry | Actea rubra | | | ▲ | ▲ | |
| Reed | Phragmites australis | | | | | ▲ |
| Reindeer moss | Cladina rangiferina | | | | ▲ | |
| Rockrose | Helianthemum bicknellii | | ▲ | | | |
| Rush | Juncus spp. | | | | | ▲ |
| Rush, beak | Rhynchospora alba | | | | | ▲ |
| Rush, spike | Eleocharis spp. | | | | | ▲ |
| Rush, wood | Luzula acuminata | | | ▲ | | |
| Sage | Artemisia spp. | | ▲ | | | |
| Scurfpea, gray | Psoralea tenuiflora | | ▲ | | | |
| Scurfpea, silverleaf | Psoralea argophylla | | ▲ | | | |
| Sea blite | Suaeda calceoliformis | | ▲ | | | ▲ |
| Sedge | Carex lasiocarpa | | | | | ▲ |
| Sedge | Carex rostrata | | | | | ▲ |
| Sedge | Cyperus spp. | | ▲ | | | ▲ |
| Sedge, blunt | Carex obtusata | ● | ▲ | | | |
| Sedge, Pennsylvania | Carex pensylvanica | | ▲ | ▲ | | |
| Shining clubmoss | Lycopodium lucidulum | | | | ▲ | |
| Shinleaf | Pyrola rotundifolia | | | | ▲ | |
| Skeleton weed | Lygodesmia juncea | | ▲ | | | |
| Skullcap | Scutellaria spp. | | | | | ▲ |
| Skunk cabbage | Symplocarpus foetidus | | | ▲ | | |
| Smartweed | Polygonum spp. | | ▲ | ▲ | ▲ | ▲ |
| Smooth rattlesnake root | Prenanthes racemosa | | ▲ | | | |
| Sneezeweed | Helenium autumnale | | ▲ | | | |
| Spearscale | Atriplex patula | | ▲ | | | |
| Sphagnum moss | Sphagnum spp. | | | | ▲ | ▲ |
| Spiderwort | Tradescantia spp. | | ▲ | | | |
| Spikemoss | Selaginella rupestris | | ▲ | | | |
| Spreading dogbane | Apocynum androsaemifolium | | ▲ | | | |

| Common name | Scientific name | State status | Prairie | Deciduous forest | Coniferous forest | Wetland, lake, river |
|---|---|---|---|---|---|---|
| Sticktight | Bidens cernua | | | ▲ | ▲ | ▲ |
| Sundew | Drosera rotundifolia | | | | ▲ | ▲ |
| Sunflower, giant | Helianthus giganteus | | ▲ | | | |
| Sunflower, Maximilian's | Helianthus maximiliani | | ▲ | | | |
| Sunflower, sawtooth | Helianthus grosseserratus | | ▲ | | | |
| Sunflower, stiff | Helianthus rigidus | | ▲ | | | |
| Sweet clover | Melilotus sp. | | ▲ | | | |
| Sweet-scented bedstraw | Galium triflorum | | | ▲ | ▲ | |
| Switchgrass | Panicum virgatum | | ▲ | | | |
| Thimbleweed | Anemone cylindrica | | ▲ | | | |
| Thistle, Canada | Cirsium arvense | | ▲ | ▲ | ▲ | |
| Thistle, Flodman's | Cirsium flodmani | | ▲ | | | |
| Tick trefoil | Desmodium glutinosum | | | ▲ | | |
| Tooth-leaved evening primrose | Calylophus serrulata | | ▲ | | | |
| Trillium, big white | Trillium grandiflorum | | | ▲ | | |
| Trillium, nodding | Trillium cernuum | | | ▲ | ▲ | |
| Trout lily, Minnesota | Erythronium propullans | ○ | | ▲ | | |
| Trout lily, white | Erythronium albidum | | | ▲ | | |
| Twin flower | Linnaea borealis | | | | ▲ | |
| Twisted stalk | Streptopus roseus | | | | ▲ | |
| Violet | Viola spp. | | ▲ | ▲ | ▲ | |
| Virginia creeper | Parthenocissus inserta | | | ▲ | | |
| Virginia strawberry | Fragaria virginiana | | | ▲ | ▲ | |
| Water cress | Nasturtium officinale | | | ▲ | | ▲ |
| Water hemlock | Cicuta bulbifera | | | | | ▲ |
| Water hemlock | Cicuta maculata | | | | | ▲ |
| Water lobelia | Lobelia dortmanna | | | | | ▲ |
| Water shield | Brasenia schreberi | | | | | ▲ |
| Water smartweed | Polygonum amphibium | | | | | ▲ |
| Water-lily, little yellow | Nuphar microphyllum | | | | | ▲ |
| Water-lily, white | Nymphaea tuberosa | | | | | ▲ |
| Water-lily, yellow | Nuphar variegatum | | | | | ▲ |
| Wheatgrass | Agropyron trachycaulum | | ▲ | | | |
| White camas | Zigadenus elegans | | ▲ | | | |
| Wild celery | Vallisneria americana | | | | | ▲ |
| Wild cucumber | Echinocystis lobata | | | ▲ | | |

| Common name | Scientific name | State status | Prairie | Deciduous forest | Coniferous forest | Wetland, lake, river |
|---|---|---|---|---|---|---|
| Wild false indigo | Baptisia alba | | ▲ | | | |
| Wild geranium | Geranium maculatum | | | ▲ | | |
| Wild grape | Vitis riparia | | | ▲ | | |
| Wild iris | Iris versicolor | | | | | ▲ |
| Wild licorice | Glycyrrhiza lepidota | | ▲ | | | |
| Wild mint | Mentha arvensis | | ▲ | | | ▲ |
| Wild quinine | Parthenium integrifolium | ○ | ▲ | ▲ | | |
| Wild rice | Zizania aquatica | | | | | ▲ |
| Wild rye | Elymus virginicus | | | ▲ | | |
| Wild sarsaparilla | Aralia nudicaulis | | | ▲ | ▲ | |
| Willow-herb | Epilobium spp. | | | | ▲ | ▲ |
| Wood betony | Pedicularis canadensis | | ▲ | | | |
| Yellow star-grass | Hypoxis hirsuta | | ▲ | | | |
| Yellow water-crowfoot | Ranunculus flabellaris | | | | | ▲ |

*Note:* ● = special concern; ◗ = threatened; ○ = endangered

*Sources:* State status data from Coffin and Pfannmuller (1988); scientific names from Ownbey and Morley (1991); common names from Aaseng et al. (1993) and Ownbey and Morley (1991).

# APPENDIX E

# MAMMALS FOUND IN MINNESOTA

| | Common name | Scientific name | State status | Prairie | Deciduous forest | Coniferous forest | Preferred habitat |
|---|---|---|---|---|---|---|---|
| Marsupials | Opossum | Didelphis virginiana | | | ▲ | | T |
| Insectivores | Masked Shrew | Sorex cinereus | | | ▲ | ▲ | T |
| | Arctic Shrew | Sorex arcticus | | | | ▲ | T W |
| | Water Shrew | Sorex palustris | | | | ▲ | L R |
| | Smoky Shrew | Sorex fumens | | | ▲ | ▲ | T W |
| | Hayden's Shrew | Sorex haydeni | | ▲ | | | T |
| | Pygmy Shrew | Microsorex hoyi | | | | ▲ | T |
| | Short-tailed Shrew | Blarina brevicauda | | ▲ | ▲ | ▲ | T |
| | Least Shrew | Cryptotis parva | ● | ▲ | ▲ | | T |
| | Eastern Mole | Scalopus aquaticus | | ▲ | ▲ | | T |
| | Star-nosed Mole | Condylura cristata | | | | ▲ | T |
| Bats | Little Brown Bat | Myotis lucifugus | | ▲ | ▲ | ▲ | T |
| | Northern Myotis | Myotis septentrionalis | ● | | ▲ | ▲ | T |
| | Silver-haired Bat | Lasionycteris noctivagans | | | ▲ | ▲ | T |
| | Eastern Pipistrelle | Pipistrellus subflavus | ● | | ▲ | | T |
| | Big Brown Bat | Eptesicus fuscus | | ▲ | ▲ | ▲ | T |
| | Red Bat | Lasiurus borealis | | ▲ | ▲ | ▲ | T |
| | Hoary Bat | Lasiurus cinereus | | ▲ | ▲ | ▲ | T |
| Lagomorphs | Eastern Cottontail | Sylvilagus floridanus | | ▲ | ▲ | ▲ | T |
| | Snowshoe Hare | Lepus americanus | | | | ▲ | T |
| | White-tailed Jackrabbit | Lepus townsendii | | ▲ | | | T |
| Rodents | Woodchuck | Marmota monax | | ▲ | ▲ | ▲ | T |
| | Eastern Chipmunk | Tamias striatus | | | ▲ | ▲ | T |
| | Least Chipmunk | Eutamias minimus | | | | ▲ | T |
| | Franklin's Ground Squirrel | Spermophilis franklinii | | ▲ | | | T |
| | Richardson's Ground Squirrel | Spermophilis richardsonii | | ▲ | | | T |

| | Common name | Scientific name | State status | Prairie | Deciduous forest | Coniferous forest | Preferred habitat |
|---|---|---|---|---|---|---|---|
| Rodents (cont.) | Thirteen-lined Ground Squirrel | Spermophilis tridecemlineatus | | ▲ | | | T |
| | Gray Squirrel | Sciurus carolinensis | | | ▲ | | T |
| | Fox Squirrel | Sciurus niger | | ▲ | ▲ | | T |
| | Red Squirrel | Tamiasciurus hudsonicus | | | ▲ | ▲ | T |
| | Northern Flying Squirrel | Glaucomys sabrinus | | | | ▲ | T |
| | Southern Flying Squirrel | Glaucomys volans | | | ▲ | | T |
| | Northern Pocket Gopher | Thomomys talpoides | ● | ▲ | | | T |
| | Plains Pocket Gopher | Geomys bursarius | | ▲ | | | T |
| | Plains Pocket Mouse | Peroganthus flavescens | | ▲ | | | T |
| | Beaver | Castor canadensis | | ▲ | ▲ | ▲ | T |
| | Western Harvest Mouse | Reithrodontomys megalotis | | ▲ | | | T |
| | Deer Mouse | Peromyscus maniculatus | | ▲ | ▲ | ▲ | T |
| | White-footed Mouse | Peromyscus leucopus | | | ▲ | ▲ | T |
| | Southern Red-backed Vole | Clethrionomys gapperi | | | | ▲ | T |
| | Heather Vole | Phenacomys intermedius | ● | | | ▲ | T |
| | Meadow Vole | Microtus pennsylvanicus | | ▲ | ▲ | ▲ | T |
| | Yellownosed (Rock) Vole | Microtus chrotorrhinus | ● | | | ▲ | T |
| | Prairie Vole | Microtus ochrogaster | ● | ▲ | | | T |
| | Woodland Vole | Microtus pinetorum | ● | | ▲ | | T |
| | Muskrat | Ondatra zibethica | | ▲ | ▲ | ▲ | W |
| | Northern Grasshopper Mouse | Onochomys leucogaster | | ▲ | | | T |
| | Southern Bog Lemming | Synaptomys cooperi | | | | ▲ | T W |
| | Northern Bog Lemming | Synaptomys borealis | ● | | | ▲ | T W |
| | Norway Rat | Rattus norvegicus | | ▲ | ▲ | ▲ | T |
| | House Mouse | Mus musculus | | ▲ | ▲ | | T |
| | Meadow Jumping Mouse | Zapus hudsonius | | ▲ | ▲ | ▲ | T |
| | Woodland Jumping Mouse | Napaeozapus insignis | | | | ▲ | T |
| | Porcupine | Erethizon dorsatum | | | | ▲ | T |
| Carnivores | Red Fox | Vulpes vulpes | | ▲ | ▲ | ▲ | T |
| | Gray Fox | Urocyon cinereoargenteus | | | ▲ | | T |
| | Coyote | Canis latrans | | ▲ | ▲ | ▲ | T |
| | Gray Wolf, Timber Wolf | Canis lupus | ▶ | | | ▲ | T |
| | Raccoon | Procyon lotor | | ▲ | ▲ | | T |
| | Black Bear | Ursus americanus | | | | ▲ | T |
| | Marten | Martes americana | ● | | | ▲ | T |
| | Fisher | Martes pennanti | | | | ▲ | T |

| | Common name | Scientific name | State status | Prairie | Deciduous forest | Coniferous forest | Preferred habitat |
|---|---|---|---|---|---|---|---|
| Carnivores (cont.) | Least Weasel | Mustela nivalis | | ▲ | ▲ | ▲ | T |
| | Ermine | Mustela erminea | | | ▲ | ▲ | T |
| | Long-tailed Weasel | Mustela frenata | | ▲ | ▲ | | T |
| | Mink | Mustela vison | | ▲ | ▲ | ▲ | W L R |
| | Wolverine | Gulo gulo | ● | | | ▲ | T |
| | Badger | Taxidea taxus | | ▲ | | | T |
| | Eastern Spotted Skunk | Spilogale putorius | ● | ▲ | ▲ | | T |
| | Striped Skunk | Mephitis mephitis | | ▲ | ▲ | ▲ | T |
| | River Otter | Lutra canadensis | | | ▲ | ▲ | L R |
| | Mountain Lion | Felis concolor | ● | | | ▲ | T |
| | Lynx | Lynx canadensis | | | | ▲ | T |
| | Bobcat | Lynx rufus | | | | ▲ | T |
| Ungulates | American Elk | Cervus elaphus | ● | ▲ | | | T |
| | White-tailed Deer | Odocoileus virginianus | | | ▲ | ▲ | T |
| | Mule Deer | Odocoileus hemionus | ● | ▲ | | | T |
| | Moose | Alces alces | | | | ▲ | T |
| | Caribou | Rangifer tarandus | ● | | | ▲ | T |
| | Pronghorn | Antilocapra americana | | ▲ | | | T |

Note: ● = special concern; ◗ = threatened; ○ = endangered; R = rivers and streams; L = lakes; W = wetlands; T = terrestrial.

Source: State status data from Coffin and Pfannmuller (1988); preferred habitat data from Hazard (1982) and Knox Jones and Birney (1988).

# BIRDS BREEDING IN MINNESOTA

| | Common name | Scientific name | State status | Prairie/ farmland | Fragmented/ disturbed | Deciduous forest | Coniferous forest | Water/ wetland |
|---|---|---|---|---|---|---|---|---|
| Loons and Grebes | Common Loon | Gavia immer | | | | | | ▲ |
| | Pied-billed Grebe | Podilymbus podiceps | | | | | | ▲ |
| | Horned Grebe | Podiceps auritus | ● | | | | | ▲ |
| | Red-necked Grebe | Podiceps grisegena | | | | | | ▲ |
| | Eared Grebe | Podiceps nigricollis | | | | | | ▲ |
| | Western Grebe | Aechmophrus occidentalis | | | | | | ▲ |
| Pelicans and Cormorants | American White Pelican | Pelecanus erythrorhynchos | ● | | | | | ▲ |
| | Double-crested Cormorant | Phalacrocorax auritus | | | | | | ▲ |
| Bitterns, Herons, and Egrets | American Bittern | Botaurus lentiginosus | ● | | | | | ▲ |
| | Least Bittern | Ixobrychus exilis | | | | | | ▲ |
| | Great Blue Heron | Ardea herodias | | | | ▲ | | ▲ |
| | Great Egret | Casmerodius albus | | | | | | ▲ |
| | Snowy Egret | Egretta thula | | | | | | ▲ |
| | Little Blue Heron | Egretta caerulea | | | | | | ▲ |
| | Cattle Egret | Bubulcus ibis | | ▲ | | | | ▲ |
| | Green Heron | Butorides striatus | | | | ▲ | | ▲ |
| | Black-crowned Night-Heron | Nycticorax nycticorax | | | | | | ▲ |
| | Yellow-crowned Night-Heron | Nycticorax violaceus | | | | ▲ | | ▲ |
| Swans, Geese, and Ducks | Trumpeter Swan | Cygnus buccinator | ▶ | | | | | ▲ |
| | Canada Goose | Branta canadensis | | | | | | ▲ |
| | Wood Duck | Aix sponsa | | | | ▲ | | ▲ |
| | Green-winged Teal | Anas crecca | | | | | | ▲ |
| | American Black Duck | Anas rubripes | | | | | | ▲ |
| | Mallard | Anas platyrhynchos | | | | | | ▲ |

| | Common name | Scientific name | State status | Prairie/ farmland | Fragmented/ disturbed | Deciduous forest | Coniferous forest | Water/ wetland |
|---|---|---|---|---|---|---|---|---|
| *Swans, Geese, and Ducks (cont.)* | Northen Pintail | *Anas acuta* | | | | | | ▲ |
| | Blue-winged Teal | *Anas discors* | | | | | | ▲ |
| | Northern Shoveler | *Anas clypeata* | | | | | | ▲ |
| | Gadwall | *Anas strepera* | | | | | | ▲ |
| | American Wigeon | *Anas americana* | | | | | | ▲ |
| | Canvasback | *Aythya valisineria* | | | | | | ▲ |
| | Redhead | *Aythya americana* | | | | | | ▲ |
| | Ring-necked Duck | *Aythya collaris* | | | | | | ▲ |
| | Lesser Scaup, Bluebill | *Aythya affinis* | | | | | | ▲ |
| | Common Goldeneye | *Bucephala clangula* | | | | | | ▲ |
| | Bufflehead | *Bucephala albeola* | | | | | | ▲ |
| | Hooded Merganser | *Lophodytes cucullatus* | | | | | | ▲ |
| | Common Merganser | *Mergus merganser* | | | | | | ▲ |
| | Red-breasted Merganser | *Mergus serrator* | | | | | | ▲ |
| | Ruddy Duck | *Oxyura jamaicensis* | | | | | | ▲ |
| *Vultures* | Turkey Vulture | *Cathartes aura* | | | ▲ | ▲ | | |
| *Ospreys, Eagles, Harriers, and Hawks* | Osprey | *Pandion haliaetus* | ● | | | | | ▲ |
| | Bald Eagle | *Haliaeetus leucocephalus* | ◗ | | | | ▲ | ▲ |
| | Northern Harrier | *Circus cyaneus* | | ▲ | | | | |
| | Sharp-shinned Hawk | *Accipiter striatus* | | | | | ▲ | |
| | Cooper's Hawk | *Accipiter cooperi* | | | | ▲ | | |
| | Northern Goshawk | *Accipiter gentilis* | | | | | ▲ | |
| | Red-shouldered Hawk | *Buteo lineatus* | ● | | | ▲ | | |
| | Broad-winged Hawk | *Buteo playtypterus* | | | | ▲ | | |
| | Swainson's Hawk | *Buteo swainsoni* | | ▲ | | | | |
| | Red-tailed Hawk | *Buteo jamaicensis* | | | ▲ | ▲ | | |
| | American Kestrel | *Falco sparverius* | | ▲ | | | | |
| | Merlin | *Falco columbarius* | | | | | ▲ | |
| | Peregrine Falcon | *Falco peregrinus* | ○ | ▲ | ▲ | | | |
| *Partridges, Pheasants, Grouse, Turkeys, and Quails* | Gray Partridge | *Perdix perdix* | | ▲ | | | | |
| | Ring-necked Pheasant | *Phaisanus colchicus* | | ▲ | | | | |
| | Spruce Grouse | *Dendragapus canadensis* | | | | | ▲ | |
| | Ruffed Grouse | *Bonasa umbellus* | | | | ▲ | ▲ | |
| | Greater Prairie-Chicken | *Tympanuchus cupido* | ● | ▲ | | | | |
| | Sharp-tailed Grouse | *Tympanuchus phasianellus* | | ▲ | | ▲ | | |
| | Wild Turkey | *Meleagris gallopavo* | | | | ▲ | | |
| | Northern Bobwhite | *Colinus virginianus* | | ▲ | ▲ | | | |

| | Common name | Scientific name | State status | Prairie/ farmland | Fragmented/ disturbed | Deciduous forest | Coniferous forest | Water/ wetland |
|---|---|---|---|---|---|---|---|---|
| Rails, Coots, and Cranes | Yellow Rail | Coturnicops noveboracensis | ● | | | | | ▲ |
| | King Rail | Rallus elegans | ● | | | | | ▲ |
| | Virginia Rail | Rallus limicola | | | | | | ▲ |
| | Sora | Porzana carolina | | | | | | ▲ |
| | Common Moorhen | Gallinula chloropus | ● | | | | | ▲ |
| | American Coot | Fulica americana | | | | | | ▲ |
| | Sandhill Crane | Grus canadensis | ● | | | | | ▲ |
| Plovers and Avocets | Piping Plover | Charadrius melodus | ○ | | | | | ▲ |
| | Killdeer | Charadrius vociferus | | ▲ | | | | |
| | American Avocet | Recurvirostra americana | | | | | | ▲ |
| Sandpipers, Godwits, Snipes, Woodcocks, and Phalaropes | Spotted Sandpiper | Actitis macularia | | | | | | ▲ |
| | Upland Sandpiper | Bartramia longicauda | ● | ▲ | | | | |
| | Marbled Godwit | Limosa fedoa | ● | ▲ | | | | |
| | Common Snipe | Gallinago gallinago | | | | | | ▲ |
| | American Woodcock | Scolopax minor | | | | ▲ | | |
| | Wilson's Phalarope | Phalaropus tricolor | ● | | | | | ▲ |
| Gulls and Terns | Franklin's Gull | Larus pipixcan | | | | | | ▲ |
| | Ring-billed Gull | Larus delawarensis | | | | | | ▲ |
| | Herring Gull | Larus argentatus | | | | | | ▲ |
| | Caspian Tern | Sterna caspia | | | | | | ▲ |
| | Common Tern | Sterna hirundo | ● | | | | | ▲ |
| | Forster's Tern | Sterna forsteri | ● | | | | | ▲ |
| | Black Tern | Childonias niger | | | | | | ▲ |
| Pigeons and Doves | Rock Dove | Columbia livia | | ▲ | | | | |
| | Mourning Dove | Zenaida macroura | | ▲ | ▲ | | | |
| Cuckoos | Black-billed Cuckoo | Coccyzus erythropthalmus | | | | ▲ | | |
| | Yellow-billed Cuckoo | Coccyzus americanus | | | | ▲ | | |
| Owls | Eastern Screech-Owl | Otus asio | | | ▲ | ▲ | | |
| | Great Horned Owl | Bubo virginianus | | | ▲ | ▲ | ▲ | |
| | Northern Hawk-Owl | Surnia ulula | | | | | ▲ | |
| | Burrowing Owl | Athene cunicularia | ○ | ▲ | | | | |
| | Barred Owl | Strix varia | | | | ▲ | | |
| | Great Gray Owl | Strix nebulosa | | | | | ▲ | |
| | Long-eared Owl | Asio otus | | | | ▲ | | |
| | Short-eared Owl | Asio flammeus | ● | ▲ | | | | |
| | Boreal Owl | Aegolius funereus | | | | | ▲ | |
| | Northern Saw-whet Owl | Aegolius acadicus | | | | | ▲ | |

| | Common name | Scientific name | State status | Prairie/ farmland | Fragmented/ disturbed | Deciduous forest | Coniferous forest | Water/ wetland |
|---|---|---|---|---|---|---|---|---|
| Goatsuckers | Common Nighthawk | Chordeiles minor | | ▲ | ▲ | | | |
| | Whip-poor-will | Caprimulgus vociferus | | | | ▲ | | |
| Swifts and Hummingbirds | Chimney Swift | Chaetura pelagica | | | ▲ | | | |
| | Ruby-throated Hummingbird | Archilochus colubris | | | ▲ | ▲ | ▲ | |
| Kingfishers | Belted Kingfisher | Ceryle alcyon | | | | | | ▲ |
| Woodpeckers | Red-headed Woodpecker | Melanerpes erythrocephalus | | | ▲ | ▲ | | |
| | Red-bellied Woodpecker | Melanerpes carolinus | | | | ▲ | | |
| | Yellow-bellied Sapsucker | Sphyrapicus varius | | | | ▲ | | |
| | Downy Woodpecker | Picoides pubescens | | | ▲ | ▲ | ▲ | |
| | Hairy Woodpecker | Picoides villosus | | | ▲ | ▲ | ▲ | |
| | Three-toed Woodpecker | Picoides tridactylus | | | | | ▲ | |
| | Black-backed Woodpecker | Picoides arcticus | | | | | ▲ | |
| | Northern Flicker | Colaptes auratus | | | ▲ | ▲ | | |
| | Pileated Woodpecker | Dryocopus pileatus | | | | ▲ | | |
| Flycatchers | Olive-sided Flycatcher | Contopus borealis | | | | | ▲ | |
| | Eastern Wood-Pewee | Contopus virens | | | | ▲ | | |
| | Western Wood-Pewee | Contopus sordidulus | | | | ▲ | | |
| | Yellow-bellied Flycatcher | Empidonax flaviventris | | | | | ▲ | |
| | Acadian Flycatcher | Empidonax virescens | | | | ▲ | | |
| | Alder Flycatcher | Empidonax alnorum | | | | | | ▲ |
| | Willow Flycatcher | Empidonax traillii | | | | | | ▲ |
| | Least Flycatcher | Empidonax minimus | | | | ▲ | | |
| | Eastern Phoebe | Sayornis nigricans | | | ▲ | ▲ | | |
| | Great Crested Flycatcher | Myiarchus critinus | | | ▲ | | | |
| | Western Kingbird | Tyrannus verticalis | | ▲ | | | | |
| | Eastern Kingbird | Tyrannus tyrannus | | ▲ | | | | |
| Larks and Swallows | Horned Lark | Eremophila alpestris | | ▲ | | | | |
| | Purple Martin | Progne subis | | ▲ | | | | ▲ |
| | Tree Swallow | Tachycineta bicolor | | | ▲ | | | ▲ |
| | Rough-winged Swallow | Stelgidopteryx serripennis | | ▲ | | | | ▲ |
| | Northern Bank Swallow | Riparia riparia | | ▲ | | | | ▲ |
| | Cliff Swallow | Hirundo pyrrhonota | | ▲ | | | | ▲ |
| | Barn Swallow | Hirundo rustica | | ▲ | | | | ▲ |

| | Common name | Scientific name | State status | Prairie/ farmland | Fragmented/ disturbed | Deciduous forest | Coniferous forest | Water/ wetland |
|---|---|---|---|---|---|---|---|---|
| Jays, Magpies, and Crows | Gray Jay | Perisoreus canadensis | | | | | ▲ | |
| | Blue Jay | Cyanactta cristata | | | ▲ | ▲ | ▲ | |
| | Black-billed Magpie | Pica pica | | | ▲ | | | |
| | American Crow | Corvus brachyrhynchos | | | ▲ | ▲ | | |
| | Common Raven | Corvus corax | | | | | ▲ | |
| Chickadees and Titmice | Black-capped Chickadee | Parus atricapillus | | | ▲ | ▲ | ▲ | |
| | Boreal Chickadee | Parus hudsonicus | | | | | ▲ | |
| | Tufted Titmouse | Parus bicolor | | | | ▲ | | |
| Nuthatches and Creepers | Red-breasted Nuthatch | Sitta canadensis | | | | | ▲ | |
| | White-breasted Nuthatch | Sitta carolinensis | | | ▲ | | | |
| | Brown Creeper | Certhia americana | | | | ▲ | ▲ | |
| Wrens | House Wren | Troglodytes aedon | | | ▲ | ▲ | | |
| | Winter Wren | Troglodytes troglodytes | | | | | ▲ | |
| | Sedge Wren | Cistothorus platensis | | | | | | ▲ |
| | Marsh Wren | Cistothorus palustris | | | | | | ▲ |
| Kinglets, Gnatcatchers, and Thrushes | Golden-crowned Kinglet | Regulus satrapa | | | | | ▲ | |
| | Ruby-crowned Kinglet | Regulus calendula | | | | | ▲ | |
| | Blue-gray Gnatcatcher | Polioptila caerulea | | | | ▲ | | |
| | Eastern Bluebird | Sialia sialis | | | ▲ | | | |
| | Veery | Catharus fuscescens | | | | ▲ | | |
| | Swainson's Thrush | Catharus ustulatus | | | | | ▲ | |
| | Hermit Thrush | Catharus guttatus | | | | | ▲ | |
| | Wood Thrush | Hylocichla mustelina | | | | ▲ | | |
| | American Robin | Turdus migratorius | | | ▲ | | | |
| Catbirds, Mockingbirds, and Thrashers | Gray Catbird | Dumetella carolinensis | | | ▲ | | | |
| | Northern Mockingbird | Mimus polyglottos | | | ▲ | | | |
| | Brown Thrasher | Toxostoma rufum | | | ▲ | | | |
| Pipets, Waxwings, and Shrikes | Sprague's Pipit | Anthus spragueii | ○ | ▲ | | | | |
| | Cedar Waxwing | Bombycilla cedrorum | | | ▲ | | | |
| | Loggerhead Shrike | Lanius ludovicianus | ▶ | ▲ | | | | |
| Starlings and Vireos | European Starling | Sturnus vulgaris | | | ▲ | | | |
| | Bell's Vireo | Vireo bellii | | | ▲ | | | |
| | Solitary Vireo | Vireo solitarius | | | | | ▲ | |
| | Yellow-throated Vireo | Vireo flavifrons | | | | ▲ | | |
| | Warbling Vireo | Vireo gilvus | | | ▲ | | | |
| | Philadelphia Vireo | Vireo philadelphicus | | | | ▲ | | |
| | Red-eyed Vireo | Vireo olivaceus | | | | ▲ | | |

306

| | Common name | Scientific name | State status | Prairie/ farmland | Fragmented/ disturbed | Deciduous forest | Coniferous forest | Water/ wetland |
|---|---|---|---|---|---|---|---|---|
| Warblers and Tanagers | Blue-winged Warbler | Vermivora pinus | | | | ▲ | | |
| | Golden-winged Warbler | Vermivora chrysoptera | | | | ▲ | | |
| | Tennessee Warbler | Vermivora peregrina | | | | | ▲ | |
| | Nashville Warbler | Vermivora ruficapilla | | | | | ▲ | |
| | Northern Parula | Parula americana | | | | | ▲ | |
| | Yellow Warbler | Dendroica petechia | | | ▲ | | | |
| | Chestnut-sided Warbler | Dendroica pensylvanica | | | | ▲ | | |
| | Magnolia Warbler | Dendroica magnolia | | | | | ▲ | |
| | Cape May Warbler | Dendroica tigrina | | | | | ▲ | |
| | Black-throated Blue Warbler | Dendroica caerulescens | | | | ▲ | | |
| | Yellow-rumped Warbler | Dendroica coronata | | | | | ▲ | |
| | Black-throated Green Warbler | Dendroica virens | | | | | ▲ | |
| | Blackburnian Warbler | Dendroica fusca | | | | | ▲ | |
| | Pine Warbler | Dendroica pinus | | | | | ▲ | |
| | Palm Warbler | Dendroica palmarum | | | | | ▲ | |
| | Bay-breasted Warbler | Dendroica castanea | | | | | ▲ | |
| | Cerulean Warbler | Dendroica cerulea | | | | ▲ | | |
| | Black-and-white Warbler | Mniotilta varia | | | | ▲ | | |
| | American Redstart | Setophaga ruticilla | | | | ▲ | | |
| | Prothonotary Warbler | Protonotaria citrea | | | | ▲ | | |
| | Ovenbird | Seiurus aurocapillus | | | | ▲ | | |
| | Northern Waterthrush | Seiurus noveboracensis | | | | ▲ | | |
| | Louisiana Waterthrush | Seiurus motacilla | ● | | | ▲ | | |
| | Kentucky Warbler | Oporornis formosus | | | | ▲ | | |
| | Connecticut Warbler | Oporornis agilis | | | | | ▲ | |
| | Mourning Warbler | Oporornis philadelphia | | | | ▲ | | |
| | Common Yellowthroat | Geothylpis trichas | | | ▲ | | | |
| | Hooded Warbler | Wilsonia citrina | | | | ▲ | | |
| | Wilson's Warbler | Wilsonia pusilla | | | | | ▲ | |
| | Canada Warbler | Wilsonia canadensis | | | | ▲ | ▲ | |
| | Yellow-breasted Chat | Icteria virens | | | ▲ | | | |
| | Scarlet Tanager | Piranga olivacea | | | | ▲ | | |
| Grosbeaks and Buntings | Northern Cardinal | Cardinalis cardinalis | | | ▲ | | | |
| | Rose-breasted Grosbeak | Pheucticus ludovicianus | | | | ▲ | | |
| | Blue Grosbeak | Guiraca caerulea | | | ▲ | | | |

| | Common name | Scientific name | State status | Prairie/ farmland | Fragmented/ disturbed | Deciduous forest | Coniferous forest | Water/ wetland |
|---|---|---|---|---|---|---|---|---|
| Grosbeaks and Buntings (cont.) | Indigo Bunting | Passerina cyanea | | | ▲ | ▲ | | |
| | Dickcissel | Spiza americana | | ▲ | | | | |
| Towhees and Sparrows | Rufous-sided Towhee | Pipilo erhythrophthalamus | | | ▲ | | | |
| | Chipping Sparrow | Spizella passerina | | | ▲ | ▲ | | |
| | Clay-colored Sparrow | Spizella pallida | | ▲ | | | | |
| | Field Sparrow | Spizella pusilla | | ▲ | | | | |
| | Vesper Sparrow | Pooecetes gramineus | | ▲ | | | | |
| | Lark Sparrow | Chondestes grammacus | | ▲ | | | | |
| | Savannah Sparrow | Passerculus sandwichensis | | ▲ | | | | |
| | Baird's Sparrow | Ammondramus bairdii | ○ | ▲ | | | | |
| | Grasshopper Sparrow | Ammondramus savannarum | | ▲ | | | | |
| | Henslow's Sparrow | Ammondramus henslowii | ● | ▲ | | | | |
| | LeConte's Sparrow | Ammodramus leconteii | | | | | | ▲ |
| | Sharp-tailed Sparrow | Ammodramus caudacutus | ● | | | | | ▲ |
| | Song Sparrow | Melospiza melodia | | | ▲ | | | |
| | Lincoln's Sparrow | Melospiza lincolnii | | | | | ▲ | |
| | Swamp Sparrow | Melospiza georgiana | | | | | | ▲ |
| | White-throated Sparrow | Zonotrichia albicollis | | | | | ▲ | |
| | Dark-eyed Junco | Junco hyemalis | | | | | ▲ | |
| Longspurs and Blackbirds | Chestnut-collared Longspur | Calcaris ornatus | ○ | ▲ | | | | |
| | Bobolink | Dolichonyx oryzivorus | | ▲ | | | | |
| | Red-winged Blackbird | Agelaius phoeniceus | | | | | | ▲ |
| | Eastern Meadowlark | Sturnella magna | | ▲ | | | | |
| | Western Meadowlark | Sturnella neglecta | | ▲ | | | | |
| | Yellow-headed Blackbird | Xanthocephalus xanthocephalus | | | | | | ▲ |
| | Rusty Blackbird | Euphagus carolinus | | | | | ▲ | |
| | Brewer's Blackbird | Euphagus cyanocephalus | | ▲ | | | | |
| | Common Grackle | Quiscalus quiscula | | | ▲ | | | |
| | Brown-headed Cowbird | Molothrus ater | | | ▲ | | | |
| | Orchard Oriole | Icterus spurius | | | ▲ | | | |
| | Northern Oriole | Icterus galbula | | | ▲ | ▲ | | |
| Finches | House Finch | Carpodacus mexicanus | | ▲ | ▲ | ▲ | ▲ | |
| | Purple Finch | Carpodacus purpureus | | | | | ▲ | |
| | Red Crossbill | Loxia curvirostra | | | | | ▲ | |
| | White-winged Crossbill | Loxia leucoptera | | | | | ▲ | |

|  | Common name | Scientific name | State status | Prairie/ farmland | Fragmented/ disturbed | Deciduous forest | Coniferous forest | Water/ wetland |
|---|---|---|---|---|---|---|---|---|
| Finches (cont.) | Pine Siskin | *Carduelis pinus* |  |  |  |  | ▲ |  |
|  | American Goldfinch | *Carduelis tristis* |  |  | ▲ |  |  |  |
|  | Evening Grosbeak | *Coccothraustes vespertinus* |  |  |  |  | ▲ |  |
|  | Pine Grosbeak | *Pinicola enucleator* |  |  |  |  | ▲ |  |
|  | House Sparrow | *Passer domesticus* |  |  | ▲ |  |  |  |

*Note:* ● = special concern; ◗ = threatened; ○ = endangered.

*Source:* State status data from Coffin and Pfannmuller (1988); nomenclature and preferred habitat data from Green (1991) and Janssen (1987).

# Amphibians and Reptiles Found in Minnesota

| | Common name | Scientific name | State status | Prairie | Deciduous forest | Coniferous forest | Preferred habitat |
|---|---|---|---|---|---|---|---|
| Turtles | Snapping Turtle | *Cheldyra serpentina* | ● | ▲ | ▲ | ▲ | L R W |
| | Painted Turtle | *Chrysemys picta* | | ▲ | ▲ | ▲ | L W |
| | Wood Turtle | *Clemmys insculpta* | ▶ | | ▲ | ▲ | R |
| | Blanding's Turtle | *Emydoidea blandingii* | ▶ | ▲ | ▲ | | L W |
| | Map Turtle | *Graptemys geographica* | | ▲ | ▲ | | R L |
| | Ouachita Map Turtle | *Graptemys ouachitensis* | | | ▲ | | R |
| | False Map Turtle | *Graptemys pseudogeographica* | | ▲ | ▲ | | R |
| | Smooth Softshell | *Apalone muticua* | | | ▲ | | R |
| | Spiny Softshell | *Apalone spiniferua* | | ▲ | ▲ | | R |
| Lizards | Six-lined Racerunner | *Cnemidophorus sexlineatus* | | ▲ | ▲ | | T |
| | Five-lined Skink | *Eumeces fasciatus* | ○ | | ▲ | | T |
| | Prairie Skink | *Eumeces septentrionalis* | | ▲ | | | T |
| Snakes | Racer | *Coluber constrictor* | ● | ▲ | ▲ | | T |
| | Timber Rattlesnake | *Crotalus horridus* | | | ▲ | | T |
| | Ringneck Snake | *Diadophis punctatus* | | | ▲ | ▲ | T |
| | Rat Snake | *Elaphe obsoleta* | ● | | ▲ | | T |
| | Fox Snake | *Elaphe vulpina* | ● | | ▲ | | T |
| | Western Hognose Snake | *Heterodon nasicus* | ● | | | ▲ | T |
| | Eastern Hognose Snake | *Heterodon platirhinos* | ● | | ▲ | | T |
| | Milk Snake | *Lampropeltis triangulum* | ● | | ▲ | | T |
| | Northern Water Snake | *Nerodia sipedon* | | | ▲ | | T L R W |
| | Smooth Green Snake | *Ophedrys vernalis* | | | ▲ | | T |
| | Gopher Snake | *Pitophis catenifer* | ● | ▲ | | | T |
| | Massasauga | *Sistrurus catenatus* | ● | | ▲ | | T W |
| | Brown Snake | *Storeria dekayi* | | | ▲ | | T |
| | Redbelly Snake | *Soreria occipitomaculata* | | ▲ | ▲ | ▲ | T |
| | Plains Garter Snake | *Thamnophis radix* | | ▲ | | | T |
| | Common Garter Snake | *Thamnophis sirtalis* | | ▲ | ▲ | ▲ | T W |
| | Lined Snake | *Tropidoclonion lineatum* | ● | ▲ | | | T |

| | Common name | Scientific name | State status | Prairie | Deciduous forest | Coniferous forest | Preferred habitat |
|---|---|---|---|---|---|---|---|
| Salamanders | Blue-spotted Salamander | Ambystoma laterale | | | ▲ | ▲ | T W |
| | Tiger Salamander | Ambystoma tigrinum | | ▲ | ▲ | | T W |
| | Mudpuppy | Necturus maculosus | | ▲ | ▲ | ▲ | L R |
| | Eastern Newt | Notophthalmus viridescens | | | ▲ | ▲ | T W |
| | Redback Salamander | Plethodon cinereus | | | ▲ | ▲ | T |
| Toads and frogs | Northern Cricket Frog | Acris crepitans | ● | ▲ | ▲ | | R W |
| | American Toad | Bufo americanus | | ▲ | ▲ | ▲ | T W |
| | Great Plains Toad | Bufo cognatus | | ▲ | | | T W |
| | Canadian Toad, Manitoba Toad | Bufo hemiophrys | | ▲ | | | T W |
| | Cope's Gray Treefrog | Hyla chrysoscelis | | ▲ | ▲ | | T W |
| | Gray Treefrog | Hyla versicolor | | | ▲ | ▲ | T W |
| | Spring Peeper | Pseudacris crucifer | | | ▲ | ▲ | T W |
| | Western Chorus Frog | Pseudacris triseriata | | ▲ | ▲ | ▲ | T W |
| | Bullfrog | Rana catesbeiana | ● | | ▲ | | L R W |
| | Green Frog | Rana clamitans | | | ▲ | ▲ | L R W |
| | Pickerel Frog | Rana palustris | ● | | ▲ | | T R W |
| | Northern Leopard Frog | Rana pipiens | | ▲ | ▲ | ▲ | T W |
| | Mink Frog | Rana septentrionalis | | | | ▲ | L W |
| | Wood Frog | Rana sylvatica | | | ▲ | ▲ | T W |

*Note:* ● = special concern; ◗ = threatened; ○ = endangered; R = rivers and streams; L = lakes; W = wetlands; T = terrestrial.

*Source:* State status data from Coffin and Pfannmuller (1988); preferred habitat data from J. Lang, University of North Dakota, personal communication; nomenclature from Oldfield and Moriarty (1994).

# APPENDIX H METRIC EQUIVALENTS OF ENGLISH WEIGHTS AND MEASURES

| English unit | Metric equivalent |
|---|---|
| parts per million (ppm) | milligrams per liter |
| parts per billion (ppb) | micrograms per liter |
| foot | 0.305 meter |
| mile | 1.609 kilometers |
| acre | 0.405 hectare |
| gallon | 3.785 liters |
| pound | 0.454 kilogram |
| °F | 0.556 (°F - 32) = °C |

# LITERATURE CITED

Aaseng, N. E., J. C. Almendinger, R. P. Dana, B. C. Delaney, H. L. Dunevitz, K. A. Rusterholz, N. P. Sather, and D. S. Wovcha. 1993. Minnesota's native vegetation: A key to natural communities. Minnesota Department of Natural Resources, Natural Heritage Program, Biological Report 20:1–III.

Ahlgren, C. E, and I. Ahlgren. 1984. *Lob trees in the wilderness.* University of Minnesota Press, Minneapolis.

Ahlgren, I. F., and C. E. Ahlgren. 1960. Ecological effects of forest fires. *Botanical Review* 26(4):483–533.

Alexander, S. C., and E. C. Alexander Jr. 1989. Residence times of Minnesota groundwater. *Journal of the Minnesota Academy of Science* 55:48–52.

Almendinger, J. E. 1988. Lake and groundwater paleohydrology: A groundwater model to explain past lake levels in west-central Minnesota. Ph.D. thesis, University of Minnesota.

Anderson, J. F., R. C. Johnson, L. A. Magnarelli, and F. W. Hyde. 1986. Involvement of birds in the epidemiology of the Lyme disease agent, *Borrelia burgdorferi. Infection and Immunity* 51:394–96.

Anderson, J. L., and D. F. Grigal. 1984. Soils and landscapes of Minnesota. University of Minnesota Agricultural Extension Service, AG-FO-2331:1–8.

Anderson, J. P., and W. J. Craig. 1984. *Growing energy crops on Minnesota's wetlands: The land use perspective.* Center for Urban and Regional Affairs, University of Minnesota. Publication No. 84-3:1–95.

Archibald, H. L. 1976. Spatial relationships of neighboring male ruffed grouse in spring. *Journal of Wildlife Management* 40: 750–60.

Atkins, A. 1984. *Harvest of grief: Grasshopper plagues and public assistance in Minnesota, 1873–78.* Minnesota Historical Society Press, St. Paul.

Baker, D. G., and R. K. Crookston. 1980. An agronomic look at the climate and weather of Minnesota. In Corn and soybean management in Minnesota, compiler R. K. Crookston, 2–25. Mimeographed.

Baker, D. G., and E. L. Kuehnast. 1978. *Precipitation normals for Minnesota: 1941–1970.* Climate of Minnesota, Part X, University of Minnesota Agricultural Experiment Station Technical Bulletin 314.

Baker, D. G., E. L. Kuehnast, and J. A. Zandlo. 1985. *Normal temperatures (1951–1980) and their application.* Climate of Minnesota, Part XV, University of Minnesota Agricultural Experiment Station Technical Bulletin AD-SB-2777.

Baker, D. G., and J. H. Strub Jr. 1963. *The agricultural and minimum-temperature-free seasons.* Climate of Minnesota, Part II, University of Minnesota Agricultural Experiment Station Technical Bulletin 245.

Baker, J. E., S. J. Eisenreich, and B. J. Eadie. 1991. Sediment trap fluxes and benthic recycling of organic carbon, polycyclic aromatic hydrocarbons, and polychlorobiphenyl congeners in Lake Superior. *Environmental Science & Technology* 25:500–509.

Batt, B. D. J., M. G. Anderson, C. D. Anderson, and F. D. Caswell. 1989. The use of prairie potholes by North American Ducks. In *Northern prairie wetlands*, ed. A. van der Valk, 204–27. Iowa State University Press, Ames.

Behrend, A. F., and J. R. Tester. 1988. Feeding ecology of the plains pocket gopher (*Geomys bursarius*) in Minnesota. *Prairie Naturalist* 20:99–107.

Bell, H. 1986. Voyage to the bottom of the inland sea. University of Minnesota Sea Grant, *Seiche* 10:1–3, 6.

————. 1989. Superior pursuit: Facts about the greatest Great Lake. Minnesota Sea Grant Extension Program, Superior Advisory Note 19.

Berg, W. E. 1992. Large mammals. In *The patterned peatlands of Minnesota*, ed. H. E. Wright Jr., B. A. Coffin, and N. Aaseng, 73–84. University of Minnesota Press, Minneapolis.

Berg, W. E., and R. A. Chesness. 1978. Ecology of coyotes in northern Minnesota. In *Coyotes—Biology, behavior and management*, ed. M. Bekoff, 229–47. Academic Press, New York.

Bird, R. D. 1961. Ecology of the aspen parkland of western Canada. Canadian Department of Agriculture, Research Branch Publication 1066.

Björkman, E. 1960. *Monotropa hypopytis* L., an epiparasite on tree roots. *Physiologia Plantarum* 13:308–27.

Borchert, J. R., and N. C. Gustafson. 1980. *Atlas of Minnesota resources and settlement.* 3d ed. Center for Urban and Regional Affairs, University of Minnesota Publication No. CURA 80-5.

Borchert, J. R., and D. P. Yaeger. 1968. *Atlas of Minnesota resources and settlement.* Minnesota State Planning Agency, St. Paul.

Botch, M. S., and V. V. Masing. 1983. Mire ecosystems in the U.S.S.R. In *Ecosystems of the World, 4B, Mires: Swamp, Bog, Fen and Moor, Regional Studies*, ed. A. J. P. Gore, 95–152. Elsevier Scientific Publishing Co., New York.

Braun, E. L. 1950. *Deciduous forests of eastern North America.* Blakiston Co., Philadelphia.

Bray, J. R., and L. A. Dudkiewicz. 1963. The composition, biomass and productivity of two *Populus* forests. *Bulletin Torrey Botanical Club* 90:298–308.

Bray, J. R., D. B. Lawrence, and L. C. Pearson. 1959. Primary production in some Minnesota terrestrial communities. *Oikos* 10:38–49.

Breckenridge, W. J. 1944. *Reptiles and amphibians of Minnesota.* University of Minnesota Press, Minneapolis.

Breckenridge, W. J., and J. R. Tester. 1961. Growth, local movements and hibernation of the Manitoba toad, *Bufo hemiophrys. Ecology* 42:637–46.

Breining, G. 1989. *Managing Minnesota's fish.* Minnesota Department of Natural Resources, Section of Fisheries, St. Paul.

Brinson, M. M., A. E. Lugo, and S. Brown. 1981. Primary productivity, decomposition and consumer activity in freshwater wetlands. *Annual Review of Ecology and Systematics* 12:123–61.

Bryson, R. A., and F. K. Hare. 1974. Climates of North America. In *Climates of North America,* ed. R. A. Bryson and F. K. Hare, 1–47. World Survey of Climatology, vol. 11. Elsevier Scientific Publishing Co., New York.

Buell, M. F., and H. F. Buell. 1959. Aspen invasion of prairie. *Bulletin Torrey Botanical Club* 86:264–65.

Buell, M. F., and J. E. Cantlon. 1951. A study of two forest stands in Minnesota with an interpretation of the prairie-forest margin. *Ecology* 32:294–316.

Buell, M. F., and V. Facey. 1960. Forest-prairie transition west of Itasca Park, Minnesota. *Bulletin Torrey Botanical Club* 87:46–58.

Buell, M. F., and W. A. Niering. 1957. Fir-spruce-birch forests in northern Minnesota. *Ecology* 38:602–10.

Burgis, M. J., and P. Morris. 1987. *The natural history of lakes.* Cambridge University Press, Cambridge, England.

Burton, T. M., and G. E. Likens. 1975. Salamander populations and biomass in the Hubbard Brook Experimental Forest, New Hampshire. *Copeia* 1975:541–46.

Bury, R. B. 1979. Population ecology of freshwater turtles. In *Turtles: perspectives and research,* ed. M. Harless and H. Morlock, 571–602. J. Wiley & Sons.

Carlander, H. B. 1954. History of fish and fishing in the Upper Mississippi River. Upper Mississippi River Conservation Commission.

Carlson, R. E. 1977. A trophic state index for lakes. *Limnology and Oceanography* 22:361–69.

Carpenter, S. R. 1989. Temporal variance in lake communities: blue-green algae and the trophic cascade. *Landscape Ecology* 3: 175–84.

Christensen, N. L. 1976. The role of carnivory in *Sarracenia flava* L. with regard to specific nutrient deficiencies. *Journal of Elisha Mitchell Science Society* 92:144–47.

Clark, J. 1979. Fresh water wetlands: Habitats for aquatic invertebrates, amphibians, reptiles, and fish. In *Wetland functions and values: The state of our understanding,* ed. P. E. Greeson, J. R. Clark, and J. E. Clark, 330–43. American Water Resources Association, Minneapolis, Minn.

Clark, J. S. 1993. Fire, climate change, and forest processes during the past 2000 years. In Elk Lake, Minnesota: Evidence for rapid climate change in the north-central United States, ed. J. P. Bradbury and W. E. Dean, 295–308. Geological Society of America, Boulder, Colo., Special Paper 276.

Clements, F. E. 1916. Plant succession: An analysis of the development of vegetation. Carnegie Institute Publication 242, Washington, D.C.

Coffin, B. A. 1988. The natural vegetation of Minnesota at the time of the Public Land Survey: 1847–1907. Biological Report No. 1, Minnesota Department of Natural Resources, St. Paul.

Coffin, B. A., and L. A. Pfannmuller. 1988. *Minnesota's endangered flora and fauna.* University of Minnesota Press, Minneapolis.

Coffman, W. P. 1978. Chironomidae. In *An introduction to the aquatic insects of North America,* ed. R. W. Merritt and K. W. Cummins, 345–76. Kendall/Hunt Publishing Co., Dubuque, Iowa.

Cohen, S. B. 1973. *Oxford world atlas.* Oxford University Press, New York.

Coker, R. E. 1954. *Streams, lakes, ponds.* University of North Carolina Press, Chapel Hill.

Colby, P. J., P. A. Ryan, D. H. Schupp, and S. L. Serns. 1987. Interactions in north-temperate lake fish communities. *Canadian Journal of Fisheries and Aquatic Sciences* 44:104–28.

Cole, G. A. 1983. *Textbook of limnology.* Waveland Press, Prospect Heights, Ill.

Connell, J. H., and R. O. Slayter. 1977. Mechanisms of succession in natural communities and their role in community stability and organization. *American Naturalist* 111:1119–44.

Conway, V. M. 1949. The bogs of central Minnesota. *Ecological Monographs* 19:173–206.

Cooper, W. S. 1935. The history of the upper Mississippi River in late Wisconsin and postglacial time. Minnesota Geological Survey Bulletin 26.

Costello, D. F. 1969. *The prairie world.* T. Y. Crowell Co., New York.

Coues, E. [1910] 1965. The expeditions of Zebulon Montgomery Pike. Vol. 1. Reprint, Ross & Haines, Minneapolis, Minn.

Coveney, M. F., G. Cronberg, M. Enell, K. Larsson, and L. Olofsson. 1977. Phytoplankton, zooplankton and bacteria—Standing crop and production relationships in a eutrophic lake. *Oikos* 29:5–21.

Cowles, H. C. 1899. The ecological relations of the vegetation on the sand dunes of Lake Michigan. *Botanical Gazette* 27: 95–117, 167–202, 281–308, 361–91.

Craig, W. J., and W. S. Anderson. 1991. Environment and the river: Maps of the Mississippi. Center for Urban and Regional Affairs, University of Minnesota, Minneapolis. CURA 91-9: 1–38.

Crow, T. R. 1978. Biomass and production in three contiguous forests in northern Wisconsin. *Ecology* 59:265–73.

Cummins, K. W. 1974. Structure and function of stream ecosystems. *Bioscience* 24:631–41.

Cummins, K. W., and M. J. Klug. 1979. Feeding ecology of stream invertebrates. *Annual Review of Ecology and Systematics* 10:147–72.

Cunningham, W. P., and B. W. Saigo. 1990. *Environmental science.* Wm. C. Brown, Publishers, Dubuque, Iowa.

Curtis, J. T. 1959. *The vegetation of Wisconsin.* University of Wisconsin Press, Madison.

Dahlman, R. C., and C. L. Kucera. 1965. Root productivity and turnover in a native prairie. *Ecology* 46:84–89.

Darnell, R. M. 1979. Impact of human modification on the dynamics of wetland systems. In *Wetland functions and values: The state of our understanding,* ed. P. E. Greeson, J. R. Clark, and J. E. Clark, 200–209. American Water Resources Association, Minneapolis, Minn.

Daubenmire, R. F. 1936. The "Big Woods" of Minnesota: Its structure, and relation to climate, fire, and soils. *Ecological Monographs* 6:233–68.

Davis, M. B. 1990. Climate change and the survival of forest species. In *The earth in transition: Patterns and processes of biotic impoverishment,* ed. G. M. Woodwell, 99–110. Cambridge University Press, Cambridge, England.

de la Cruz, A. A. 1979. Production and transport of detritus in wetlands. In *Wetland functions and values: The state of our understanding,* ed. P. E. Greeson, J. R. Clark, and J. E. Clark, 162–74. American Water Resources Association, Minneapolis, Minn.

Dexter, M. H., comp. 1993. Status of wildlife populations, fall 1993 and 1981–1992 hunting and trapping harvest statistics. Section of Wildlife, Minnesota Department of Natural Resources, St. Paul. Unpublished report.

Douglas, P. 1990. *Prairie skies.* Voyageur Press, Stillwater, Minn.

Dubbe, D. R., E. G. Garver, and D. C. Pratt. 1988. Production of cattail (*Typha* spp.) biomass in Minnesota, USA. *Biomass* 17:79–104.

Dunlap, K. D., and J. W. Lang. 1990. Offspring sex ratio varies with maternal size in the common garter snake, *Thamnophis sirtalis. Copeia* 1990:568–70.

Dunevitz, H. L., and K. Bolin. 1990. Combating alien plants. *Minnesota Volunteer* 53(312):17, 22–26.

Dunevitz, H. L., and A. E. Epp. 1995. Natural communities and rare species of Rice County, Minnesota. 1:75,000; 27.5 x 40 in.; colored. Minnesota Department of Natural Resources, Minnesota County Biological Survey Map Series No. 8.

Eddy, S. 1966. Minnesota and the Dakotas. In *Limnology in North America,* ed. D. G. Frey, 301–15. University of Wisconsin Press, Madison.

Eddy, S., and J. C. Underhill. 1974. *Northern fishes.* 3d ed. University of Minnesota Press, Minneapolis.

Eleutarius, L. N., and S. B. Jones Jr. 1969. A floristic and ecological study of pitcher plant bogs in south Mississippi. *Rhodora* 71:29–34.

Elson, J. A. 1983. Lake Agassiz—Discovery and a century of research. In *Glacial Lake Agassiz ,* ed. J. T. Teller and L. Clayton, 21–41. Geological Association of Canada, Special Paper 26.

Ensor, K. L., D. D. Helwig, and L. C. Wemmer. 1992. Mercury and lead in Minnesota common loons (*Gavia immer*). Water Quality Division, Minnesota Pollution Control Agency, St. Paul.

Errington, P. L. 1957. *Of men and marshes.* Macmillan Co., New York.

———. 1963. *Muskrat populations.* Iowa State University Press, Ames.

Ewert, M. A., and C. E. Nelson. 1991. Sex determination in turtles: Diverse patterns and some possible adaptive values. *Copeia* 1991:50–69.

Ewing, J. 1924. Plant successions of the brush-prairie in northwestern Minnesota. *Ecology* 12:238–66.

Faaborg, J. 1988. *Ornithology—An ecological approach.* Prentice Hall, Englewood Cliffs, N.J.

Fandrei, G., C. Mockovak, and G. Reetz. 1990. Protecting Minnesota's waters: The land-use connection. Water Quality Division, Minnesota Pollution Control Agency, St. Paul.

Farrar, D., and C. L. Johnson-Groh. 1991. A new prairie moonwort (*Botrychium* subgenus *Botrychium*) from northwestern Minnesota. *American Fern Journal* 81:1–6.

Fashingbauer, B. A. 1965. The elk in Minnesota. In Big game in Minnesota, ed. J. B. Moyle. Minnesota Department of Conservation Technical Bulletin 9:99–132.

Fashingbauer, B. A., A. C. Hodson, and W. H. Marshall. 1957. The interrelations of a forest tent caterpillar, outbreak, song birds and DDT application. *Flicker* 29:132–47.

Featherstonhaugh, G. W. [1847] 1970. A canoe voyage up the Minnay Sotor. Reprint, Minnesota Historical Society, St. Paul.

Fisher, S. G. 1983. Succession in streams. In *Stream Ecology,* ed. J. R. Barnes and G. W. Minshall, 7–27. Plenum Press, New York.

Fisher, S. G., and G. E. Likens. 1973. Energy flow in Bear Brook, New Hampshire: An integrative approach to stream ecosystem metabolism. *Ecological Monographs* 43:421–39.

Flaccus, E., and L. F. Ohmann. 1964. Old-growth northern hardwood forests in northeastern Minnesota. *Ecology* 45: 448–59.

Fowells, H. A. 1965. Silvics of forest trees of the United States. U.S. Department of Agriculture, Forest Service, Agricultural Handbook 271.

French, D. W., and W. C. Stienstra. 1975. Oak wilt disease. University of Minnesota Agricultural Extension Service, Extension Folder 310.

French, D. W., W. C. Stienstra, and D. M. Noetzel. 1974. The Dutch elm disease. University of Minnesota Agricultural Extension Service, Extension Folder 211.

Frissell, S. S. Jr. 1973. The importance of fire as a natural ecological factor in Itasca State Park, Minnesota. *Journal of Quaternary Research* 3:397–407.

Fritts, S. H. 1980. The timber wolf in Minnesota. Proceedings of symposium on mammalian ecology and habitat management in Minnesota, 70. Mimeographed.

Gale, M. R., and D. F. Grigal. 1987. Vertical root distributions of northern tree species in relation to successional status. *Canadian Journal of Forest Research* 17:829–34.

Galli, J. 1991. Laying bat myths to rest. *Minnesota Volunteer* 54(317):17–23.

Garland, H. 1928. *A son of the Middle Border.* Macmillan, New York.

Gilmer, D. S., I. J. Ball, L. M. Cowardin, J. H. Reichman, and J. R. Tester. 1975. Habitat use and home ranges of mallards

314

breeding in Minnesota. *Journal of Wildlife Management* 39: 781–89.

Glaser, P. H. 1987. The ecology of patterned boreal peatlands of northern Minnesota: A community profile. U.S. Fish and Wildlife Service Report 85(7.14).

Glaser, P. H., and J. A. Janssens. 1986. Raised bogs in eastern North America: Transitions in landforms and gross stratigraphy. *Canadian Journal of Botany* 64:395–415.

Glaser, P. H., G. A. Wheeler, E. Gorham, and H. E. Wright Jr. 1981. The patterned mires of the Red Lake peatland, northern Minnesota: Vegetation, water chemistry and landforms. *Journal of Ecology* 69:575–99.

Glenn-Lewin, D. C., L. A. Johnson, T. W. Jurik, A. Akey, M. Leoschke, and T. Rosburg. 1990. Fire in central North American grasslands: Vegetative reproduction, seed germination, and seedling establishment. In *Fire in North American tallgrass prairies,* ed. S. L. Collins and L. L. Wallace, 28–45. University of Oklahoma Press, Norman.

Glenn-Lewin, D. C., R. K. Peet, and T. T. Veblen, eds. 1992. *Plant succession: Theory and prediction.* Chapman and Hall, London.

Gorham, E. 1990. Biotic impoverishment in northern peatlands. In *The Earth in Transition,* ed. G. M. Woodwell, 65–98. Cambridge University Press, Cambridge.

Gorham, E., W. E. Dean, and J. E. Sanger. 1983. The chemical composition of lakes in the north-central United States. *Limnology and Oceanography* 28:287–301.

Graedel, T. E., and P. J. Crutzen. 1989. The changing atmosphere. *Scientific American* 261(3):58–68.

Graham, A., and F. Graham. 1976. *The milkweed and its world of animals.* Doubleday & Co., Garden City, N.Y.

Green, J. C. 1991. A landscape classification for breeding birds in Minnesota: An approach to describing regional biodiversity. *Loon* 63:80–91.

Green, R. G., and C. A. Evans. 1940. Studies on a population cycle of snowshoe hares on the Lake Alexander area. I. Gross annual censuses, 1932–1939. *Journal of Wildlife Management* 4: 220–38.

Gregory, S. V. 1983. Plant-herbivore interactions in stream ecosystems. In *Stream Ecology,* ed. J. R. Barnes and G. W. Minshall, 157–89. Plenum Press, New York.

Grigal, D. F. 1985. *Sphagnum* production in forested bogs of northern Minnesota. *Canadian Journal of Botany* 63:1204–7.

Grigal, D. F., C. G. Buttleman, and L. K. Kernik. 1985. Biomass and productivity of the woody strata of forested bogs in northern Minnesota. *Canadian Journal of Botany* 63: 2416–24.

Grigal, D. F., and L. F. Ohmann. 1975. Classification, description, and dynamics of upland plant communities within a Minnesota wilderness area. *Ecological Monographs* 45:389–407.

Grime, J. P. 1979. *Plant strategies and vegetation processes.* Wiley, Chichester, England.

Grimm, E. C. 1983. Chronology and dynamics of vegetation change in the prairie-woodland region of southern Minnesota, USA. *New Phytologist* 93:311–50.

———. 1984. Fire and other factors controlling the Big Woods

vegetation of Minnesota in the mid-nineteenth century. *Ecological Monographs* 54:291–311.

Gullion, G. W. 1990. Forest-wildlife interactions. In *Introduction to forest science,* ed. R. A. Young and R. L. Giese, 349–83. J. Wiley, New York.

Haarstad, J. 1990. The Acrididae of Minnesota. In Summaries of wildlife research findings, ed. B. Joselyn, 121–23. Section of Wildlife, Minnesota Department of Natural Resources, St. Paul. Unpublished report.

Haas, C. H. 1974. Habitat selection, reproduction, and movements in female spruce grouse. Ph.D. thesis, University of Minnesota.

Habicht, G. S., G. Beck, and J. L. Benach. 1987. Lyme disease. *Scientific American* 257:78–83.

Hadley, E. B. 1970. Net productivity and burning responses of native eastern North Dakota prairie communities. *American Midland Naturalist* 84:121–35.

Hahn, J. T., and W. B. Smith. 1987. Minnesota's forest statistics, 1987: An inventory update. U.S. Department of Agriculture, Forest Service, General Technical Report NC-118:1–44.

Hall, J. S., R. J. Cloutier, and D. R. Griffin. 1957. Longevity records and notes on tooth wear of bats. *Journal of Mammalogy* 38:407–9.

Hansen, H. L., V. Kurmis, and D. D. Ness. 1974. The ecology of upland forest communities and implications for management in Itasca State Park, Minnesota. University of Minnesota Agricultural Experiment Station, Forestry Series 16, Technical Bulletin 298:1–44.

Hanson, H. C. 1965. *The giant Canada goose.* Southern Illinois University Press, Carbondale.

Hanson, S. R., P. A. Renard, N. A. Kirsch, and J. W. Enblom. 1984. Biological survey of the Otter Tail River. Minnesota Department of Natural Resources, Division of Fish and Wildlife, Special Publication 137:1–101.

Hardin, G. C. 1960. The competitive exclusion principle. *Science* 131:1292–97.

Harrington, F. H., and L. D. Mech. 1979. Wolf howling and its role in territory maintenance. *Behaviour* 68:207–49.

Harris, S. W., and W. H. Marshall. 1963. Ecology of water-level manipulations on a northern marsh. *Ecology* 44:331–43.

Hart, D. D. 1983. The importance of competitive interactions within stream populations and communities. In *Stream Ecology,* ed. J. R. Barnes and G. W. Minshall, 99–136. Plenum Press, New York.

Hassinger, R. 1987. Minnesota's favorite game fish. *Minnesota Volunteer* 50(292):27–31.

Hazard, E. B. 1982. *The mammals of Minnesota.* University of Minnesota Press, Minneapolis.

Heeden, S. G. 1970. The ecology and life history of the mink frog, *Rana septentrionalis.* Ph.D. thesis, University of Minnesota.

Heinselman, M. L. 1970. Landscape evolution, peatland types, and the environment in the Lake Agassiz Peatlands Natural Area, Minnesota. *Ecological Monographs* 40:235–61.

———. 1973. Fire in the virgin forest of the Boundary Waters Canoe Area, Minnesota. *Journal of Quaternary Research* 3:329–82.

_____ 1974. Interpretation of Francis J. Marschner's map of the original vegetation of Minnesota. Text on reverse of The Original Vegetation of Minnesota—map by F. J. Marschner, 1930. Published in color by U.S. Department of Agriculture, Forest Service, North Central Forest Experiment Station, through U.S. Government Printing Office, Washington, D.C.

_____. 1981. Fire and succession in the conifer forests of northern North America. In *Forest succession*, ed. D. C. West, H. H. Shugart, and D. B. Botkin, 374–405. Springer-Verlag, New York.

Heiskary, S. A., and J. Lindbloom. 1993. Water quality trends in Minnesota. Division of Water Quality, Minnesota Pollution Control Agency, St. Paul. Mimeographed.

Heiskary, S. A., and C. B. Wilson. 1989. The regional nature of lake water quality across Minnesota: An analysis for improving resource management. *Journal of the Minnesota Academy of Science* 55:71–77.

_____. 1990. Minnesota lake water quality assessment report. 2d ed. Division of Water Quality, Minnesota Pollution Control Agency, St. Paul. Mimeographed.

Held, J. W., and J. J. Peterka. 1974. Age, growth, and food habits of the fathead minnow, *Pimephales promelas,* in North Dakota saline lakes. *Transactions of American Fisheries Society* 103:743–56.

Henderson, C. 1979. The occurrence, distribution, legal status, and utilization of reptiles and amphibians in Minnesota. Minnesota Department of Natural Resources. Unpublished mimeographed report.

_____. 1980. Nongame wildlife program 1979 summary. Proceedings of symposium on mammalian ecology and habitat management in Minnesota, 14–19. Mimeographed.

_____. 1981a. Breeding birds in Minnesota, 1975–1979: Abundance, distribution, and diversity. *Loon* 53:11–49.

_____. 1981b. Conservation of nongame wildlife on midwestern wetlands. In Wetland values and management, ed. B. Richardson, 27–34. Proceedings Midwest Conference, Minnesota Water Planning Board, St. Paul.

Herrick, C. L. 1892. *The mammals of Minnesota.* Geological and Natural History Survey of Minnesota, Bulletin 7:1–299. Johnson, Smith and Harrison, State Printers, Minneapolis.

Higgins, K. 1986. Interpretation and compendium of historical fire accounts in the northern Great Plains. U.S. Fish and Wildlife Service, Resource Publication 161.

Hoskinson, R. L., and L. D. Mech. 1976. White-tailed deer migration and its role in wolf predation. *Journal of Wildlife Management* 40:429–41.

Hothem, R. L., R. W. DeHaven, and S. D. Fairaizl. 1988. Bird damage to sunflower in North Dakota, South Dakota, and Minnesota, 1979–1981. U.S. Fish and Wildlife Service, Fisheries and Wildlife Technical Report 15.

Huempfner, R. A., and J. R. Tester. 1988. Winter arboreal feeding behavior of ruffed grouse in east-central Minnesota. In *Adaptive strategies and population ecology of northern grouse*, ed. A. T. Bergerud and M. W. Gratson, 122–57. University of Minnesota Press, Minneapolis.

Huenecke, H. A., A. B. Erickson, and W. H. Marshall. 1958.

Marsh gases in muskrat houses in winter. *Journal of Wildlife Management* 22:240–45.

Huff, D. E. 1973. A preliminary study of ruffed grouse–aspen nutrient relationships. Ph.D. thesis, University of Minnesota.

Huston, M., and T. Smith. 1987. Plant succession: Life history and competition. *American Naturalist* 130:168–98.

Hynes, H. B. N. 1963. Imported organic matter and secondary productivity in streams. *International Congress Zoology* 16:324–29.

_____. 1970. *The ecology of running water.* University of Toronto Press, Toronto.

Idstrom, J. M. 1965. The moose in Minnesota. In Big game in Minnesota, ed. J. B. Moyle. Minnesota Department of Conservation Technical Bulletin 9:57–98.

Irving, F. G. 1970. Field instruction in prescribed burning techniques at the University of Minnesota. Tall Timbers Fire Ecology Conference, Tall Timbers Research Station, Tallahassee, Fla. 10:323–31.

Irving, W. 1835. *A tour on the prairies.* John Murray, London.

Jaakko Pöyry Consulting, Inc. 1992. Biodiversity. A technical paper for a generic environmental impact statement—timber harvesting and forest management in Minnesota. Jaakko Pöyry Consulting, Inc., Tarrytown, N.Y.

Jackson, H. H. T. 1961. *Mammals of Wisconsin.* University of Wisconsin Press, Madison.

Jakes, P. J. 1980. The fourth Minnesota forest inventory: Area. U.S. Department of Agriculture, Forest Service, Resource Bulletin NC-54:1–37.

Janssen, R. B. 1987. *Birds in Minnesota.* University of Minnesota Press, Minneapolis.

Jenks, A. E. 1898. The wild rice gatherers of the upper lakes. Smithsonian Institution Bureau of American Ethnology, Annual Report 19:1019–1131.

Johnson, D. H., and M. D. Schwartz. 1993. The Conservation Reserve Program and grassland birds. *Conservation Biology* 7:934–37.

Johnson, R. C. 1985. Lyme disease: A new, rapidly spreading spirochetosis. *Norden News* 60(2):14–17.

Johnson, R. G., and S. A. Temple. 1986. Assessing habitat quality for birds nesting in fragmented tallgrass prairie. In *Wildlife 2000*, ed. J. Verner, M. L. Morrison, and C. J. Ralph, 245–49. University of Wisconsin Press, Madison.

Johnson, W. T., and H. H. Lyon. 1976. *Insects that feed on trees and shrubs.* Cornell University Press, Ithaca, N.Y.

Johnston, C. A., and R. J. Naiman. 1990. The use of a geographic information system to analyze long-term landscape alteration by beaver. *Landscape Ecology* 4:5–19.

Jones, E. 1962. *The Minnesota, forgotten river.* Holt, Rinehart and Winston, New York.

Kadlec, R. H., and J. A. Kadlec. 1979. Wetlands and water quality. In *Wetland functions and values: The state of our understanding*, ed. P. E. Greeson, J. R. Clark, and J. E. Clark, 436–56. American Water Resources Association, Minneapolis, Minn.

Kalin, O. T. 1976. Distribution, relative abundance, and species richness of small mammals in Minnesota, with an analysis of some structural characteristics of habitats as factors influencing species richness. Ph.D. thesis, University of Minnesota.

Kantrud, H. A., J. B. Millar, and A. G. van der Valk. 1989. Vegetation of wetlands of the prairie pothole region. In *Northern prairie wetlands*, ed. A. van der Valk, 132–87. Iowa State University Press, Ames.

Karns, D. R. 1984. Toxic bog water in northern Minnesota peatlands: Ecological and evolutionary consequences for breeding amphibians. Ph.D. thesis, University of Minnesota.

_____. 1992. Amphibians and reptiles. In *The patterned peatlands of Minnesota*, ed. H. E. Wright Jr., B. A. Coffin, and N. Aaseng, 131–50. University of Minnesota Press, Minneapolis.

Karns, P. D. 1976. *Pneumostrongylus tenuis* in deer in Minnesota and implications for moose. *Journal of Wildife Management* 31: 299–303.

Keeling, C. D., R. B. Bacastow, A. F. Carter, S. C. Piper, T. P. Whorf, M. Heimann, W. G. Mook, and H. Roeloffzen. 1989. A three-dimensional model of atmospheric $CO_2$ transport based on observed winds: 1. Analysis of observational data. In Geophysical Monograph 55, ed. D. H. Peterson, 165–236. American Geophysical Union, Washington, D.C.

Keith, L. B., and L. A. Windberg. 1978. A demographic analysis of the snowshoe hare cycle. *Wildlife Monographs* 58:6–70.

Kelleher, K. E., and J. R. Tester. 1969. Homing and survival in the Manitoba toad, *Bufo hemiophrys*, in Minnesota. *Ecology* 50: 1040–48.

Kirsch, N. A., S. A. Hanson, P. A. Renard, and J. W. Enblom. 1985. Biological survey of the Minnesota River. Minnesota Department of Natural Resources, Division of Fish and Wildlife, Special Publication 139:1–85.

Knox Jones, J. Jr., and E. C. Birney. 1988. *Handbook of mammals of the north-central states.* University of Minnesota Press, Minneapolis.

Kornas, J. 1972. Corresponding taxa and their ecological background in the forests of temperate Eurasia and North America. In *Taxonomy, phytogeography and evolution*, ed. D. H. Valentine, 37–59. Academic Press, New York.

Kreuger, C. C., and T. F. Waters. 1983. Annual production of macroinvertebrates in three streams of different water quality. *Ecology* 64:840–50.

Kroodsma, D. E. 1979. Habitat values for nongame wetland birds. In *Wetland functions and values: The state of our understanding*, ed. P. E. Greeson, J. R. Clark, and J. E. Clark, 320–29. American Water Resources Association, Minneapolis, Minn.

Kucera, C. L., R. C. Dahlman, and M. R. Koelling. 1967. Total net productivity and turnover on an energy basis for tallgrass prairie. *Ecology* 48:536–41.

Kuehnast, E. L. 1972. Climate of Minnesota. National Oceanic and Atmospheric Administration, Climatography of the United States No. 60-21.

Kuehnast, E. L., D. G. Baker, and J. A. Zandlo. 1982. *Duration and depth of snow cover.* Climate of Minnesota, Part XIII, University of Minnesota Agricultural Experiment Station Technical Bulletin 333.

Lammers, R. K. T. 1976. Plant and insect communities in a Minnesota wetland. Ph.D. thesis, University of Minnesota.

Lane, B. 1990. My night life with the boreal owl. *Minnesota Volunteer* 43(312):10–13.

Lang, J. W. 1971. Overwintering of three species of snakes in northwestern Minnesota. Master's thesis, University of North Dakota, Grand Forks.

Larsen, J. A. 1980. *The boreal ecosystem.* Academic Press, New York.

Larson, A. M. 1989. Foreign fish invades Lake Superior. *Minnesota Volunteer* 52(306):26–29.

Lass, W. E. 1984. Minnesota—An American Siberia? *Minnesota History* 49:149–55.

Lawrence, D. B. 1972. The Arboretum's reed marsh in historical perspective. Landscape Arboretum, University of Minnesota Agricutural Experiment Station Miscellaneous Report 111: 24–27.

Leatherberry, E. C. 1991. Forest statistics for Minnesota's Central Hardwood Unit. U.S. Department of Agriculture, Forest Service, Resource Bulletin NC-135:1–46.

Leisman, G. A. 1953. The rate of organic matter accumulation on the sedge mat zones of bogs in the Itasca State Park region of Minnesota. *Ecology* 34:81–101.

Likens, G. E., F. H. Borman, R. S. Pierce, J. S. Eaton, and N. M. Johnson. 1977. *Biogeochemistry of a forested ecosystem.* Springer-Verlag, Berlin.

Lindeman, R. L. 1942. The trophic-dynamic aspect of ecology. *Ecology* 23:399–418.

Lowery, G. H. 1974. *The mammals of Louisiana and its adjacent waters.* Louisiana State University Press, Baton Rouge.

Lydecker, R. 1976. The edge of the Arrowhead. University of Minnesota, Duluth, Minnesota Marine Advisory Service.

Manabe, S., and R. T. Wetherald. 1987. Large-scale changes of soil wetness induced by an increase in atmospheric carbon dioxide. *Journal Atmospheric Sciences* 44:1211–35.

Manweiler, J. 1938. Minnesota's big bog. *Minnesota Conservationist* 63:12–13, 24–25, 28.

Marschner, F. J. 1974. The original vegetation of Minnesota (map). U.S. Department of Agriculture, Forest Service. North Central Forest Experiment Station, St. Paul, Minn. Redraft of the original 1930 edition.

McAndrews, J. H. 1966. Postglacial history of prairie, savanna, and forest in northwestern Minnesota. *Torrey Botanical Club, Memoir* 23(2):1–72.

McIntyre, J. W. 1988. *The common loon, spirit of northern lakes.* University of Minnesota Press, Minneapolis.

Mech. L. D. 1970. *The wolf: The ecology and behavior of an endangered species.* Doubleday, New York.

Mech, L. D., D. M. Barnes, and J. R. Tester. 1968. Seasonal weight changes, mortality, and population structure of raccoons in Minnesota. *Journal of Mammalogy* 49:63–73.

Mech, L. D., and P. D. Karns. 1977. Role of the wolf in a deer decline in the Superior National Forest. U.S. Department of Agriculture, Forest Service, Research Paper NC-148:1–23.

Mellichamp, L. 1978. Botanical history of CP, 1. Sarraceniaceae. *Carnivorous Plant Newsletter* 7(2):56–59.

Melquist, W. E., and M. G. Hornocker. 1983. Ecology of river otters in west central Idaho. *Wildlife Monographs* 83:1–60.

Merrell, D. J. 1977. Life history of the leopard frog *Rana pipiens*,

317

in Minnesota. Bell Museum of Natural History, University of Minnesota, Occasional Papers 15:1–23.

Miller, G. T. Jr. 1992. *Living in the environment.* 7th ed. Wadsworth Publishing Co., Belmont, Calif.

Miller, M., and L. Pfannmuller. 1991. Bald eagle population in Minnesota. *Loon* 63:268–71.

Minnesota Department of Natural Resources. 1982. *Inventory of peat resources, Aitkin County, Minnesota.* Minnesota Department of Natural Resources, Hibbing.

Minnesota Department of Natural Resources. 1988. Sixteen year study of Minnesota flash floods. State climatology office and University of Minnesota Soil Science Department.

Minnesota Department of Natural Resources. 1990. Minnesota's outdoor legacy: Strategies for the '90's. Minnesota Department of Natural Resources, St. Paul.

Minnesota Environmental Quality Board. 1988. Minnesota Environmental Quality, Trends in Resource Conditions and Current Issues. Minnesota Environmental Quality Board, St. Paul.

Minnesota Pollution Control Agency. 1992. Minnesota water quality: Water years 1990–1991. Minnesota Pollution Control Agency, St. Paul.

———— 1994. Minnesota River assessment project report, Executive Summary. Minnesota Pollution Control Agency, St. Paul.

Moore, C. T. 1972. Man and fires in the central North American grassland 1535–1890: A documentary historical geography. Ph.D. thesis, University of California, Los Angeles.

Moore, P. D., and D. J. Bellamy. 1974. *Peatlands.* Springer-Verlag, New York.

Morris, R. F., D. R. Redmond, A. B. Vincent, E. L. Howie, and D. W. Hudson. 1958. The numerical response of avian and mammalian predators during a gradation of the spruce budworm. *Ecology* 39:487–94.

Moss, B. 1988. *Ecology of fresh waters.* 2d ed. Blackwell Scientific Publications, Oxford.

Moyle, J. B. 1944. Wild rice in Minnesota. *Journal of Wildlife Management* 8:177–84.

———— 1945. Some chemical factors influencing the distribution of aquatic plants in Minnesota. *American Midland Naturalist* 34:402–20.

————. 1956. Relationships between the chemistry of Minnesota surface waters and wildlife management. *Journal of Wildlife Management* 20:303–20.

————. 1972. Mercury levels in Minnesota fish, 1970–1971. Special Publication 97, Minnesota Department of Natural Resources, St. Paul.

Moyle, J. B., and J. H. Kuehn. 1964. Carp, a sometimes villain. In *Waterfowl tomorrow,* ed. J. P. Linduska, 635–42. U.S. Fish and Wildlife Service, Washington, D.C.

Moyle, J. B., and P. Krueger. 1964. Wild rice in Minnesota. *Conservation Volunteer* 27(158):30–37.

Moyle, J. B., and E. W. Moyle. 1977. *Northland wild flowers.* University of Minnesota Press, Minneapolis.

Murkin, H. R. 1989. The basis for food chains in prairie wetlands. In *Northern prairie wetlands,* ed. A. van der Valk, 317–338. Iowa State University Press, Ames.

Murray, P. 1991. Forest statistics for Minnesota's Northern Pine Unit. U.S. Department of Agriculture, Forest Service, Resource Bulletin, NC-131.

Naiman, R. J., J. M. Melillo, M. A. Lock, T. E. Ford, and S. R. Rice. 1987. Longitudinal patterns of ecosystem processes and community structure in a subarctic river continuum. *Ecology* 68:1139–56.

Neel, J. K. Jr. 1985. *A northern prairie stream.* University of North Dakota Press, Grand Forks.

Neely, R. K., and J. L. Baker. 1989. Nitrogen and phosphorus dynamics and the fate of agricultural runoff. In *Northern prairie wetlands,* ed. A. van der Valk, 92–131. Iowa State University Press, Ames.

Nelson, U. C. 1947. Woodland caribou in Minnesota. *Journal of Wildlife Management* 11:283–84.

Neimi, G. J., and J. M. Hannowski. 1992. Birds populations. In *The patterned peatlands of Minnesota,* ed. H. E. Wright Jr., B. A. Coffin, and N. Aaseng, 111–29. University of Minnesota Press, Minneapolis.

Nichols, C. W. 1939. The Northampton Colony and Chanhassen. *Minnesota History* 20:140–45.

Noble, M. G. 1979. The origin of *Populus deltoides* and *Salix interior* zones on point bars along the Minnesota River. *American Midland Naturalist* 102:59–67.

Noon, B. R., V. P. Bingman, and J. P. Noon. 1979. The effect of changes in habitat on northern hardwood forest bird communities. In Management of north central and northeastern forests for nongame birds, ed. R. M. DeGraaf and K. E. Evans. U.S. Department of Agriculture, Forest Service, General Technical Report NC-51:33–48.

Nordquist, G. E. 1992. Small mammals. In *The patterned peatlands of Minnesota,* ed. H. E. Wright Jr., B. A. Coffin, and N. Aaseng, 85–110. University of Minnesota Press, Minneapolis.

Odum, E. P. 1969. The strategy of ecosystem development. *Science* 164:262–70.

Oechel, W. C., and W. T. Lawrence. 1985. Taiga. In *Physiological ecology of North American plant communities,* ed. B. F. Chabot and H. A. Mooney, 66–94. Chapman and Hall, New York.

Ohmann, L. F., and D. F. Grigal. 1979. Early revegetation and nutrient dynamics following the 1971 Little Sioux forest fire in northeastern Minnesota. *Forest Science Monograph* 21:1–80.

————. 1985. Biomass distribution of unmanaged upland forests in Minnesota. *Forest Ecology and Management* 13:205–22.

Ojakangas, R. W., and C. L. Matsch. 1982. *Minnesota's geology.* University of Minnesota Press, Minneapolis.

Oldfield, B., and J. J. Moriarty. 1994. *Amphibians and reptiles native to Minnesota.* University of Minnesota Press, Minneapolis.

Olson, S. T., and W. H. Marshall. 1952. The common loon in Minnesota. Minnesota Museum of Natural History, Occasional Paper No. 5. University of Minnesota Press, Minneapolis.

Omernik, J. M. 1987. Ecoregions of the conterminous United States. *Annals Association American Geographers* 77:118–25.

Ostry, M. E., and T. H. Nicholls. 1979. Eastern dwarf mistletoe on black spruce. U.S. Department of Agriculture, Forest Service, Forest Insect and Disease Leaflet 158.

Otis, D. L., and C. M. Kilburn. 1988. Influence of environmental factors on blackbird damage to sunflower. U.S. Fish and Wildlife Service Technical Report 16.

Ovington, J. D., D. Heitkamp, and D. B. Lawrence. 1963. Plant biomass and productivity of prairie, savanna, oakwood and maize field ecosystems in central Minnesota. *Ecology* 44: 52–63.

Ownbey, G. B., and T. Morley. 1991. *Vascular plants of Minnesota*. University of Minnesota Press, Minneapolis.

Palmer, W. C. 1965. Meteorological drought. U.S. Department of Commerce, Weather Bureau Research Paper No. 45.

Pastor, J., B. Dewey, R. J. Naiman, P. F. McInnes, and Y. Cohen. 1993. Moose browsing and soil fertility in the boreal forests of Isle Royale National Park. *Ecology* 74:467–80.

Pastor, J., R. J. Naiman, B. Dewey, and P. McInnes. 1988. Moose, microbes, and the boreal forest. *BioScience* 38:770–76.

Peek, J. M., D. L. Urich, and R. J. Mackie. 1976. Moose habitat selection and relationships to forest management in northeastern Minnesota. *Wildlife Monographs* 48:1–65.

Peet, M., R. Anderson, and M. S. Adams. 1975. Effect of fire on big bluestem production. *American Midland Naturalist* 94: 15–26.

Penko, J. M. 1985. Ecological studies of *Typha* in Minnesota: *Typha*-insect interactions, and the productivity of floating stands. Master's thesis, University of Minnesota.

Perry, H. R. Jr. 1982. Muskrats. In *Wild mammals of North America: Biology, management, and economics*. ed. J. A. Chapman and G. A. Feldhomer, 282–325. The Johns Hopkins University Press, Baltimore, Md.

Peterson, D. L., and F. Hennagir. 1980. Minnesota live bait industry assessment study. National Marine Fisheries Service, Completion Report 3-261-R. Minnesota Department of Natural Resources, St. Paul. Unpublished report.

Peterson, R. L. 1955. *North American moose*. University of Toronto Press, Toronto.

Petraborg, W. H., and D. W. Burcalow. 1965. The white-tailed deer in Minnesota. In Big game in Minnesota, ed. J. B. Moyle, 11–48. Minnesota Department of Conservation, Technical Bulletin 9.

Pfannmuller, L. A., and B. A. Coffin. 1989. The uncommon ones: Minnesota's endangered plants and animals. Section of Wildlife, Minnesota Department of Natural Resources, St. Paul.

Pietz, P. J., and J. R. Tester. 1982. Habitat selection by sympatric spruce and ruffed grouse in north central Minnesota. *Journal of Wildlife Management* 46:391–403.

———. 1983. Habitat selection by snowshoe hares in north-central Minnesota. *Journal of Wildlife Management* 47:686–96.

Polunin, N. V. C. 1984. The decomposition of emergent macrophytes in fresh water. *Advances in Ecological Research* 14:115–66.

Porcher, E. 1988. Ground water contamination susceptibility in Minnesota. Minnesota Pollution Control Agency, St. Paul.

Porter, K. G. 1977. The plant-animal interface in freshwater ecosystems. *American Scientist* 65:159–70.

Postel, S., and C. Flavin. 1991. Reshaping the global economy. In *State of the world 1991*, ed. L. R. Brown, 170–88. W. W. Norton Co., New York.

Reader, R. J. 1978. Primary production in northern bog marshes. In *Freshwater wetlands: Ecological processes and management potential*, ed. R. E. Good, D. F. Whigham, and R. L. Simpson, 53–62. Academic Press, New York.

Reid, G. K., and R. D. Wood. 1976. *Ecology of inland waters and estuaries*. 2d ed. D. Van Nostrand Co., New York.

Reinartz, J. A., J. W. Popp, and M. A. Kuchenreuther. 1987. Purple loosestrife (*Lythrum salicaria*): Its status in Wisconsin and control methods. Field Station Bulletin, University of Wisconsin, Milwaukee 20:25–35.

Reiners, W. A. 1972. Structure and energetics of three Minnesota forests. *Ecological Monographs* 42:71–94.

Rexrode, C. O., and H. D. Brown. 1983. Oak wilt. U.S. Department of Agriculture, Forest Service, Forest Insect and Disease Leaflet 29.

Richardson, C. J. 1979. Primary productivity values in fresh water wetlands. In *Wetland functions and values: The state of our understanding*, ed. P. E. Greeson, J. R. Clark, and J. E. Clark, 131–145. American Water Resources Association, Minneapolis, Minn.

Roberts, T. S. 1932. *The birds of Minnesota*. University of Minnesota Press, Minneapolis.

Rock, C., G. Howse, and L. Wright. 1993. Minnesota agriculture statistics 1993. Minnesota Department of Agriculture, St. Paul.

Rogers, L. L. 1976. Effects of mast and berry crop failures on survival, growth and reproductive success of black bears. *Transactions of North American Wildlife Resources Conference* 41:431–38.

Rogers, R. S. 1981. Mature mesophytic hardwood forest: Community transitions, by layer, from east-central Minnesota to southeastern Michigan. *Ecology* 62:1634–47.

Rolvag, O. E. 1927. *Giants in the earth*. Harper and Row, New York.

Rongstad, O. J., and J. R. Tester. 1969. Movements and habitat use of white-tailed deer in Minnesota. *Journal of Wildlife Management* 33:366–79.

———. 1971. Behavior and maternal relations of young snowshoe hares. *Journal of Wildlife Management* 35:338–46.

Ross, B. A., J. R. Tester, and W. J. Breckenridge. 1968. Ecology of Mima-type mounds in northwestern Minnesota. *Ecology* 49: 172–77.

Ross, M. J., and J. D. Winter. 1981. Winter movements of four fish species near a thermal plume in northern Minnesota. *Transactions of American Fisheries Society* 110:14–18.

Ross, P. Jr., and D. Crews. 1977. Influence of the seminal plug on mating behavior in the garter snake. *Nature* 2657:344–45.

Rusterholz, K. 1990. Minnesota's old growth forests. *Minnesota Forests* 3:12–16.

Sakai, A., and C. J. Weiser. 1973. Freezing resistance of trees in North America with reference to tree regions. *Ecology* 54:118–26.

Sansome, C. J. 1983. *Minnesota underfoot*. Voyageur Press, Bloomington, Minn.

319

Sargeant, A. B., and D. W. Warner. 1972. Movements and denning habits of a badger. *Journal of Mammalogy* 53:207–10.

Schindler, D. W., and G. W. Comita. 1972. The dependence of primary production upon physical and chemical factors in a small, senescing lake, including the effects of complete winter oxygen depletion. *Archives für Hydrobiologie* 69:413–51.

Schmid, W. D. 1982. Survival of frogs in low temperature. *Science* 215:697–98.

Schorger, A. W. 1955. *The passenger pigeon—Its natural history and extinction.* University of Wisconsin Press, Madison.

Schwartz, G. M., and G. A. Thiel. 1954. Minnesota's rocks and waters. Minnesota Geological Survey Bulletin 37, University of Minnesota Press, Minneapolis.

Searle, R. N., and M. E. Heitlinger. 1980. Prairies, woods, and islands. The Nature Conservancy, Minnesota Chapter.

Shamess, J. J., G. G. C. Robinson, and L. G. Goldsborough. 1985. The structure and comparison of periphytic and planktonic algal communities in two eutrophic prairie lakes. *Archives für Hydrobiologie* 103:99–116.

Shapiro, J., and D. I. Wright. 1984. Lake restoration by biomanipulation: Round Lake, Minnesota, the first two years. *Freshwater Biology* 14:371–83.

Shaw, S. P., and C. G. Fredine. 1956. Wetlands of the United States. Circular 39, U.S. Fish and Wildlife Service, Washington, D.C.

Skaggs, R. H., and D. G. Baker. 1989. Temperature change in eastern Minnesota. *Journal of Climate* 2:629–30.

Smeins, F. E., and D. E. Olsen. 1970. Species composition and production of a native northwestern Minnesota tallgrass prairie. *American Midland Naturalist* 84:398–410.

Smith, C. S., and J. W. Barko. 1990. Ecology of Eurasian watermilfoil. *Journal of Aquatic Plant Management* 28:55–64.

Smith, L. L. Jr., and J. B. Moyle. 1944. A biological survey and fishery management plan for the streams of the Lake Superior North Shore watershed. Minnesota Department of Conservation, Division of Game and Fish, Technical Bulletin No. 1.

Smith, R. L. 1990. *Ecology and field biology.* Harper Collins Publishers, New York.

Soil Survey Staff. 1975. *Soil Taxonomy: A basic system of classification for making and interpreting soil surveys.* U.S. Department of Agriculture, Soil Conservation Service, Agriculture Handbook No. 436. U.S. Government Printing Office, Washington, D.C.

Soper, E. K. 1919. The peat deposits of Minnesota. *Minnesota Geological Survey Bulletin* 16:1–261.

Spurr, S. H. 1954. The forests of Itasca in the nineteenth century as related to fire. *Ecology* 35:21–35.

Stearns, F. 1988. The changing forests of the Lake States. In *The Lake States forest,* ed., W. E. Shands, 25–35. The Conservation Foundation, Lake States Forestry Alliance, St. Paul, Minn.

Stefan, H. G., and M. J. Hanson. 1981. Phosphorus recycling in five shallow lakes. ASCE Environmental Engineering Division 107:714–30.

Stewart, R. E., and H. A. Kantrud. 1971. Classification of natural ponds and lakes in the glaciated prairie region. U.S. Fish and Wildlife Service Research Publication 92.

Strong, P. I. V., and R. Baker. 1991. An estimate of Minnesota's summer population of adult common loons. Minnesota Department of Natural Resources Biological Report No. 37, St. Paul, Minn.

Stuckey, R. L. 1980. Distributional history of *Lythrum salicaria* (purple loosestrife) in North America. *Bartonia* 47:3–21.

Surber, T., and T. S. Roberts. 1932. *The mammals of Minnesota.* Minnesota Department of Conservation Bulletin.

Svejcar, T. J. 1990. Response of *Andropogon gerardii* to fire in the tallgrass prairie. In *Fire in North American tallgrass prairies,* ed. S. L. Collins and L. L. Wallace, 19–27. University of Oklahoma Press, Norman.

Swain, A. 1973. A history of fire and vegetation in northeastern Minnesota as recorded in lake sediments. *Quaternary Research* 3:383–96.

Swain, E. B., D. R. Engstrom, M. E. Brigham, T. A. Henning, and P. L. Brezonik. 1992. Increasing rates of atmospheric mercury deposition in midcontinental North America. *Science* 257:784–87.

Swain, E. B., and D. D. Helwig. 1989. Mercury in fish from northeastern Minnesota lakes: Historical trends, environmental correlates, and potential sources. *Journal of Minnesota Academy of Science* 55:103–9.

Swain, E. B., R. L. Strassman, D. C. Bock, and C. J. Twaroski. 1993. Lake Superior's North Shore streams: Sensitivity to acid depositions. Research Summary, Minnesota Pollution Control Agency, Air Quality Division, St. Paul. Mimeograph.

Swanson, E. B. 1940. The use and conservation of Minnesota game 1850–1900. Ph.D. thesis, University of Minnesota.

Swanson, G. 1945. The American elk in Minnesota. *Conservation Volunteer* 1(2):4–7.

Swanson, G. A. 1977. Diet food selection by Anatinae on a waste-stabilization system. *Journal of Wildlife Management* 41:226–31.

Swanson, G. A., and H. F. Duebbert. 1989. Wetland habitats of waterfowl in the prairie pothole region. In *Northern prairie wetlands,* ed. A. van der Valk, 228–67. Iowa State University Press, Ames.

Swihart, R. K., and R. H. Yahner. 1982. Habitat features influencing use of farmstead shelterbelts by the eastern cottontail (*Sylvilagus floridanus*). *American Midland Naturalist* 107:411–14.

Tester, J. R. 1963. Techniques for studying movements of vertebrates in the field. In *Radioecology,* ed. by V. Schultz and A. W. Klement Jr., 445–50. Reinhold Publ. Corp., New York.

———. 1965. Effects of a controlled burn on small mammals in a Minnesota oak savanna. *American Midland Naturalist* 74:240–43.

———. 1987. Seasonal changes in activity rhythms of free-ranging animals. *Canadian Field Naturalist* 101:13–21.

———. 1989. Effects of fire frequency on oak savanna. *Bulletin Torrey Botannical Club* 116:134–44.

Tester, J. R., and W. J. Breckenridge. 1964a. Population dynamics of the Manitoba toad, *Bufo hemiophrys,* in northwestern Minnesota. *Ecology* 45:592–601.

_____. 1964b. Winter behavior patterns of *Bufo hemiophrys* in northwestern Minnesota. Annals of Academy of Science Fennicae, Series A, Biologica IV 71/31: 423–31.

Tester, J. R., and J. Figala. 1990. Effects of biological and environmental factors on activity rhythms of wild animals. In *Chronobiology: Its role in clinical medicine, general biology, and agriculture*, Part B, ed. D. K. Hayes, J. E. Pauly, and R. J. Reiter, 809–19. Wiley-Liss, New York.

Tester, J. R., and W. H. Marshall. 1961. A study of certain plant and animal interrelations on a native prairie in northwestern Minnesota. Minnesota Museum Natural History, Occasional Papers No. 8.

_____. 1962. Minnesota prairie management techniques and their wildlife implications. *Transactions of North American Wildlife Conference* 27:267–87.

Thienemann, A. 1925. Die Binnengewasser Mitteleuropas: eine Limnologische Einführung. *Die Binnengewasser* 1:1–255.

Thompson, J. S. Jr., and K. L. Sather. 1970. The giant Canada returns to Heron Lake. In Home Grown Honkers, ed. H. H. Dill and F. B. Lee, 7–17. U.S. Fish and Wildlife Service.

Tilman, D. 1978. Cherries, ants and tent caterpillars: Timing of nectar production in relation to susceptibility of caterpillars to ant predation. *Ecology* 59:686–92.

Tilman, D., and J. A. Downing. 1994. Biodiversity and stability in grasslands. *Nature* 367:363–65.

Titus, J., and L. VanDruff. 1981. Response of the common loon to recreational pressure in the Boundary Waters Canoe Area, northeastern Minnesota. *Wildlife Monograph* 79:1–59.

Twaroski, C. J., J. D. Thornton, R. L. Strassman, and P. L. Brezonik. 1989. Susceptibility of northern Minnesota lakes to acid deposition impacts. *Journal of Minnesota Academy of Science* 55:95–102.

United States Department of Commerce (NOAA). 1993. Climatological Data Minnesota, Annual Summary. Environmental Data Service, National Oceanic and Atmospheric Administration, National Climatic Data Center, Asheville, N.C.

Upham, W. 1884. *Flora of Minnesota*. Geological and Natural History Survey of Minnesota, Part VI of the Annual Report of Progress for the Year 1983. Johnson, Smith and Harrison, Minneapolis.

Urquhart, F. A., and N. R. Urquhart. 1976. The overwintering site of the eastern population of the monarch butterfly (*Danaus plexippus*: Danaidae) in southern Mexico. *Journal of the Lepidopterists' Society* 30:153–58.

Usinger, R. L. 1967. *The life of rivers and streams*. McGraw-Hill Book Co., New York.

Van Cleve, K., C. T. Dyrness, L. A. Viereck, J. Fox, F. S. Chapin III, and W. C. Oechel. 1983. Taiga ecosystems in interior Alaska. *BioScience* 33:39–44.

van der Schalie, H., and A. van der Schalie. 1950. The mussels of the Mississippi River. *American Midland Naturalist* 44:448–66.

van der Valk, A. G. 1981. Succession in wetlands: A Gleasonian approach. *Ecology* 62:688–96.

van der Valk, A. G., and C. B. Davis. 1978. Primary production of prairie glacial marshes. In *Freshwater wetlands: Ecological processes and management potential*, ed. R. E. Good, D. F. Whigham, and R. L. Simpson, 21–37. Academic Press, New York.

Vankat, J. L. 1979. *The natural vegetation of North America*. John Wiley & Sons, New York.

Vannote, R. L., G. W. Minshall, K. W. Cummins, J. R. Sedell, and C. E. Cushing. 1980. The river continuum concept. *Canadian Journal of Fisheries and Aquatic Science* 37:130–37.

Vogl, R. J. 1964. Vegetational history of Crex Meadows, a prairie savannah in northwestern Wisconsin. *American Midland Naturalist* 789:487–95.

Vogt, R. C. 1981. *Natural history of amphibians and reptiles of Wisconsin*. Milwaukee Public Museum, Milwaukee.

Voss, E. G. 1985. Michigan flora, Part II, Dicots (Saururaceae-Cornaceae). Cranbrook Institute Science, Bloomfield Hills, Mich.

Warner, D. W., and D. Wells. 1980. Bird population structure and seasonal habitat use as indicators of environment quality of peatlands. Minnesota Department of Natural Resources, Minnesota Peat Program, St. Paul.

Waters, T. F. 1972. The drift of stream insects. *Annual Review of Entomology* 17:253–72.

_____. 1977. *The streams and rivers of Minnesota*. University of Minnesota Press, Minneapolis.

_____. 1986. The benthos of three Minnesota streams tributary to Lake Superior and its relationship to acid precipitation. University of Minnesota Agricultural Experiment Station, Miscellaneous Publication 40.

Waters, T. F., and J. C. Hokenstrom. 1980. Annual production and drift of the stream amphipod *Gammarus pseudolimnaeus* in Valley Creek, Minnesota. *Limnology and Oceanography* 25:700–710.

Webb, T. III. 1989. The spectrum of temporal climatic variability: Current estimates and the need for global and regional time series. In Global changes of the past, ed. R. S. Bradley, 61–81. University Corporation for Atmospheric Research, Office for Interdisciplinary Earth Studies, Boulder, Colo.

Webb, T., III, P. J. Bartlein, and J. E. Kutzbach. 1987. Climatic change in eastern North America during the past 18,000 years; comparisons of pollen data with model results. In *North America and adjacent oceans during the last deglaciation*, ed. W. F. Ruddiman and H. E. Wright Jr., 447–62. Geological Society of America, Boulder, Colo.

Wedin, D. A., and D. Tilman. 1992. Nitrogen cycling, plant competition, and the stability of tallgrass prairie. Proceedings of North American Prairie Conference 12:5–8.

Weller, M. W. 1979. Wetland habitats. In *Wetland functions and values: The state of our understanding*, ed. P. E. Greeson, J. R. Clark, and J. E. Clark, 210–34. American Water Resources Association, Minneapolis, Minn.

_____. 1987. *Freshwater marshes: Ecology and wildlife management*. University of Minnesota Press, Minneapolis.

Weller, M. W., and C. E. Spatcher. 1965. Role of habitat in the distribution and abundance of marsh birds. Iowa State University of Agriculture and Home Economics, Experiment Station Special Report 43.

Wendt, K. M. 1984. A guide to Minnesota prairies. Natural Heritage Program, Minnesota Department of Natural Resources, St. Paul.

Wheeler, G. A., E. J. Cushing, E. Gorham, T. Morley, and G. B. Ownbey. 1992. A major floristic boundary in Minnesota: An analysis of 280 taxa occurring in the western and southern portions of the state. *Canadian Journal of Botany* 70:319–33.

Wheeler, G. A., R. P. Dana, and C. Converse. 1991. Contribution to the vascular (and moss) flora of the Great Plains: A floristic survey of six counties in western Minnesota. *Michigan Botanist* 30(3):75–129.

Wheeler, G. A., P. H. Glaser, E. Gorham, C. M. Wetmore, F. D. Bowers, and J. A. Janssens. 1983. Contributions to the flora of the Red Lake Peatland, northern Minnesota, with special attention to *Carex*. *American Midland Naturalist* 110:62–96

Whitaker, J. O. Jr., and L. L. Schmeltz. 1973. Food and external parasites of *Sorex palustris* and food of *Sorex cinereus* from St. Louis County, Minnesota. *Journal of Mammalogy* 54:283–85.

White, A. 1983. The effects of thirteen years of annual prescribed burning on a *Quercus ellipsoidalis* community in Minnesota. *Ecology* 64:1081–85.

Whittaker, R. H., and G. E. Likens. 1975. The biosphere and man. In *The primary production of the biosphere*, ed. H. Leith and R. H. Whittaker, 305–28. Springer-Verlag, New York.

Willard, D. E. 1979. Support for birds and mammals. Abstract of paper presented at National Symposium on Wetlands, 7–9 November 1978, at Lake Buena Vista, Fla.

Winchell, N. H. 1884. Geology of Minnesota. Vol. I of the Final Report, Geological and Natural History Survey of Minnesota. Johnson, Smith & Harrison, State Printers, Minneapolis.

Winkler, J. A., R. A. Skaggs, and D. G. Baker. 1981. Effect of temperature adjustments on the Minneapolis–St. Paul urban heat island. *Journal of Applied Meteorology* 20:1295–1300.

Woodward, F. I. 1987. *Climate and plant distribution*. Cambridge University Press, Cambridge, England.

Wovcha, D. S., B. C. Delaney, and G. E. Nordquist. 1995. *Minnesota's St. Croix River valley and Anoka sandplain: A guide to native habitats*. Published for the Minnesota Department of Natural Resources by the University of Minnesota Press, Minneapolis.

Wright, H. E. Jr. 1972. Quaternary history of Minnesota. In *Geology of Minnesota: A centennial volume*, ed. P. K. Sims and G. B. Morey, 515–47. Minnesota Geological Survey.

———. 1984. Red Lake Peatland: Its past and patterns. Bell Museum of Natural History, University of Minnesota, *Imprint* I (Summer):1–3.

———. 1989. Origin and developmental history of Minnesota lakes. *Journal of Minnesota Academy of Science* 55:26–31.

Wright, H. E. Jr., B. A. Coffin, and N. E. Aaseng. 1992. *The patterned peatlands of Minnesota*. University of Minnesota Press, Minneapolis.

Wright, L. 1990. Minnesota agriculture statistics 1990. Minnesota Agricultural Statistics Service, Minnesota Department of Agriculture, St. Paul.

Wrubleski, D. A. 1984. Chironomid (Diptera:Chironomidae) species composition, emergence phenologies, and relative abundances in the Delta Marsh, Manitoba, Canada. Master's thesis, University of Manitoba, Winnipeg.

Yahner, R. H. 1982. Microhabitat use by small mammals in farmstead shelterbelts. *Journal of Mammalogy* 63:440–45.

———. 1983. Seasonal dynamics, habitat relationships, and management of avifauna in farmstead shelterbelts. *Journal of Wildlife Management* 47:85–110.

Zinnel, K. C. 1992. Behavior of free-ranging pocket gophers. Ph.D. thesis, University of Minnesota.

Zinnel, K. C., and J. R. Tester. 1992. Effects of plains pocket gophers on root biomass and soil nitrogen. Proceedings of North American Prairie Conference 12:43–46.

Zumberge, J. H. 1952. The lakes of Minnesota, their origin and classification. Minnesota Geological Survey Bulletin 35:1–99. University of Minnesota Press, Minneapolis.

# PERMISSIONS

Figures 1.1, 1.11, 1.16, 4.8, 6.1, 7.11, 7.13, 7.14, 7.17, 8.19, 9.13, 9.16, and 10.1 and jacket photo: **Blacklock Nature Photography,** by permission

Figures 1.2, 1.3, 1.4, 1.8, 1.15, 1.21, 2.2, 3.1, 3.3, 4.5, 6.5, 6.6, 6.10, 7.4, 7.15, 8.6, 8.7, 10.3, 10.5, and 10.9: **Richard Hamilton Smith,** by permission

Figures 1.5, 1.7, 2.1, 2.10, 2.11, 3.6, 3.13, 3.17, 4.1, 4.3, 4.9, 4.14, 5.1, 5.2, 5.5, 5.12, 5.15, 7.1, 7.22, 8.1, 8.14, 9.1, 9.4, 9.5, 9.6, 9.17, 10.2, 10.7, and 10.8 and title page photo: **Firth PhotoBank, Minneapolis,** by permission

Figure 1.6: modified, by permission, from **J. L. Vankat,** *Natural vegetation of North America,* © 1979, John Wiley & Sons, Inc.

Figures 1.9 and 1.10: modified, by permission, from **R. W. Ojakangas and C. L. Matsch,** *Minnesota's geology,* 1982, University of Minnesota Press

Figure 1.12: modified, by permission, from **J. A. Elson,** "Lake Agassiz—Discovery and a century of research," in J. T. Teller and L. Clayton, eds., *Glacial Lake Agassiz,* 1983, Geological Association of Canada, Special Paper 26

Figure 1.13: modified, by permission, from **H. E. Wright Jr.,** "Physiography of Minnesota," in P. K. Sims and G. B. Morey, eds., *Geology of Minnesota,* 1972, Minnesota Geological Survey

Figure 1.14: modified, by permission, from **J. R. Borchert and D. P. Yaeger,** *Atlas of Minnesota resources and settlement,* 1968, Minnesota State Planning Agency

Figure 1.17: **H. E. Wright Jr.,** by permission

Figures 1.18, 8.9, 8.10, and 8.13: **Robert Megard,** by permission

Figure 1.19: **E. J. Cushing,** by permission

Figure 1.20: modified, by permission, from **J. H. McAndrews,** "Postglacial history of prairie, savanna, and forest in northwestern Minnesota," 1966, *Torrey Botanical Club, Memoir 22* (2)

Figures 1.24, 3.10, and 3.11: modified, by permission, from **G. T. Miller Jr.,** *Living in the environment,* 7th edition, © 1992, Wadsworth Publishing Co.

Figure 1.25: modified, by permission, from **B. A. Coffin,** The natural vegetation of Minnesota at the time of the Public Land Survey, Minnesota Department of Natural Resources, 1988

Figure 2.3: modified, by permission, from **D. G. Baker et al.,** *Normal temperatures (1951-1980) and their application,* 1985,

Climate of Minnesota, Part XV, University of Minnesota Agricultural Experiment Station Technical Bulletin AD-SB-2777

Figures 2.4 and 2.8: modified, by permission, from **S. B. Cohen,** *Oxford world atlas,* 1973, Oxford University Press

Figure 2.5: modified, by permission, from **D. G. Baker and J. H. Strub Jr.,** *The agricultural and minimum-temperature-free seasons,* 1963, Climate of Minnesota, Part II, University of Minnesota Agricultural Experiment Station Technical Bulletin 245

Figure 2.9: modified, by permission, from **D. G. Baker and E. L. Kuehnast,** *Precipitation normals for Minnesota: 1941-1970,* 1978, Climate of Minnesota, Part X, University of Minnesota Agricultural Experiment Station Technical Bulletin 314

Figure 2.12: modified, by permission, from Annual 1951–1980 normal snowfall, 1988, **Minnesota Department of Natural Resources**

Figures 2.14 and 2.15: modified, by permission, *Star Tribune,* Minneapolis-St. Paul

Figure 3.2: **David Tilman,** by permission

Figures 3.4, 3.8, 5.13, 5.16, 5.17, 6.2, 6.7, 6.9, 6.16, 7.5, and 10.10: © **Jim Brandenburg,** by permission

Figure 3.7: **Cedar Creek Natural History Area,** by permission

Figures 3.9 and 3.12: modified, by permission, from **W. P. Cunningham and B. W. Saigo,** *Environmental science,* 2d ed., © 1992, Wm. C. Brown Communications, Inc.

Figure 3.15: modified, by permission, from **W. P. Cunningham and B. W. Saigo,** *Environmental science,* © 1990, Wm. C. Brown Communications, Inc.

Figures 3.16, 4.18, 6.15, 8.16, and 9.11: **Barney Oldfield,** by permission

Figures 4.2, 4.6, 4.7, 4.16, 5.3, 5.4, 5.7, 5.8, 5.14, 6.4, 7.20, 8.3, 8.20, and 8.22: **James LaVigne,** by permission

Figures 4.10, 6.8, 7.19, and 7.21: **Minnesota Department of Natural Resources,** by permission

Figure 4.12: modified, by permission, from **J. R. Tester,** "Effects of fire frequency on oak savanna," 1989, *Bulletin Torrey Botanical Club* 116

Figure 4.13: modified, by permission, from **O. J. Rongstad and J. R. Tester,** "Movements and habitat use of white-tailed deer in Minnesota," 1969, *Journal of Wildlife Management* 33 (2), © The Wildlife Society

Figure 4.17: **Artis Orjala**, by permission

Figures 4.19, 5.21, 6.4, 8.11, 8.12, 9.8, and 9.15: **Donald L. Rubbelke**, by permission

Figure 4.20: **Gilbert Ahlstrand**, by permission

Figure 4.21: modified, by permission, from **H. L. Dunevitz and A. E. Epp**, Natural communities and rare species of Rice County, Minnesota, 1995, Minnesota Department of Natural Resources

Figure 5.6: **Jon Ross**, by permission

Figure 5.18: modified, by permission, from **G. W. Gullion**, in R. A. Young and R. L. Giese, eds., *Introduction to forest science*, 1990, John Wiley & Sons, Inc.

Figures 5.20, 8.17, 8.18, and 9.12: **Thomas D. Mangelsen/Images of Nature**, by permission

Figure 5.22: modified, by permission, from **F. Stearns**, in W. E. Shands, ed., *The Lake States forest*, 1988, The Conservation Foundation

Figures 6.12 and 8.15: **William. D. Schmid**, by permission

Figure 6.14: modified, by permission, from **J. R. Tester and W. J. Breckenridge**, "Winter behavior patterns of *Bufo hemiophrys* in northwestern Minnesota," 1964, Annals of Academy of Science Fennicae, Series A, Biologica IV 71/31, Finnish Academy of Science and Letters

Figure 6.17: modified, by permission, from an unpublished map, 1990, **Minnesota Department of Natural Resources**

Figures 7.2 and 10.4: **U.S. Fish and Wildlife Service**, by permission

Figure 7.6: modified, by permission, from **J. B. Moyle**, "Relationships between the chemistry of Minnesota surface waters and wildlife management," 1956, *Journal of Wildlife Management* 20 (3), © The Wildlife Society

Figures 7.7 and 7.8: modified, by permission, from **M. W. Weller**, *Freshwater marshes*, 1987, University of Minnesota Press

Figure 7.12: modified, by permission, from **B. D. J. Batt et al.,** "The use of prairie potholes by North American ducks," 1989, in A. van der Valk, ed., *Northern prairie wetlands*, Iowa State University Press

Figure 7.18: modified, by permission, from *Inventory of peat resources, Aitkin County, Minnesota*, 1982, **Minnesota Department of Natural Resources**

Figure 7.24: modified, by permission, from a map adapted by **A. E. Epp** of the Minnesota Department of Natural Resources, Natural Heritage and Nongame Research Program, from The Original Vegetation of Minnesota, a map compiled in 1930 by F. J. Marschner from the U.S. General Land Office Survey Notes and published in 1974 under the direction of M. L. Heinselman of the U. S. Forest Service; the water layers are Tiger data, and the peatland Scientific and Natural Area boundaries were digitized by the Natural Heritage and Nongame Research Program, Minnesota Department of Natural Resources

Figure 8.2: modified, by permission, from **J. M. Omernik**, "Ecoregions of the conterminous United States," 1987, *Annals Association American Geographers* 77, Basil Blackwell Inc.

Figure 8.4: modified, by permission, from **R. L. Smith**, *Ecology and field biology*, 1990, HarperCollins Publishers

Figure 8.21: modified, by permission, from **G. Breining**, *Managing Minnesota's fish*, 1989, Minnesota Department of Natural Resources

Figure 9.2: **Barb LaVigne**, by permission

Figure 9.3: modified, by permission, from **G. M. Schwartz and G. A. Thiel**, "Minnesota's rocks and waters," 1954, Minnesota Geological Survey Bulletin 37, University of Minnesota Press

Figure 9.10: modified, by permission, from **R. L. Vannote et al.,** "The river continuum concept," 1980, *Canadian Journal of Fisheries and Aquatic Science* 37, National Research Council Canada

# INDEX

*John R. Tester* is professor emeritus in the Department of Ecology, Evolution, and Behavior at the University of Minnesota, where he served on the graduate faculties in ecology, conservation biology, and wildlife conservation. Dr. Tester is a Fellow of the American Association for the Advancement of Science, an Honorary Research Fellow of Aberdeen University, Scotland, and a member of numerous scientific societies. His recent research has focused on the use of fire in managing savanna habitat, studies of effects of pocket gophers on community succession, studies of population dynamics of wild horses in the western United States, and design of management strategies for restoration of Minnesota pine forests. He has published more than 100 papers in scientific books and journals.

*Mary Keirstead* is a freelance editor and writer specializing in the fields of natural and environmental sciences. From 1979 to 1991, she worked as an editor, writer, and environmental planner for the Minnesota Department of Natural Resources.